Achieving the Paris Climate Agreement Goals

Sven Teske

Editor

Achieving the Paris Climate Agreement Goals

Part 2: Science-based Target Setting
for the Finance industry — Net-Zero
Sectoral 1.5°C Pathways for Real
Economy Sectors

 Springer

Editor
Sven Teske
Institute for Sustainable Futures
University of Technology Sydney
Sydney, NSW, Australia

ISBN 978-3-030-99179-1   ISBN 978-3-030-99177-7   (eBook)
https://doi.org/10.1007/978-3-030-99177-7

This Springer imprint is published by the registered company Springer Nature Switzerland AG
The registered company address is: Gewerbestrasse 11, 6330 Cham, Switzerland

v

For the next generation.
For my son Travis.

Foreword

Climate change is the defining threat of the twenty-first century. It is at the centre of our global risk landscape, affecting global societies and economies through extreme weather events, food and water crises, sea-level rise, and large-scale migration.

This decade is decisive—it is far later than hoped, but not too late to avoid the worst consequences for mankind and our planet. Climate scientists, through the Intergovernmental Panel on Climate Change (IPCC), have detailed the strategies necessary to limit global warming to 1.5 °C above the average pre-industrial temperature level by 2100.

The COP26 summit in Glasgow brought parties together to accelerate action towards the goals of the Paris Agreement and the UN Framework Convention on Climate Change. 'Real economy' actors, cities, regions, businesses, investors, and institutions of higher education, responsible for nearly 25% of global CO_2 emissions and over 50% of global GDP, made commitments to net-zero greenhouse gas emissions under the UNFCCC's Race to Zero. This is critical mass signalling to governments that non-state actors are already united in meeting the Paris Agreement's goals and in creating a more inclusive and resilient economy.

The pre-COP26 commitments predicted a 2.7 °C warming pathway. It is somewhat encouraging that in Glasgow at COP26, with the revised Nationally Determined Contributions for 2030 and the announcement of new net-zero targets and pledges, limiting global warming to 2 °C or even lower became achievable.

A major announcement at COP26 was the pledge of the Glasgow Financial Alliance for Net Zero (GFANZ)—a global coalition of over 450 finance firms across 45 countries, jointly managing US$130 trillion—to structure their financial activities to achieve net-zero emissions by 2050. This pledge is indicative of the scale and awareness of the global private sector.

Three UN-convened financial alliances, which are part of the GFANZ—the Net-Zero Asset Owner Alliance (NZAOA), the Net-Zero Banking Alliance (NZBA), and the Net-Zero Insurance Alliance (NZIA)—have committed to immediate action in aligning their investment, lending, and underwriting portfolios with *a pathway to limit the global temperature rise to 1.5 °C, with no or only limited overshoot*, consistent with the findings of the IPCC Special Report.

If current emission levels are maintained, the remaining carbon budget required to limit global warming to 1.5 °C will be exhausted in less than a decade. Unless emissions are urgently and severely limited, the world will overshoot this carbon budget, and therefore exceed 1.5 °C warming. Overshoot scenarios come with a high risk of failing to reach the 1.5 °C target altogether.

To achieve the steep decarbonization of the global economy, all actors require information on how the transition can be achieved. Policies must be adjusted to support a transitioning economy, green technologies must be scaled up, energy efficiency must increase, and consumer behaviour must change.

The United Nations Framework Convention on Climate (UNFCCC) Marrakech Partnership for Global Climate Action works to accelerate the implementation of the Paris Agreement by enabling collaborations between governments and cities, regions, businesses, and investors. In November 2020, it launched the *Climate Action Pathways*, which outlines sectorial visions for a 1.5 °C climate-resilient world. These pathways provide an overview of the actions and milestones required for the transformation of systems within sectors. They are supported and enhanced by the growing body of sectorial decarbonization pathways developed by the scientific community and others, built on industry intelligence. One such effort, a collaboration between the scientific community and in consultation with investors, is contained within this book.

In this book, Dr. Sven Teske and his research team provide data points for sectorial pathways on a low/no-overshoot basis. These pathways do not rely on carbon removal technologies but instead build on the rapid deployment of renewable energy and the preservation of natural carbon sinks. These detailed roadmaps provide highly ambitious information on the routes for various sectors and businesses. They also inform financial institutions of what they must require of their clients or investees to ensure that they participate in the journey to net-zero emissions by 2050.

This book provides a detailed analysis of 12 industry sectors, their interconnections, and their potential decarbonization in the short and longer terms. This assessment may be the first to translate a global energy system model into 12 financial sectors, and to report the *Scope 1, 2,* and *3* interconnections and therefore the final responsibilities for greenhouse gas emissions. This approach allows investors and actors in the real economy to engage with a common map and work together with all stakeholders towards change.

We must make use of all the intelligence at our disposal to move this critical mass of actors towards the finish line in the race to achieve net-zero greenhouse gas emissions.

UK   Nigel Topping

Acknowledgement

The authors thank the experts, asset owners, and other stakeholders who provided peer review and input during the research between May 2020 and November 2021. In particular, the authors thank Elke Pfeiffer (NZAOA UNPRI) and Jes Andrews (UNEPFI) for their input, guidance, support, and collaborative spirit throughout this project. We also acknowledge and thank the researchers involved in the development of the One Earth Climate Model on which this study builds.

This research has been supported and financed in parts by the UN-convened Net-Zero Asset Owner Alliance, the Rockefeller Foundation, and the European Climate Foundation (ECF). The ECF stresses that responsibility for the information and views set out in this research lies with the authors. None of the founders can be held responsible for any use which may be made of the information contained or expressed therein. A special thank you to Dr. Anna Irimisch of ECF for suggestions and support.

Furthermore, we would like to thank Deutsche Gesellschaft für Internationale Zusammenarbeit (GIZ) GmbH, the Transformative Urban Mobility Initiative (TUMI), and the Federal Ministry for Economic Cooperation and Development (BMZ) who financed the development of the global and regional transport pathways which have been the basis for the 1.5 °C pathways for transport (Chap. 8). Thank you in particular to the GIZ team Daniel Ernesto Moser, Marvin Stolz, and Rohan Shailesh Modi.

The authors would like to thank the One Earth, a philanthropic organization working to accelerate collective action to limit global average temperature rise to 1.5 °C. Especially the One Earth Climate team Karl Burkart, Justin Winters, Edward Bell, and Edith Espejo for ongoing support. Furthermore, we would like to thank the former Leonardo DiCaprio Foundation, which funded the initial research between July 2017 and February 2019.

This project has been supported by numerous people since the book project started with *Achieving the Paris Climate Agreement Goals (Part 1)* in July 2017 and our thanks go to each of them. The ongoing support was key and kept all researchers highly motivated.

Special thanks to Anna Leidreiter, Anna Skowron, and Naemie Dubbles of the World Future Council (https://www.worldfuturecouncil.org/), Dr. Joachim Fuenfgelt of Bread for the World (https://www.brot-fuer-die-welt.de/en/bread-for-the-world), and Stefan Schurig of F20—Foundations 20 (http://www.foundations-20.org/) who provided initial support to make this project possible. Finally, we would like to thank Greenpeace International and Greenpeace Germany for their ongoing support of the Energy [R]evolution energy scenario research series between 2004 and 2015 which resulted in the development of the long-term energy scenario model, the basis for the One Earth Climate Model.

Executive Summary

Abstract: To put this research project into context, a short introduction to the status of the climate debate is given. The methodology of the research is presented in brief, followed by the socio-economic assumptions made and key technological parameters used. The storylines of the energy demand projections for the 12 industry and service sectors analysed are described. The supply side of the sectorial pathways for power, heat, and fuels is documented. Finally, the main results are presented in terms of the final and primary energy demands, including energy- and non-energy-related greenhouse gas emissions. Key conclusions are drawn by sector, and policy recommendations are offered.

Introduction

Extreme weather events, such as extreme rainfall and floods, cyclones, and bushfires, have increased in frequency. Australia experienced the worst bushfire season on record between September 2019 and March 2020—known as the *Black Summer* (Cook et al., 2021). In June 2020, the Arctic region of Siberia experienced a heat wave with temperatures of up to 38 °C and wildfires covering almost 1 million hectares. The World Meteorological Organization (WMO) recognized this as a new Arctic temperature record.

Time is running out. In August 2021, the Sixth Assessment Report (AR6) of the United Nations Intergovernmental Panel on Climate Change (IPCC) was published. The First Assessment Report was launched in 1990 and underlined the importance of climate change as a challenge with global consequences that required international co-operation. Thirty years later, the IPCC states unequivocally that the world is already in the middle of climate change. The UN Secretary-General António Guterres said the Working Group's report was nothing less than '*a code red for humanity. The alarm bells are deafening, and the evidence is irrefutable*'.

Our first book laid out global and regional 100% renewable energy scenarios with non-energy GHG pathways for +1.5 °C or +2 °C warming scenarios and compared them with a reference case. Those scenarios were calculated under the leadership of the Institute for Sustainable Futures (ISF) at the University of Technology Sydney (UTS) in close co-operation with the German Aerospace Center (DLR) and the University of Melbourne, Australia. The energy scenario model used for that project became known as the *One Earth Climate Model* (OECM) in 2020 during the numerous debates that followed the book launch in February 2019.

The Second Book Focuses on Sectorial Pathways and Provides Key Performance Indicators (KPIs) for Industry Sectors to Limit the Global Temperature Increase to 1.5 °C

The book documents all the steps in the scenario development and provides a detailed analysis of the main assumptions and scenario narratives. The results of the OECM 1.5 °C pathways for 12 industry and service sectors include the total remaining carbon budget and the *Scope 1, 2,* and *3* emissions for each sector.

Science-Based Industry Greenhouse Gas (GHG) Targets— Defining the Challenge

The UN-convened *Net Zero Asset Owners Alliance* is a *Program for Responsible Investment* and a United Nations Environment Programme Finance Initiative (UNEP FI)-supported initiative. The members of the Alliance have committed to transitioning their investment portfolios to net-zero GHG emissions by 2050, consistent with a maximum global temperature rise of 1.5 °C above pre-industrial levels. This requires intermediate targets to be established for 5-yearly intervals, and regular reporting on progress.

Outlining the Task—Trend Reversals Until 2025

The global economy must decarbonize the energy system entirely within the next 30 years—in one generation. In historical terms, this means breaking the connections between population growth, steady economic development fuelled by energy, and the increase in CO_2 emissions of the past 120 years, and reversing those trends within the next 5 years. Between 2025 and 2030, global energy-related CO_2 must peak and start to decline to zero by 2050.

Figure 1 shows the historic development of the global population, GDP, energy demand, and the resulting annual CO_2 emissions between 1950 and 2020 on the left side, and the projected trajectory until 2050 on the right side. Based on the projected

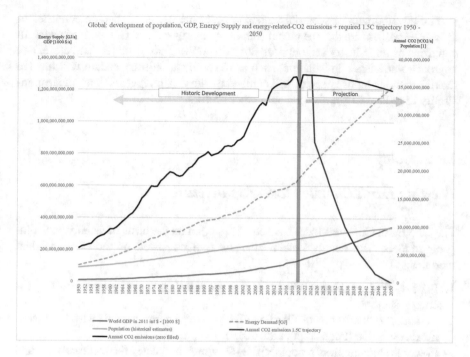

Fig. 1 Global development of key parameters

population and economic growth until 2050, and under the assumed historical trends of the past two decades, with an annual decline of 1% in both the energy and emissions intensities, the global energy demand will double, whereas CO_2 emissions will remain at around current levels.

The OECM does not question the development of the population or the global economy projected by international organizations but focuses on technical measures to increase energy efficiencies and decarbonize the energy supply by a transition to renewable energies to achieve the 1.5 °C decarbonization trajectory (marked with the red line). This will require a bottom-up assessment of the energy demand combined with an alternative energy supply concept for power, heating, and transport, which are documented in the following chapters of this book.

Science-Based Target Setting

The latest available scientific information is IPCC's Sixth Assessment Report *Climate Change 2021: The Physical Science Basis*. According to the IPCC definition, 67% likelihood is 'good', whereas 50% likelihood is 'fair'. The OECM aims to limit the global mean temperature rise to 1.5 °C with 'good' likelihood. Therefore, the 'science-based target' for the OECM 1.5 °C pathway in terms of the global carbon budget between 2020 and 2050 is set to 400 Gt CO_2.

The development of sectorial targets to meet the needs of specific countries or industries will ensure that the global sum of all energy-related CO_2 emissions for all countries or all industry sectors does not exceed the global budget. Therefore, any approach undertaken in isolation, such as for a single industry sector, will involve the risk that one industry sector will demand a high CO_2 budget and push the responsibility to reduce CO_2 emissions onto other sectors.

Methodology

The One Earth Climate Model Architecture

The One Earth Climate Model has been developed on the basis of established computer models. The energy system analysis tool consisted of three independent modules:

1. Energy System Model (EM): a mathematical accounting system for the energy sector
2. Transport scenario model TRAEM (TRAnsport Energy Model) with high technical resolution (Pagenkopf et al., 2019)
3. Power system analysis model [R]E 24/7, which simulates the electricity system on an hourly basis and at geographic resolution to assess the requirements for infrastructure, such as grid connections between different regions and electricity storage types, depending on the demand profile and power-generation characteristics of the system

The advanced One Earth Climate Model, OECM 2.0, merges the energy system model EM, the transport model TRAEM, and the power system model [R]E 24/7 into one MATLAB-based energy system. The Global Industry Classification System (GICS) was used to define sub-areas of the economy. The global finance industry must increasingly undertake mandatory climate change stress tests for GICS-classified industry sectors in order to develop energy and emissions benchmarks to implement the Paris climate protection agreement. This requires very high technical resolution for the calculation and projection of future energy demands and the supplies of electricity, (process) heat, and fuels that are necessary for the steel and chemical industries. An energy model with high technical resolution must be able to calculate the energy demand based on either projections of the sector-specific gross domestic product (GDP) or market forecasts of material flows, such as the demand for steel, aluminium, or cement in tonnes per year.

The MATLAB model has an object-oriented structure and two modules—to calculate demand and supply—that can be operated independently of each other. Therefore, an energy demand analysis independent of the specific supply options or the development of a supply concept based on demand from an external source is possible.

The High-Efficiency Buildings Model (HEB)

The HEB model was originally developed in 2012 to calculate the energy demand and CO_2 emissions of the residential and tertiary building sectors until 2050 under three different scenarios (Urge-Vorsatz et al., 2012). Since then, the model has been developed and updated several times. With the latest update, the model calculates the energy demand under four scenarios until 2060, based on the most recent data for macroeconomic indicators and technological development. This model is novel in its methodology compared with earlier global energy analyses and reflects an emerging paradigm—the performance-oriented approach to the energy analysis of buildings. Unlike component-oriented methods, a systemic perspective is taken: the performance of whole systems (e.g., whole buildings) is studied and these performance values are used as the input in the scenarios. The model calculates the overall energy performance levels of buildings, regardless of the measures applied to achieve them. It also captures the diversity of solutions required in each region by including region-specific assumptions about advanced and sub-optimal technology mixes. The elaborated model uses a bottom-up approach, because it includes rather detailed technological information for one sector of the economy. However, it also exploits certain macroeconomic (GDP) and socio-demographic data (population, urbanization rate, floor area per capita, etc.). The key output of the HEB model is floor area projections for different types of residential and tertiary buildings in different regions and their member states, the total energy consumption of residential and tertiary buildings, the energy consumption for heating and cooling, the energy consumption for hot water energy, the total CO_2 emissions, the CO_2 emissions for heating and cooling, and the CO_2 emissions for hot water energy.

Integration of HEB Results into the OECM

To capture the complexity of regional and global building demand projection, both in terms of data availability and high technical resolution, the HEB was used to develop four bottom-up demand scenarios. The HEB was developed by the Central European University (CEU) of Budapest under the scientific leadership of Prof Dr. Diana Uerge-Vorsatz.

Classification Systems for Setting Net-Zero Targets for Industries

Investment decisions, such as the decarbonization targets for the finance industry, are highly complex processes. In November 2020, the European Central Bank published a Guide on Climate-related and Environmental Risks, which maps a detailed process

Table 1 GICS: 11 main industries

10	Energy
15	Materials
20	Industrials
25	Consumer Discretionary
30	Consumer Staples
35	Health Care
40	Financials
45	Information Technology
50	Communication Services
55	Utilities
60	Real Estate

for undertaking 'climate stress tests' for investment portfolios. To achieve the Paris Climate Agreement goals in the global finance industry, decarbonization targets and benchmarks for individual industry sectors are required. This opens up a whole new research area for energy modelling because although decarbonization pathways have been developed for countries, regions, and communities, few have been developed for industry sectors. The One Earth Climate Model (OECM) is an integrated assessment model for climate and energy pathways that focuses on 1.5 °C scenarios and has been further improved to meet this need. To develop energy scenarios for industry sectors classified under the Global Industry Classification Standard (GICS), the technological resolution of the OECM required significant improvement. Furthermore, all demand and supply calculations had to be broken down into industry sectors before the individual pathways could be developed (Table 1).

The GICS has four classification levels, and includes 11 sectors, 24 industry groups, 69 industries, and 158 sub-industries. The 11 GICS sectors are: energy, materials, industrials, consumer discretionary, consumer staples, health care, financials, information technology, communication services, utilities, and real estate.

Sectorial Energy Scenarios for Industry and Services Provide Key Performance Indicators for Investors

The finance industry requires sectorial energy scenarios for the industry and service sectors to set sector-specific decarbonization targets. Increasingly, investment decisions of international and national banks, insurance companies, and investor groups are driven by key performance indicators (KPIs) not only for profitability but also with regard to the embedded GHG emissions of a company. For asset managers, it has become increasingly important to have access to detailed information about GHG emissions, e.g., whether or not a steel manufacturer is on a decarbonization trajectory. The emissions must be further divided according to the responsibility for those emissions. This is done by calculating so-called Scopes 1, 2, and 3.

Methodologies for Calculating Scopes 1, 2, and 3

Reporting corporate GHG emissions is important, and the focus is no longer only on direct energy-related CO_2 emissions but includes the other GHGs emitted by industries. These increasingly include the indirect emissions that occur in supply chains. The Greenhouse Gas Protocol, a global corporate GHG accounting and reporting standard, distinguishes between three 'scopes':

- Scope 1—emissions are direct emissions from owned or controlled sources.
- Scope 2—emissions are indirect emissions from the generation of purchased energy.
- Scope 3—emissions are all the indirect emissions (not included in Scope 2) that occur in the value chain of the reporting company, including both upstream and downstream emissions.

The OECM model focuses on the development of 1.5 °C net-zero pathways for industry sectors classified under the GICS, for countries or regions or at the global level. Emissions-calculating methodologies for entity-level Scope 3 require bottom-up entity-level data to arrive at exact figures. Therefore, data availability and accounting systems for whole industry sectors on a regional or global level present significant challenges.

Therefore, the Scope 3 calculation methodology was simplified for country-, region-, and global-level calculations and to avoid double counting. The OECM reports only emissions directly related to the economic activities classified by the GICS. Furthermore, the industries are broken down into three categories: Primary Class, Secondary Class, and End-use Activity Class.

Table 2 shows a schematic representation of the OECM *Scope 1, 2*, and *3* calculation methods according to GICS class, which are used to avoid double counting. The sum of *Scopes 1, 2*, and *3* for each of the three categories is equal to the actual emissions.

Double counting can be avoided by defining a primary class for the primary energy industry, a secondary class for the supply utilities, and an end-use class for all the economic activities that use the energy from the primary- and secondary-class companies. The separation of all emissions by the defined industry categories—such as GICS—also streamlines the accounting and reporting systems. The volume of data required is reduced and reporting is considerably simplified under the OECM methodology.

For a specific industry sector to achieve the global targets of a 1.5 °C temperature increase and net-zero emissions by 2050 under the Paris Agreement requires that all its business activities are with other sectors that are also committed to a 1.5 °C increase and net-zero emissions.

Table 2 Schematic representation of OECM Scopes 1, 2, and 3 according to GICS classes, to avoid double counting

	Primary Class			Secondary Class			End-use Activity Class		
	GICS 10 *Energy*			GICS 55 *Utilities*			All Other Industries & Services		
	Scope 1	Scope 2	Scope 3	Scope 1	Scope 2	Scope 3	Scope 1	Scope 2	Scope 3
CO_2									
CH_2 AFOLU									
CH_4									
N_2O									
CFCs									
Total GHG	Sum of *Scopes 1, 2, & 3* equals total emissions			Sum of *Scopes 1, 2, & 3* equals total emissions			Sum of *Scopes 1, 2, & 3* equals total emissions		

Decarbonization Pathways for Industries

The global gross domestic product (GDP) in 2019 was US$87.8 trillion, 3% of which came from agriculture, 26% from industry, 15% from manufacturing, and the remaining 65% from services.

Chemical Industry

The economic development of the global chemical industry is significantly more complex than that of the aluminium and steel industries. The product range of the chemical industry is diverse, and the material flow approach used for aluminium and steel is very data intensive, and is therefore beyond the scope of this research. The chemical industry produces materials for almost all parts of the economy— from mining to services—and it is therefore intrinsically connected to overall economic development. Consequently, a GDP-based approach has been used to develop the energy demand projections for the chemical industry over the next three decades.

Projection of the Chemical Industry Energy Intensity

The energy demands for the five sub-sectors—pharmaceuticals, agricultural chemicals, inorganic chemicals and consumer products, manufactured fibres and synthetic rubber, and the petrochemical industry—were calculated with the energy intensities, which are based on the IEA Energy Efficiency extended database and our own research. The energy intensities for primary feedstocks were also considered in estimating the efficiency trajectories of the different sub-sectors. An increase in the efficiency of primary feedstock production of 1% per year over the entire modelling period is required to achieve the assumed efficiency gains for all sub-sectors. However, inadequate data are available to calculate the specific energy intensities of the chemical industry, and no detailed breakdown of the electricity and process heat temperature levels is available in public databases. Therefore, our estimates should be seen as approximate values and more research, in co-operation with the chemical industry, is required. However, the energy requirements of the entire chemical industry are precisely known and were taken from the IEA statistics Advanced Energy Balances. The energy requirements of the sub-sectors were determined on the basis of market shares and GDP and in discussions with representatives of the chemical industry—specifically members of the Net-Zero Asset Owner Alliance and the Strategic Approach to International Chemicals Management of the United Nations Environmental Program (SAICM UNEP).

Projection of the Energy Demand for the Chemical Industry

The projections of the economic development and energy intensities of an industry yield the overall global energy demand projection for that industry. In another step, the share of electricity required to generate thermal process heat has been estimated. Table 3 shows the calculated electricity demand and the process heat demand by temperature level for the chemical industry sub-sectors.

Cement Industry

Cement is the second most-consumed substance in the world after water and is a central component of the built environment—from civil infrastructure projects and power generation plants to residential houses. Typically made from raw materials such as limestone, sand, clay, shale and chalk, cement acts as a binder between aggregates in the formation of concrete. Cement manufacture is a resource- and emissions-intensive process, and is associated with around 7% of total global CO_2 emissions, according to the Intergovernmental Panel on Climate Change.

Beyond the mining of the raw materials, there are five main steps in the cement production process:

Table 3 Projected electricity and process heat demand for the chemical industry to 2050

Sub-sector	Units	2019	2025	2030	2035	2040	2050
Chemical Industries							
Chemical Industry—Electricity Demand by Sub-sector							
Pharmaceutical Industry	[PJ/a]	1431	1652	1873	2118	2341	2799
	[TWh/a]	398	459	520	588	650	778
Agricultural Chemicals	[PJ/a]	782	899	1019	1152	1274	1523
	[TWh/a]	217	250	283	320	354	423
Inorganic Chemicals and Consumer Products	[PJ/a]	1447	1663	1884	2131	2355	2817
	[TWh/a]	402	462	523	592	654	782
Manufactured Fibres & Synthetic Rubber	[PJ/a]	273	314	356	403	445	532
	[TWh/a]	76	87	99	112	124	148
Bulk Petrochemicals & Intermediates, Plastic Resins	[PJ/a]	1450	1649	1849	2070	2264	2651
	[TWh/a]	403	458	514	575	629	736
Total Chemical Industry	[PJ/a]	5384	6178	6981	7874	8678	10,323
	[TWh/a]	1496	1716	1939	2187	2411	2867
Heat Demand	[PJ/a]	12,163	15,949	18,024	20,329	22,406	26,653
Heat Share:	[%]	56%	56%	56%	56%	56%	56%
Heat demand <100 °C	[PJ/a]	2196	2879	3254	3670	4044	4811
Heat demand 100–500 °C	[PJ/a]	2722	3570	4034	4550	5015	5965
Heat demand 500–1000 °C	[PJ/a]	5813	7623	8615	9716	10,709	12,739
Heat demand >1000 °C	[PJ/a]	1432	1878	2122	2394	2638	3138

1. Raw Material Preparation—This stage involves the crushing or grinding, clas-sification, mixing, and storage of raw materials and additives. This is an electricity-intensive production step requiring between 25–35 kilowatt hours (kWh) per tonne of raw material.
2. Fuel Preparation—This phase involves optimizing the size and moisture content of the fuel for the pyroprocessing system of the kiln.
3. Clinker Production—The production of clinker involves the transformation of raw materials (predominantly limestone) into clinker (lime), the basic compo-nent of cement. This is achieved by heating the raw materials to temperatures >1450 °C in large rotary kilns. Clinker production is the most energy-intensive stage of the cement-manufacturing process, accounting for >90% of the total energy used in the cement industry.
4. Clinker Cooling—After the clinker is discharged from the kiln, it is cooled rapidly.
5. Finish Grinding—After cooling, the clinker is crushed and mixed with other materials (gypsum, fly ash, ground granulated blast-furnace slag, and fine lime-stone) to produce the final product, cement.

Global Cement Production and Energy Intensity Projections

Table 4 summarizes the assumptions of the 1.5 °C OECM cement industry pathway in terms of the projected volume of global cement production, the development of energy intensities for the relevant processes, and the process emissions per tonne of clinker produced. These assumptions are similar, to a large extent, to those made for the IEA Technology Roadmap—Low-Carbon Transition in the cement industry projections.

Projections of the Cement Industry Energy Demand

Table 5 shows the calculated electricity and process heat demand developments based on the documented assumptions. The breakdown by temperature level is based on the five cement production steps required and their shares of the overall energy demand. No detailed statistical documentation of the exact breakdown of the process heat demand by temperature level and quantity is available.

Aluminium Industry

Aluminium is among the most important building and construction materials glob-ally. To understand the opportunities and challenges facing the industry, the global flow of aluminium metal must be considered. Since 1880, an estimated 1.5 billion

Table 4 Assumed global cement market development and production energy intensities

Parameter	Units	2019	2025	2030	2035	2040	2050
			Projection				
Cement—production volume	[Mt/a]	4200	4448	4595	4739	4883	5094
Cement—variation compared with 2019	[%]	0	6%	9%	13%	16%	21%
Clinker							
Clinker—production volume	[Mt/a]	2730	2869	2941	3000	3076	3056
Clinker—variation compared with 2019	[%]	0	5%	8%	10%	13%	12%
Clinker/cement ratio	[%]	65.0%	64.5%	64.0%	63.3%	63.0%	60.0%
Energy intensities							
Thermal Energy Intensity—per tonne of clinker	[GJ/t]	3.5	3.4	3.3	3.25	3.2	3.1
Variation compared with 2019	[%]	0	−3%	−6%	−7%	−9%	−11%
Cement Production—electricity intensity	[kWh/t]	116	90	87	85	83	79
Variation compared with 2019	[%]	0	−22%	−25%	−27%	−28%	−32%
Thermal Energy Intensity—per tonne of cement	[GJ/t]	2.33	2.27	2.20	2.12	2.07	2.01
Variation compared with 2019	[%]	0	−2%	−5%	−9%	−11%	−14%
Process emissions							
Process emissions (calcination process)	[tCO$_2$/t clinker)	0.40	0.40	0.37	0.34	0.30	0.24

Table 5 Projected electricity and process heat demands for the cement industry

Parameter	Units		2019	2025	2030	2035	2040	2050
				Projection				
Energy Demand—Limestone mining	[PJ/a]		510	526	618	724	829	1034
Energy Demand—Clinker production	[PJ/a]		9555	9753	9705	9749	9844	9475
Energy Demand—Cement production	[PJ/a]		11,530	11,550	11,552	11,517	11,546	11,670
Electricity Demand—Cement production	[PJ/a]		1754	1441	1439	1450	1459	1449
	[TWh/a]		487	400	400	403	405	402
Heat Demand (energy used)	[PJ/a]		7213	7514	7516	7483	7497	7597
Heat share (final energy):	[%]		81%	88%	88%	87%	87%	88%
Heat demand <100 °C	[PJ/a]	5%	346	361	361	359	360	365
Heat demand 100–500 °C	[PJ/a]	2%	146	152	152	152	152	154
Heat demand 500–1000 °C	[PJ/a]	30%	2189	2280	2281	2271	2275	2305
Heat demand >1000 °C	[PJ/a]	63%	4532	4721	4722	4701	4710	4773

tonnes of aluminium have been produced worldwide, and about 75% of the aluminium produced is in productive use. In 2019, 36% of aluminium was located in buildings, 25% in electrical cables and machinery, and 30% in transport applications. Aluminium can be recycled, but the availability of scrap is limited by the high proportion of aluminium in use.

Bauxite Production

Primary aluminium production requires bauxite. Bauxite ore occurs in the top-soils of tropical and sub-tropical regions, such as Africa, the Caribbean, South America, and Australia. The largest producers/miners of bauxite include Australia, China, and Guinea. Australia supplies 30% of global bauxite production.

Aluminium Production Processes

An analysis of current and future aluminium production processes is required to understand the decarbonization opportunities within each process.

Primary Aluminium Production Involves the Following Processes (Excluding Mining)

1. Refining bauxite to produce alumina.
2. Smelting: is the process of refining alumina to pure aluminium metal. An electrical reduction line is formed by connecting several electrolysis cells in series. Electrolysis separates alumina into aluminium metal at the cathode and oxygen gas at the anode.

In the *secondary production of aluminium (aluminium recycling process)*, the process of refining the raw material (bauxite) to alumina is not required. Instead, scrap aluminium is re-melted and refined. Therefore, the energy consumption for this process is much lower than for its primary production.

Projection of the Aluminium Industry Energy Demand

Due to the assumed increase in the share of recycled aluminium in global production and the reduced energy intensity per tonne of aluminium produced, a decoupling of the increases in production and energy demand is possible. Between 2019 and 2050, global aluminium production is projected to increase by 75%, whereas the overall energy demand will increase by only 12% (Table 6). Due to the already high electrification rates in the aluminium industry—which are projected to increase further—and the decarbonization of the electricity supply based on renewable power generation, the aluminium industry can halve its specific CO_2 emissions by 2035.

Global Steel Industry

Steel is an important material for engineering and the construction sector world-wide, and it is also used for everyday appliances at the domestic and industrial levels. About 52% of steel usage is for buildings and infrastructure: 16% is used for mechanical equipment, such as construction cranes and heavy machinery; 12% is used for automotive vehicles (road transport); 10% is used for metal products, including tools; 5% is used for other means of transport, including cargo ships, aeroplanes, and two-wheeler vehicles; 3% is used for electrical equipment; and 2% is used for domestic appliances, such as white goods.

Technological Overview of Steel Production

On average, 20 GJ of energy is consumed to produce one tonne of crude steel globally. There are two routes by which steel is produced. Primary or crude steel is produced by the coal- or natural-gas-based blast furnace–basic oxygen furnace (BF–BOF) route, in which iron ore is reduced at very high temperatures in a blast furnace. The iron ore is melted to a liquefied form, and then oxidized and rolled. Coal or natural gas is required to generate high temperatures of up to 1650 °C.

In the secondary production route, scrap steel is melted in electric arc furnaces (EAFs). The EAF route has the lowest emissions intensities. In the EAF (gas-fuelled) process, scrap is usually blended at a rate of about 10% with direct reduced iron. A more energy-efficient pathway for primary production is to use scrap steel with ore-based inputs in BF–BOF production, usually at a rate of 15–20% scrap (Table 7).

Projection of the Steel Industry Energy Demand and CO_2 Emissions

The assumed division between primary and secondary production rates and the assumed production process technologies are key to the energy demand projections. Whereas secondary steel production requires significantly more electricity per tonne, its demand for high-temperature process heat is significantly lower (Table 8).

Table 6 Projected electricity and process heat demands for the aluminium industry to 2050

Sub-sector	Units	2019	2025	2030	2035	2040	2050
			Projection				
Total Electricity Demand—Aluminium Industry	[PJ/a]	3694	3860	3924	3982	4035	4125
Total Electricity Demand—Aluminium Industry (including re-melting)	[TWh/a]	1026	1048	1066	1082	1097	1123
Electricity Demand—Primary Aluminium	[TWh/a]	1005	1027	1040	1051	1062	1079
Electricity Demand—Secondary Aluminium (excluding re-melting)	[TWh/a]	21	21	26	31	36	44
Total Process Heat Demand—Aluminium Industry	[PJ/a]	3110	2581	2590	2597	2601	2601
Process Heat Demand—Primary Aluminium	[PJ/a]	3079	2556	2559	2560	2558	2549
Process Heat Demand—Secondary Aluminium	[PJ/a]	31	25	31	37	42	52
Heat demand <100 °C	[PJ/a]	261	216	217	218	218	218
Heat demand 100–500 °C	[PJ/a]	48	40	40	40	40	40
Heat demand 500–1000 °C	[PJ/a]	569	472	474	475	476	476
Heat demand >1000 °C	[PJ/a]	2232	1852	1859	1864	1867	1867

Table 7 Assumed market and energy intensity developments for the global steel industry according to the production process

Parameter	Units	2019	2025	2030	2035	2040	2050
			Projection				
Global Iron Ore Production—estimates based on steel growth projections	[Mt/a]	2339	2377	2511	2676	2851	3289
Global: Annual Production Volume—iron and steel industry	[Mt/a]	1869.6	1904.2	2018.4	2159.7	2310.9	2695.4
Calculated Annual Growth Rate for Global Steel Market	[%/a]		0.95%	1.13%	1.31%	1.31%	1.48%
Development of Production Structures (Primary and Secondary)							
PRIMARY Steel Production	[%]	65%	63%	61%	59%	56%	52%
SECONDARY Steel Production (share of scrap)	[%]	35%	37.2%	39.3%	41.5%	43.7%	48%
Share of electricity in PRIMARY steel production	[%]	2%	2%	2%	2%	2%	2%
Share of electricity in SECONDARY steel production	[%]	91%	91%	91%	91%	91%	91%
Energy Intensities							
Energy Intensity for Iron Ore Mining	[PJ/Mt]	0.069	0.067	0.066	0.064	0.062	0.059
Global: Average Energy Intensity for Steel Production	[GJ/t]	18.6	12.81	12.4	12.2	12.0	11.4

(Continued)

Table 7 (Continued)

Parameter	Units	2019	2025	2030	2035	2040	2050
			Projection				
Global Range: Average Energy Intensity for PRIMARY Steel Production	[GJ/t]	21	16	16	16	16	16
Global Range: Average Energy Intensity for SECONDARY Steel Production	[GJ/t]	9.1	8.26	7.65	7.55	7.45	7
Primary Steel Production—Electricity demand	[GJ/t]	0.42	0.31	0.31	0.31	0.31	0.31
Primary Steel Production—Process heat demand	[GJ/t]	15.57	11.51	11.51	11.51	11.51	11.51
Secondary Steel Production—Electricity demand	[GJ/t]	8.28	7.52	6.96	6.87	6.78	6.37
Secondary Steel Production—Process heat demand	[GJ/t]	6.75	6.13	5.68	5.61	5.53	5.20
Electricity Intensities							
Electricity Intensity—PRIMARY steel production	[TWh/Mt]	0.12	0.09	0.09	0.09	0.09	0.09
Electricity Intensity—SECONDARY steel production	[TWh/Mt]	2.30	2.09	1.93	1.91	1.88	1.77
Development of Process-related Emissions							
Specific Process Emissions—Assumption in the OECM for the global average	[tCO$_2$/t crude steel]	1.06	0.92	0.60	0.37	0.23	0.08
Basic Oxygen Furnace (BOF)—production share	[%]	65%	58%	35%	20%	10%	0%
Basic Oxygen Furnace (BOF)—emission factor	[tCO$_2$/t steel]	1.46	1.46	1.46	1.46	1.46	1.46
Open Hearth Furnace (OHF)—production share	[%]	5%	3.0%	2.5%	1.0%	1.0%	0%
Open Hearth Furnace (OHF)—emission factor	[tCO$_2$/t steel]	1.72	1.72	1.72	1.72	1.72	1.72
Electric Arc Furnace (EAF)—production share	[%]	30%	40%	63%	79%	89%	100%
Electric Arc Furnace (EAF)—emission factor	[tCO$_2$/t steel]	0.08	0.08	0.08	0.08	0.08	0.08

Table 8 Projected electricity and process heat demands for the steel industry to 2050

Sub-sector	Units	2019	2025	2030	2035	2040	2050
			Projection				
Steel Industry							
Total Electricity Demand—Iron & Steel Industry	[PJ/a]	4559	5691	5906	6550	7245	8676
Total Electricity Demand—Iron & Steel Industry	[TWh/a]	1266	1581	1641	1819	2012	2410
Electricity Demand—Primary steel	[TWh/a]	83	103	105	109	112	121
Electricity Demand—Secondary steel	[TWh/a]	1184	1478	1535	1711	1900	2289
Total Process Heat Demand—Iron & Steel Industry (final energy)	[PJ/a]	17,451	18,146	18,639	19,603	20,604	22,900
Process Heat Demand—Primary steel	[PJ/a]	13,269	13,797	14,120	14,569	15,011	16,163
Process Heat Demand—Secondary steel	[PJ/a]	4183	4349	4518	5034	5593	6738
Heat Demand	[PJ/a]	13,060	18,146	18,639	19,603	20,604	22,900
Heat Share:	[%]	74%	76%	76%	75%	74%	73%
Heat demand <100 °C	[PJ/a]	595	2341	2405	2529	2658	2955
Heat demand 100–500 °C	[PJ/a]	211	336	345	363	382	424
Heat demand 500–1000 °C	[PJ/a]	2489	5038	5175	5442	5720	6358
Heat demand >1000 °C	[PJ/a]	9765	10,431	10,714	11,268	11,844	13,164

Furthermore, as the share of primary steel is reduced with higher recycling rates, the energy demand for iron-ore mining (volumes) that is required will decrease.

Textile and Leather Industry

The international fashion industry is estimated to be worth US$2.4 trillion, and the textile and leather industry constitutes a large proportion of it (valued at US$818.19 billion in 2020). 'Textiles' refers to natural and synthetic materials used in the manufacture of clothing (including finished garments and ready-to-wear clothing), furniture and furnishings, automotive accessories, and decorative items. Therefore, the textile industry spans activities related to the design, manufacture, distribution, and sale of yarn, cloth, and clothing. The textile and leather industry has close links with the agricultural and chemical industries. Agricultural output provides the raw materials for the textile industry in the form of natural fibres; similarly, the chemical industry outputs are used as synthetic raw materials in the textile industry.

Table 9 Projected economic development and energy intensities of the textile and leather industry

Parameter	Units	2019	2025	2030	2035	2040	2050
			Projection				
Textile Industries—Economic Value	[bn $GDP]	1275	1632	1927	2270	2614	3392
Variation compared with 2019	[%]	0%	28%	51%	78%	105%	166%
Leather Industry—Economic Value	[bn $GDP]	252	323	381	449	516	670
Variation compared with 2019	[%]	0%	28%	51%	78%	105%	166%
Total Textile & Leather Value	[bn $GDP]	1527	1955	2308	2719	3130	4062
Variation compared with 2019	[%]	0%	28%	51%	78%	105%	166%
Textile & Leather—Sector share (global/total GDP)	[%]	1.2%	1.2%	1.2%	1.2%	1.2%	1.2%
Textile Industry—Energy Intensities							
Textile Mills	[MJ/$GDP]	4.5	4.4	4.3	4.2	4.1	3.9
Textile Products Mills	[MJ/$GDP]	4.6	4.5	4.4	4.2	4.1	3.9
Clothing Industries	[MJ/$GDP]	0.9	0.8	0.8	0.8	0.8	0.7
Textile Industry—average energy intensity	[MJ/$GDP]	2.0	1.9	1.9	1.8	1.8	1.7
Variation compared with 2019	[%]	0%	−2%	−5%	−7%	−10%	−14%
Leather Industry—Energy Intensities							
Leather and Allied Products Industries	[MJ/$GDP]	1.49	1.45	1.42	1.38	1.35	1.28
Variation compared with 2019	[%/a]	0%	−2%	−5%	−7%	−10%	−14%

Projections for the Global Textile and Leather Industry: Production and Energy Intensities

Table 9 shows the assumed economic development and energy intensities for the textile and leather industry used to calculate the 1.5 °C OECM pathway. The energy intensities per product volume (e.g., in tonnes per year) are not available, so the energy demand is calculated as a product of the assumed economic development in $GDP and the average energy units required per dollar. This simplification is necessary because the level of detail in the available energy demand data for the textile and leather industry on the global level does not allow a more exact approach. Textile mills have a significantly higher energy intensity than the clothing industry, which manufactures the clothing in downstream processes. The assumed average energy intensities for both the textile and leather sections of the industry are estimated on the basis of the overall energy demand for both industries according to the IEA World Energy Statistics and the GDP shares.

Table 10 Projected electricity and process heat demands for the textile and leather industry to 2050

Parameter	Units	2019	2025	2030	2035	2040	2050
			Projections				
Energy Demand—Textile Industry	[PJ/a]	2474	3134	3607	4143	4650	5737
Variation compared with 2019	[%]	0%	27%	46%	67%	88%	132%
Energy Demand—Leather Industry	[PJ/a]	425	469	539	620	696	858
Variation compared with 2019	[%]	0%	10%	27%	46%	64%	102%
Energy Demand—Textile & Leather Industry	[PJ/a]	2899	3603	4146	4763	5346	6595
Variation compared with 2019	[%]	0%	24%	43%	64%	84%	128%
Electricity Demand—Textile & Leather Industries	[PJ/a]	1277	1569	1805	2074	2328	2872
	[TWh/a]	355	436	501	576	647	798
Heat Demand	[PJ/a]	2899	3603	4146	4763	5346	6595
Heat share:	[%]	56%	56%	56%	56%	56%	56%
Heat demand <100 °C	[PJ/a]	1622	2034	2341	2689	3018	3723
Heat demand 100–500 °C	[PJ/a]	0	0	0	0	0	0
Heat demand 500–1000 °C	[PJ/a]	0	0	0	0	0	0
Heat demand >1000 °C	[PJ/a]	0	0	0	0	0	0

Projection of the Textile and Leather Industry Energy Demand and CO_2 Emissions

Analogous to the previous industry energy and emissions projections, Table 10 shows the results for the textile and leather industry. All values are calculated on the basis of the documented assumptions. Based on the production processes typical of the industry, it is assumed that the process heat demand does not exceed the temperature level of 100 °C. The 1.5 °C OECM pathway requires that the global textile and leather industry decarbonizes the required energy demand entirely by 2050, whereas a reduction by almost 50% seems achievable by 2030.

Decarbonization Pathways for Services

The *Service* sector contributes 65% of the global GDP (US$56.9 trillion in 2019). In this analysis, we use the IEA World Energy Balances as the basis for the energy statistics, which define three main sub-sectors: *Industry*, *Transport*, and *Other Sectors*. Although *Industry* and *Transport* overlap with corresponding GICS classification used for the 1.5 °C OECM sectorial pathways, to a large extent, the *Service* sector is scattered across several GICS sectors and the IEA *Other Sectors* and *Industry* groups. In this section, we describe four service sectors that supply essential goods:

1. Agriculture and food processing
2. Forestry and wood products
3. Fisheries
4. Water utilities

The combined share of the global energy demand of these sectors is about 7.5%, which is relatively minor. Although their energy demand is low and their current energy-related CO_2 emissions contribute only 6% to global CO_2 emissions, their non-energy GHG emissions are significant. Agriculture and forestry are among the main emitters of non-energy CO_2, methane (CH_4), and nitrous oxide (N_2O)— referred to in climate science as *AFOLU* (agriculture, forestry, and other land uses) emissions.

Global Agriculture and Food Sector

The *Agriculture & Food* sector is an essential economic sector that contributes to food security, livelihoods, and well-being. Valued at US$3.5 trillion, agriculture, forestry, and fisheries (AFF) accounted for 4% of the global GDP in 2019, with the largest contributions from China and India. The value added in agriculture alone was US$0.2 trillion. Value is also added in some of the manufacturing sectors supported by AFF. In 2018, the manufacture of food and beverages contributed S$1.5 trillion and the manufacture of tobacco products contributed U&S$167 billion.

Energy Demand Projection for the Global Agriculture and Food Sector

Although energy is an important input to agriculture, the sector accounts for only 2.2% of the total final energy consumption globally, with oil and oil products meeting most of this demand. Generally, as agriculture is industrialized, this energy consumption increases. In regions where most agricultural systems are industrialized, efficiency gains may have plateaued (in the USA, after a peak in 2006) and the sectorial final energy consumption may even have decreased (in EU, 10.8% reduction since 1998).

However, the global food system is estimated to account for almost one-third of the world's total final energy demand. In high-GDP countries, approximately 25% of the total sectorial energy is consumed behind the farm-gate (in agriculture, including in fisheries): 45% in food processing and distribution, and 30% in retail, preparation, and cooking. In low-GDP countries, a smaller share is spent on the farm and a greater share on cooking.

The estimated global population growth is based on UN population projections and will decrease evenly from about 1% per year in 2020 to 0.5% per year in 2050.

Table 11 Energy demand projection for agriculture and food processing

Parameter	Unit	2019	2025	2030	2035	2040	2050
			Projection				
Agriculture							
Energy Demand – Agriculture	[PJ/a]	7803	8655	9297	9967	10,442	11,221
Agriculture: Electricity Demand	[PJ/a]	2450	2873	3087	3309	3467	3725
	[TWh/a]		681	798	857	919	963
Agriculture: Heat & Fuels Demand	[PJ/a]	5352	5781	6210	6658	6975	7496
Food processing							
Energy Demand – Food Processing	[PJ/a]	6071	6381	7498	8795	10,079	12,549
Food Processing: Electricity Demand	[PJ/a]	1931	2000	2349	2755	3156	3932
	[TWh/a]	536	556	653	765	877	1092
Food Processing: Heat & Fuels Demand	[PJ/a]	4140	4381	5149	6040	6923	8617
Agriculture & Food Processing							
Energy Demand – Agriculture & Food Processing	[PJ/a]	13,873	15,036	16,795	18,762	20,520	23,770
Agriculture & Food Processing: Electricity Demand	[PJ/a]	4382	4873	5436	6064	6622	7657
	[TWh/a]	1217	1354	1510	1684	1840	2127
Agriculture & Food Processing: Heat & Fuels Demand	[PJ/a]	9492	10,162	11,359	12,698	13,898	16,113

The food production volumes for each product will develop accordingly. No dietary or life-style changes are assumed in estimating the future energy demand of the agriculture and food-processing sector. In addition to food for human consumption, agricultural products are also required for animal feed.

The majority of the energy demand is estimated to be for fuel for agricultural machinery, such as tractors and harvesters, whereas 30% of the energy is electricity. Efficiency gains are assumed to be higher in the agriculture sector—0.8–1% per year—than in the food-processing industry.

Table 11 shows the calculated energy demand broken down according to the electricity, heat, and fuel requirements for the agriculture and food-processing sector.

Global Forestry and Wood Sector

Forestry contributes to food security, livelihoods, and well-being, supports terrestrial ecosystems and biodiversity, and provides (human)-life-sustaining ecosystem services, and forests act as carbon sinks. Value is also added by some of the manufacturing sectors supported by forestry. In 2018, wood and wood products contributed US$183 billion, and paper and paper products contributed US$324 billion to the global economy. Together with agricultural manufacturing, this is about 18% of the value added in total manufacturing globally.

Globally, 30% of all forests are used for production. Of this 30%, about 1.15 billion ha of forest are primarily used for the production of wood and non-wood forest products, and another 749 million ha are designated for multiple uses. In contrast, only 10% is allocated for biodiversity conversation, although more than half of all forests have management plans.

The energy demand for forestry was calculated both as the energy intensity multiplied by the global GDP for this sector, as shown in Table 12, and by subtracting the calculated energy for agriculture from the combined energy demand for agriculture and forestry provided by IEA. This dual calculation of the energy intensity for forestry was confirmed again with data from the literature (Table 13).

Global Fisheries Sector

About 7% of total protein intake globally is from seafood. Over 200 million tonnes of fish and seafood are produced annually. According to the Organisation for Economic Co-operation and Development (OECD), the fisheries industry employs over 10% of the world's population. Whereas the overall food fish consumption expanded by 122% between 1990 and 2018, the global capture fisheries—fish that are caught from natural environments with various fishing methods—only grew by 14%. The main increase in fish 'production' was in aquaculture, the output of which increased five-fold. However, the percentage of fish stocks caught in the open ocean within biologically sustainable levels decreased from 90% in 1909 to only 65.8% in 2018. The economic (first sale) value of the global fishing industry in 2018 was estimated at US$401 billion, of which US$250 billion was from aquaculture production.

Although the fishing industry plays a significant role in the food supply and economic income of a large part of the global coastal population, its share of the global energy demand is minor, at <0.1% of the global energy demand. However, in the OECM, we developed a specific scenario for fisheries because of their importance for small island states. Subsistence fishing is a key economic pillar of island nations in the Pacific, the Indian Ocean, and the Caribbean. Over the past decades, large fishing vessels have disputed the traditional fishing grounds of local indigenous people.

Among the most unsustainable fishing methods is bottom trawling by large vessels, which accounts for about one-quarter of the global fish catch. Traditional artisanal fishing boats, which are either entirely unpowered or powered by small outboard engines, cannot compete with industrial fishing vessels. Increasing fuel costs make it increasingly uneconomic for fishermen, because fuel costs often exceed the income from fishing. Moreover, most island states still rely on expensive diesel generators to provide electricity for households and cooling equipment for food preservation.

The economic value of the fishery industry is assumed retain its current global GDP share of 0.2% and to increase, according to the growth projection for global GDP, from US$272 billion in 2019 to over US$700 billion in 2050. However, the

Table 12 Global economic development of the forestry, wood, and wood products industry

Parameter	Units	2019	2025	2030	2035	2040	2050
			Projection				
Forestry Industry—Economic Value	[bn $]	155	187	221	261	300	390
Wood Industry—Economic Value	[bn $]	143	183	216	255	293	381
Pulp & Paper Industry—Economic Value	[bn $]	117	150	177	209	240	312
Round wood	[million m³]	3969	3993	4013	4033	4053	4094
Variation compared with 2019	[%]	0.0%	0.6%	1.1%	1.6%	2.1%	3.1%
Sawn wood	[million m³]	489	492	494	497	499	504
Variation compared with 2019	[%]	0.0%	0.6%	1.1%	1.6%	2.1%	3.1%
Pulp for paper	[million tonnes]	194	195	196	197	198	200
Variation compared with 2019	[%]	0.0%	0.6%	1.1%	1.6%	2.1%	3.1%
Paper and paperboard	[million tonnes]	404	429	446	461	475	499
Variation compared with 2019	[%]	0%	6%	10%	14%	18%	24%

Table 13 Energy demand for the forestry and wood products industry

Energy Demand	Unit	2019	2025	2030	2035	2040	2050
			Projection				
Forestry							
Energy Demand – Forestry	[PJ/a]	832	923	992	1063	1114	1197
Forestry: Electricity Demand	[PJ/a]	74	5	11	22	44	176
	[TWh/a]	20	2	3	6	12	49
Forestry: Heat & Fuels Demand	[PJ/a]	759	918	981	1041	1070	1021
Wood & Wood Products							
Energy Demand – Wood & Paper	[PJ/a]	7039	7791	8737	9779	10,695	13,330
Wood & Paper: Electricity Demand	[PJ/a]	2165	2259	2534	2836	3102	3866
	[TWh/a]	602	628	704	788	862	1074
Wood & paper: Heat & Fuels demand	[PJ/a]	4873	5532	6204	6943	7593	9464
Forestry & Wood Products							
Total Energy Demand	[PJ/a]	7871	8715	9729	10,842	11,809	14,526
Electricity	[PJ/a]	2239	2265	2545	2858	3146	4042
	[TWh/a]	622	629	707	794	874	1123
Heat & Fuels	[PJ/a]	5632	6450	7184	7984	8663	10,484

proportions of marine fishing, aquaculture, and inland fishing will change significantly in favour of aquaculture. Table 14 shows all the key assumptions used to calculate the 1.5 °C pathway for fisheries.

The projected development of fish production, in million tonnes per year, is certainly arguable and no forecasts of the fish production volumes over the next 30 years are available. Therefore, it is assumed that the volumes of wild fish catch and fish from aquaculture will plateau at the 2020 levels, whereas the market value will

Table 14 Key assumptions for the energy demand projections of the global fisheries industry

Parameter	Units	2019	2025	2030	2035	2040	2050
			Projection				
Fishing (Marine)—Economic Value	[bn $]	194	317	315	313	320	346
Fishing (Aquaculture)—Economic Value	[bn $]	65	150	157	185	267	346
Fishing (Inland)—Economic Value	[bn $]	13	17	20	23	27	35
Fishing—Total Economic Value	[bn $]	272	483	492	521	614	727
Total Volume—Fish Consumption	[million tonnes]	159	159	159	159	160	160
	[%]	0%	0%	0%	1%	1%	1%
Marine Landings	[million tonnes]	47	47	47	47	46	46
Variation compared with 2019	[%]	0%	0%	0%	−1%	−1%	−1%
Aquaculture	[million tonnes]	106	107	107	107	107	108
Variation compared with 2019	[%]	0%	1%	1%	1%	1%	2%
Inland Fisheries	[million tons]	6	6	6	6	6	6
Variation compared with 2019	[%]	0%	1%	1%	1%	1%	2%
Fishing Fleet—Number of vessels: powered	[million]	2.07	2.26	2.33	2.40	2.47	2.62
Unpowered	[million]	1.16	1.16	1.16	1.16	1.16	1.16
Powered artisanal	[million]	1.63	1.81	1.91	2.02	2.13	2.36
Powered, industrial (incl. aquaculture)	[million]	0.43	0.45	0.42	0.38	0.35	0.26
Fishing Fleet—Total motor power	[GW]	144	154	151	147	144	135
Artisanal motor power	[GW]	57	63	67	71	74	83
Industrial motor power	[GW]	87	90	84	77	69	52
Catch per unit effort (CPUE)—Energy Units	[PJ/Mt fish]	6	6	6	6	6	6

steadily increase. The rationale behind this is that marine fishing will be unable to increase the volume of catch, whereas the costs and economic value per tonne of fish will continue to increase. The catch per unit effort (CPUE)—the amount of energy per tonne—is assumed to remain stable. In this case, the longer distances and sailing times required to catch one tonne of fish can be compensated by the increased energy efficiency of fishing vessels.

The 1.5 °C OECM pathway for the fishing industry suggests moving away from large-scale fish trawlers towards a more decentralized fleet of fishing boats.

In terms of the fishing vessel fleet, 2.07 million vessels were registered in 2019: 1.16 million were unpowered, 1.63 million were powered artisanal vessels, and 0.43 million were industrial vessels. The overall motor power of the global fishing fleet is estimated have a capacity of 144 GW, 87GW of which is from industrial vessels. The 1.5 °C pathways assumes that the power artisanal fishing vessels will steadily increase in number at the expense of industrial vessels, which will lose market shares by volume in a stable fish market.

Table 15 Projected energy demand for global fisheries industry

Parameter	Units	2019	2025	2030	2035	2040	2050
			Projection				
Energy Demand—Fisheries	[PJ/a]	300	309	315	327	349	483
Variation compared with 2019	[%]	0%	3%	5%	9%	16%	61%
Fuel Demand—Fishing Fleet	[PJ/a]	272	276	276	276	276	277
Variation compared with 2019	[%]	0%	1%	1%	2%	2%	2%
Electricity & Synthetic Fuel Demand of Fishing Fleet	[PJ/a]	27	32	37	50	72	205
	[TWh/a]	8	9	10	14	20	57

Table 15 shows the resulting energy demands under the documented assumptions. However, the available data on the energy demand of fishing vessels is sparse and the results are estimates. More research is required to develop more-detailed scenarios for and around the fishing industry, their vessels, and electrification concepts for artisanal fishing boats.

Overview of the Global Water Utilities Sector

Water is important for basically every process that supports human life on Earth. Potable drinking water of high quality is therefore a basic requirement for the health of humans, the environment, and an intact economy. Therefore, the economic value of water utilities is far beyond the monetary value of this industry. Although the projection of future energy demands for various sectors in this analysis is based on economic values, the energy demand projections for water utilities must be based on production volumes.

According to the OECD, 70% of all water abstracted is used for agriculture. Whereas freshwater dominates the total water extracted, desalination plants are an important parameter because their consumption of energy is high. However, water extraction by desalination plants constitutes only 0.2% of global water extraction. Globally, about one-third of all countries, representing 80% of the global population, are connected to sewerage treatment plants. Table 16 shows the assumed quantities of global water withdrawn—broken down by usage sector—which form the basis for the energy demand projections for water utilities.

Projections of the Energy Demand and CO$_2$ Emission for Water Utilities

The projected global energy demand for water utilities was calculated with the documented assumed global quantities of water required and energy intensities (Table 17). However, the main GHG emissions from water utilities do not originate

Table 16 Assumed quantities of global water withdrawn, used to predict the energy demands for water utilities

Parameter	Units	2019	2025	2030	2035	2040	2050
			Projection				
Water withdrawal—total	[billion m³]	4134	4388	4608	4838	5080	5601
Variation compared to 2019	[%]	0%	6%	11%	17%	23%	35%
Of which is saltwater	[billion m³]	11	11	11	12	12	13
Saltwater share (of total water withdrawal)	[%]	0.3%	0.3%	0.2%	0.2%	0.2%	0.2%
Agricultural water	[billion m³]	2956	3138	3295	3459	3632	4005
Variation compared with 2019	[%]	0%	6%	11%	17%	23%	35%
Municipal water	[billion m³]	475	505	530	556	584	644
Variation compared with 2019	[%]	0%	6%	11%	17%	23%	35%
Industrial water	[billion m³]	703	746	783	822	864	952
Variation compared with 2019	[%]	0%	6%	11%	17%	23%	35%

Table 17 Projected global energy demand for water utilities

Energy Demand	Units	2019	2025	2030	2035	2040	2050
			Projection				
Water Utilities: Total Energy Demand	[PJ/a]	5358	5284	5510	5745	5992	6518
Variation compared with 2019	[%]	0%	−1%	3%	7%	12%	22%
Water Utilities: Process Heat Energy Demand	[PJ/a]	2143	2098	2164	2232	2303	2451
Variation compared with 2019	[%]	0%	−2%	1%	4%	7%	14%
Water Utilities: Electricity Demand	[PJ/a]	3215	3186	3346	3513	3688	4066
	[TWh/a]	893	885	929	976	1025	1130
Variation compared with 2019	[%]	0%	−1%	4%	9%	15%	26%

from energy-related CO_2, but from methane and N_2O (or 'laughing gas'), which have significant greenhouse potential.

Decarbonization Pathways for Buildings

The *Buildings* sector is responsible for 39% of process-related GHG emissions globally and for almost 32% of the global final energy demand, making the *Buildings* sector pivotal in reducing the global energy demand and climate change. With the increasing rates of population growth and urbanization, the building stock is projected to double in developing regions by 2050, so reducing the global energy demand will become challenging. Together with these challenges, new building stocks in developing regions will simultaneously provide opportunities for energy-efficient construction, which could substantially reduce the global energy demand. In developed regions, opportunities to reduce the energy demand will predominantly involve renovating the existing building stock.

To develop detailed energy demand projections for the regional and global *Buildings* sectors, the *High-Efficiency Buildings Model* (HEB) was used. The HEB is based on a bottom-up approach and includes rather detailed technological information for the building sector. The model is based on socio-economic data, including population growth rates, urbanization rates, and floor areas per capita. The HEB model uses four different scenarios to understand the dynamics of energy use and to explore the potential of the buildings sector to mitigate climate change by exploiting various opportunities. The four scenarios are:

1. *Deep Efficiency Scenario*: The *Deep Efficiency* scenario demonstrates the potential utility of state-of-the-art construction and retrofitting technologies, which can substantially reduce the energy consumption of the buildings sector and therefore CO_2 emissions, while also providing full thermal comfort in buildings. In this scenario, exemplary building practices are implemented worldwide for both new and renovated buildings.
2. *Moderate Efficiency Scenario*: The *Moderate Efficiency* scenario incorporates present policy initiatives, particularly the implementation of the Energy Building Performance Directive (EPBD) in the EU and building codes for new buildings in other regions.
3. *Frozen Efficiency Scenario*: This scenario assumes that the energy performance of new and retrofitted buildings does not improve relative to the baseline. Retrofitted buildings will consume around 10% less energy for space heating and cooling than standard existing buildings, whereas most new buildings will have a lower level of energy performance than that in the *Moderate Efficiency* scenario because of their lower compliance with building codes.
4. *Nearly Net Zero Scenario*: The last scenario models the potential of deploying 'nearly net zero energy buildings' (buildings that can produce as much energy locally through the utilization of renewables as they consume, on annual balance) around the world. It differs from the other three scenarios in that it not only calculates the energy consumption, but already incorporates the local energy supply to arrive at the final energy demand. In other aspects, it uses the same parameters as the *Deep Efficiency* scenario.

Final Energy Use for Space Heating and Cooling under the HEB Scenarios

The final energy use for space heating and cooling will largely depend upon the calculated floor area. After the floor area is calculated for each region, the thermal energy use is calculated. Like the floor area calculations, thermal energy use is calculated for the four different scenarios.

Key regions, such as China, EU-27, and India, consume most of the global energy, so it is important to know how the building sectors in these regions will perform under different scenarios. Regions such as the USA and EU-27 have much

greater potential to reduce space-heating- and space-cooling-related energy use
with the help of best practices.

1.5 °C OECM Pathway for Buildings

Based on the results of the detailed HEB model analysis, the *Deep Efficiency* scenario was chosen for commercial buildings and the *Moderate Efficiency* scenario for
residential buildings. These scenarios were chosen after stakeholder consultation
with representatives of the respective industries, members of the Carbon Risk Real
Estate Monitor (CRREM), the Net-Zero Asset Owner Alliance, and academia. To
integrate the buildings sector into the 1.5 °C pathway as part of the OECM, consistent with all other industry and service sectors and the transport sector, the selection
of one specific pathway for the buildings sector as a whole was necessary.

Table 18 shows the assumed development of floor space for residential and commercial buildings, which was taken from the HEB analysis and the projected economic development of the construction sector. The increase in the construction
industry is based on the overall global GDP, developed as documented in Chap. 2,
and is therefore not directly related to the HEB floor space projections. The direct
link between both parameters was beyond the scope of this analysis and is therefore
highlighted as a potential source of error.

Table 19 shows the calculated annual energy demand for residential and commercial buildings and for the construction industry. The energy demand consists of
the energy required for space heating and cooling ('heating energy') and the electricity demand, which includes all electrical applications in the buildings but
excludes electricity for heating and cooling. This separation is necessary to harmo-
nize the input data from the HEB (which do not include electricity for household
applications such as washing machines, etc.) with the OECM.

The electricity demand for residential buildings is based on the bottom-up analysis of households documented in Sect. 3.1.2. The electricity demand for the service

Table 18 OECM—Global buildings: projected floor space and economic value of construction

Parameter	Units	2019	2025	2030	2035	2040	2045	2050
			Projection					
Residential Buildings	[billion m²]	184	196	211	226	241	255	269
Residential Buildings—variation compared with 2019	[%]	0	6%	15%	23%	31%	38%	46%
Commercial Buildings	[billion m²]	101	112	130	146	163	177	192
Commercial Buildings— variation compared with 2019	[%]	0	12%	29%	46%	62%	76%	91%
Construction: Residential and Commercial Building— Economic value	[bn $GDP]	2149	2699	3186	3753	4321	4964	5607
Variation compared with 2019	[%]	0	26%	48%	75%	101%	131%	161%

Table 19 OECM—Global buildings: Calculated annual energy demand for residential and commercial buildings and construction

Parameter	Units	2019	2025	2030	2035	2040	2045	2050
			Projection					
Residential Buildings: Total Energy Demand	[PJ/a]	82,565	77,724	77,039	75,274	75,199	66,944	63,147
Variation compared with 2019	[%]	0%	−6%	−7%	−9%	−9%	−19%	−24%
Residential Buildings: Heat Energy Demand	[PJ/a]	60,417	54,746	56,056	56,739	56,983	56,677	55,989
Variation compared with 2019	[%]	0%	−9%	−7%	−6%	−6%	−6%	−7%
Residential Buildings: Electricity Demand	[PJ/a]	22,148	22,979	20,983	18,536	18,216	10,268	7,158
Variation compared with 2019	[%]	0%	4%	−5%	−16%	−18%	−54%	−68%
Commercial Buildings: Total Energy Demand	[PJ/a]	34,567	40,609	44,311	42,549	39,315	35,991	31,676
Variation compared with 2019	[%]	0%	17%	28%	23%	14%	4%	−8%
Commercial Buildings: Heat Energy Demand	[PJ/a]	28,432	34,736	38,346	36,482	33,137	29,690	25,243
Variation compared with 2019	[%]	0%	22%	35%	28%	17%	4%	−11%
Commercial Buildings: Electricity Demand	[PJ/a]	2921	2686	2619	2554	2490	2428	2367
Variation compared with 2019	[%]	0%	−8%	−10%	−13%	−15%	−17%	−19%
Construction of Residential & Commercial Building: Energy Demand	[PJ/a]	1505	1531	1798	2108	2415	2719	3010
Variation compared with 2019	[%]	0%	2%	20%	40%	60%	81%	100%

sector is based on a break down of electricity and heating in 2019 across all service sectors. The future values until 2050 are based on the projections for the analysed service and industry sectors.

Decarbonization Pathways for Transport

The transport sector consumed 28% of the final global energy demand in 2019 and its decarbonization potential is therefore among the most important of all industries. Given its size and diversity, not only with regard to different transport modes and technologies, but also regional differences, it is also one of the most challenging sectors. In 2019, transport consumed 78% of the total oil demand globally. Therefore, the transition from oil to electric drives and to synthetic fuels and biofuels is key to achieving the goals of the Paris Climate Agreement. The rapid uptake of electric mobility, combined with a renewable power supply, is the single most important measure to be taken to remain within the carbon budget of the 1.5 °C pathway.

As a result of the restricted mobility imposed to stop spread of the COVID-19 virus, the global pandemic led to a significant reduction in the oil demand, especially for road transport and aviation, which are responsible for nearly 60% of oil use. The global oil demand is estimated to have dropped by 8% in 2020. At the time of writing, the global pandemic is still on-going, although travel restrictions have been relaxed in many countries, increasing in the transport demand relative to that in 2020. In our transport demand projections, we assume that the demand will continue to increase to pre-pandemic levels by 2025.

The majority of all passenger transport—in terms of overall kilometres—is by road. However, international freight transport is more strongly dominated by rail and shipping, which account for 45% of all tonne–kilometre. The high efficiency of rail and shipping means that their share of the global transport energy demand is small relative to the share of global tonnage transported.

Shipping and Aviation: Dominated by Combustion Engines for Decades to Come

Navigation will probably remain predominantly powered by internal combustion engines (ICEs) in the next few decades. Therefore, we did not model the electrification of freight vessels. However, pilot projects using diesel hybrids, batteries, and fuel cells are in preparation. We assumed the same increase in the share of bio- and synthetic fuels over time as in the road and rail sectors.

In aviation, energy efficiency can be improved by measures such as winglets, advanced composite-based lightweight structures, powertrain hybridization, and enhanced air traffic management systems. We project a 1% annual increase in efficiency on a per passenger–kilometre (pkm) basis and a 1% annual increase in efficiency on a per tonne–kilometre (tkm) basis (Tables 20 and 21).

A key target for the global transport sector is the introduction of incentives for people to drive smaller cars and use new, more-efficient vehicle concepts. It is also vital to shift transport use to efficient modes, such as rail, light rail, and buses, especially in large expanding metropolitan areas. Furthermore, the 1.5 °C scenario cannot be implemented without behavioural changes. It is not enough to simply exchange vehicle technologies, but the transport demand must be reduced in terms of the kilometres travelled and by an increase in 'non-energy' travel modes, such as cycling and walking (Table 22).

The proportion of battery electric vehicles (BEVs) among all passenger cars and light commercial vehicles in use is projected to be between 8% and 15% by 2030. This will require a massive build-up of battery production capacity in the coming years. New car sales will already be dominated by battery electric passenger vehicles in 2030 under the 1.5 °C scenario. However, with an assumed average lifetime of 15 years for ICE passenger cars, the existing car fleet will still predominantly use ICEs. Under the assumption that new ICE passenger cars and buses will not be

Table 20 Aviation—energy demand and supply

Parameter	Unit	2019	2025	2030	2035	2040	2050
			Projection				
Air Freight: Energy Intensity	[MJ/tkm]	32.2	29.1	27.2	26.5	25.8	25.2
Air-Passenger: Energy Intensity	[MJ/pkm]	5.8	4.8	4.5	4.4	4.3	4.2
Air Freight: Energy Demand	[PJ/a]	1445	911	809	712	595	430
Air Passenger: Energy Demand	[PJ/a]	13,004	8195	7279	6410	5359	3866
Air Freight Fuel: Fossil	[PJ/a]	580.5	892.4	740.0	284.9	59.5	0.0
Air Freight Fuel: Renewable & Synthetic Fuels	[PJ/a]	0.0	18.2	68.7	427.3	535.9	429.5
Air Freight Fuel: Renewables share	[%]	0%	2%	9%	60%	90%	100%
Air Freight electricity: Fossil	[PJ/a]	0	0	0	0	0	0
Air Freight electricity: Renewables	[PJ/a]	0	0	0	0	0	0
Air Freight electricity: Renewables share	[%]	0	0	0	0	0	0
Air Passenger Fuel: Fossil	[PJ/a]	5224	8031	6660	2564	536	0
Air Passenger Fuel: Renewable & Synthetic Fuels	[PJ/a]	0	164	619	3846	4823	3866
Air Passenger Fuel: Renewables share	[%]	0%	2%	9%	60%	90%	100%

Table 21 Shipping—energy demand and supply

Parameter	Unit	2019	2025	2030	2035	2040	2050
			Projection				
Shipping Freight: Energy Intensity	[MJ/tkm]	0.2	0.2	0.2	0.2	0.2	0.2
Shipping-Passenger: Energy Intensity	[MJ/pkm]	0.1	0.1	0.1	0.1	0.1	0.1
Shipping Freight: Energy Demand	[PJ/a]	11,067	10,659	11,023	11,121	11,233	11,554
Shipping Passenger: Energy Demand	[PJ/a]	833	802	830	837	846	870
Shipping Freight Fuel: Fossil	[PJ/a]	2270	10,425	7441	7507	1460	0
Shipping Freight Fuel: Renewable & Synthetic Fuels	[PJ/a]	11	235	3582	3614	9773	11,554
Shipping Freight Fuel: Renewables share	[%]	0%	2%	33%	33%	87%	100%
Shipping Passenger Fuel: Fossil	[PJ/a]	171	785	560	565	110	0
Shipping Passenger Fuel: Renewable & Synthetic Fuels	[PJ/a]	1	18	270	272	736	870
Shipping Passenger Fuel: Renewables share	[%]	0%	2%	33%	33%	87%	100%
Shipping Passenger electricity: Fossil	[PJ/a]	0	0	0	0	0	0
Shipping Passenger electricity: Renewables	[PJ/a]	0	0	0	0	0	0
Shipping Passenger electricity: Renewables share	[%]	0%	0%	0%	0%	0%	0%

Table 22 Road transport—energy demand and supply

Parameter	Unit	2019	2025	2030	2035	2040	2050
			Projection				
Road Freight: Energy Intensity	[MJ/tkm]	1.33	1.17	1.11	0.86	0.79	0.71
Road-Passenger: Energy Intensity	[MJ/pkm]	1.47	1.17	1.07	0.73	0.65	0.58
Road Freight: Energy Demand	[PJ/a]	38,598	28,937	26,027	19,137	16,736	11,058
Road Passenger: Energy Demand	[PJ/a]	53,302	50,113	39,315	23,445	19,000	13,787
Road Freight Fuel: Fossil	[PJ/a]	36,898	26,621	23,513	8670	3787	0
Road Freight Fuel: Renewable, Electric & Synthetic Fuels	[PJ/a]	1700	2317	2514	10,467	12,949	11,058
Road Freight Fuel: Renewables share	[%]	4.4%	8.0%	9.7%	54.7%	77.4%	100.0%
Road Freight electricity: Fossil	[PJ/a]	77	260	166	629	326	0
Road Freight electricity: Renewables	[PJ/a]	25	282	478	4883	6081	5928
Road Freight electricity share	[%]	0.3%	1.9%	2.5%	28.8%	38.3%	53.6%
Road Passenger Fuel: Fossil	[PJ/a]	50,954	46,485	34,491	8481	4043	0
Road Passenger Fuel: Renewable, Electric & Synthetic Fuels	[PJ/a]	2348	3628	4825	14,963	14,957	13,787
Road Passenger Fuel: Renewables share	[%]	4.4%	7.2%	12.3%	63.8%	78.7%	100.0%
Road Passenger electricity: Fossil	[PJ/a]	119	783	1154	8168	8148	6783
Road Passenger electricity: Renewables	[PJ/a]	22	88	98	512	477	338

produced after 2030, BEVs will dominate the passenger vehicle fleet of 2050 under the 1.5 °C scenario. OECD countries and China are assumed to lead the development of BEVs and therefore to have the highest shares, whereas Africa and Latin America are expected to have the lowest BEV shares. Fuel-cell-powered passenger vehicles are projected to play a significantly smaller role than BEVs and will only be used for larger vehicles, such as SUVs and buses.

Transition of the Energy Industry to (Net)-Zero Emissions

To reduce emissions to zero in line with a 1.5 °C increase in global temperature, the use of coal, oil, and gas must be phased out by at least 56% by 2030. However, current climate debates have not involved an open discussion of the orderly withdrawal from the coal, oil, and gas industries. Instead, the political debate about coal, oil, and gas has continued to focus on supply and price security, neglecting the fact that mitigating climate change is only possible when fossil fuels are phased out.

The primary energy demand analysis—and therefore the projections for the primary energy industry and possible future operation strategies—is the product of the energy demand projections for all end-use sectors and the energy supply concept. The challenge for the primary energy industry is to supply energy services for sustained economic development and a growing global population while remaining within the global carbon budget to limit the global temperature rise to 1.5 °C.

The trajectories for oil, gas, and coal depend on how quickly an alternative energy supply can be built up and how energy consumption can be reduced technically and/or by behavioural changes. The OECM 1.5 °C pathway represents such a trajectory and is based on a detailed bottom-up sectorial demand and supply analysis. However, for the primary energy industry, it is important to assess whether or not new oil, gas, or coal extraction projects are required to meet the demand, even under an ambitious fossil-fuel phase-out scenario.

A scenario designated the *Existing International Production Trajectory* ('No Expansion') was developed and modelled, specifically to understand what global fossil fuel production will look like under the following assumptions:

- No new fossil fuel projects are developed
- Existing fossil fuel production projects stop producing once the resource at the existing site is exhausted, and no new mines are dug or wells drilled in the surrounding field
- Production at existing projects declines at standard industry rates:

  - Coal: – 2% per year
  - Oil: – 4% per year onshore and 6% per year offshore
  - Gas: – 4% per year on- and offshore

The *No Expansion* scenario was compared with the OECM 1.5 °C pathways for coal, oil, and gas to understand whether security of supply is possible under the immediate implementation of a 'stop exploration' policy.

The decline rates for oil, gas, and coal that would result from the implementation of the 1.5 °C pathway and the assumed annual production decline rates for oil, gas, and coal are compared in Table 23.

Our analysis shows that even with no expansion of fossil fuel production, the current productions levels—especially for coal—will exhaust the carbon budget associated with the 1.5 °C target before 2030. Without the active phase-out of fossil-fuel production, production will significantly surpass what can be produced under a 1.5 °C scenario by 2025 onwards, for all fossil-fuel types.

Power and Gas Utilities

Throughout the description of the OECM 1.5 °C pathway, the increased electrification of the transport and heating sectors is the overarching scenario narrative, and runs across all sectors. Increased electrification will lead to 'sector coupling', with

Table 23 Decline rates required to remain within the 1.5 °C carbon budget versus the production decline rates under 'no expansion'

	Average annual decline rate required to remain within the 1.5 °C carbon budget (67%)		Typical industry production decline rates (global average)
	2021–2030	2030–2050	2021–
Coal	−9.5%	−5%	−2%
Gas: onshore & offshore	−3.5%	−9%	−4%
Oil: onshore	−8.5%	−6%	−4%
Oil: offshore			−6%

the interconnection of the heating and transport sectors with the electricity sector. The sectors are still largely separate at the time of writing. However, the interconnection of these sectors offers significant advantages in terms of the management of the energy demand and the management of generation with storage technologies. The synergies of sector coupling in terms of the infrastructural changes required to transition to 100% renewable energy systems are well documented in the literature.

The OECM 1.5 °C pathway will lead to an annual increase in electricity generation from about 26,000 TWh in 2019 to 76,000 TWh by 2050, which will require a significant increase in renewable generation capacity (Table 24). Although there is clear agreement that the global electricity demand will increase, the predictions of how this electricity will be generated are very different. Despite the significant growth in renewable power generation during the last decade, short-term projections from the IEA still expect that fossil-fuel-based power generation will continue to grow.

The changes in gas utilities under the OECM 1.5 °C scenario are more profound than those for power utilities, because the main product—natural gas—will be phased out globally by 2050. However, the OECM acknowledges the significant value of the existing gas infrastructure and recommends that the gas distribution network be re-purposed to utilize it for the future decarbonized energy supply. According to the Global Energy Monitor, 900,757 km of natural gas long-distance transmission pipelines were in operation globally at the end of 2020. Research has shown that there are no fundamental technical barriers to the conversion of natural gas pipelines for the transport of pure hydrogen.

Table 25 shows the development of the demand and supply of natural-gas-derived electricity for the global utilities sector under the OECM 1.5 °C pathway. Global renewable electricity generation will increase significantly, by a factor of 10. The projected transition of gas utilities to the distribution of hydrogen and synthetic fuels will represent 50% of their sales by 2045. Therefore, the transition is assumed to have a lead time of about 10 years for the implementation of the required technical and regulatory changes.

Based on the OECM 1.5 °C decarbonization pathway, we propose a horizontal integration of all three sub-sectors, to combine the core areas of expertise and avoid

Table 24 Renewable power, heat capacities, and energy demand for hydrogen and synthetic fuel production under the 1.5 °C scenario

Parameter	Units	2019	2025	2030	2035	2040	2050
Solar Photovoltaic (roof-top + utility scale)	[GW$_{electric}$]	537	4197	8212	14,093	15,658	16,950
Solar Photovoltaic (utility-scale share, 25% of total capacity)	[GW$_{electric}$]	134	1049	2053	3523	3915	4238
Concentrated Solar Power	[GW$_{electric}$]	5	113	657	1979	2770	3603
Solar Thermal and Solar District Heating Plants	[GW$_{thermal}$]	388	2463	4087	5402	6173	8154
Onshore Wind	[GW$_{electric}$]	617	1350	2528	4393	5733	7620
Offshore Wind	[GW$_{electric}$]	0	233	451	934	1293	2024
Hydro Power Plants	[GW$_{electric}$]	1569	1419	1576	1726	1830	1980
Ocean Energy	[GW$_{electric}$]	1	44	91	262	379	701
Bio-energy Power Plants	[GW$_{electric}$]	77	198	174	200	200	231
Bio-energy Co-Gen Plants	[GW$_{electric}$]	49	111	175	304	520	668
Bio District Heating Plants	[GW$_{thermal}$]	5221	6586	8924	6262	5476	3817
Geo Energy Power Plants	[GW$_{electric}$]	12	37	92	165	267	441
Geo Energy Co-Gen Plants	[GW$_{electric}$]	1	1	6	8	10	17
Gas Power Plant for H$_2$ Conversion	[GW$_{electric}$]	0	9	56	243	375	650
Gas Power Co-Gen for H$_2$ Conversion	[GW$_{electric}$]	0	0	0	32	70	199
Fuel Cell & Synthetic Fuel Co-Gen Plants	[GW$_{electric}$]	0	0	0	32	70	199
Nuclear Power Plants	[GW$_{electric}$]	429	322	232	141	43	0
Industrial/District Heat Pumps + Electrical Process Heat	[GW$_{thermal}$]	157	2223	3302	7461	8909	11,060
Hydrogen Fuel Production— Electricity demand	[TWh$_{electric}$/a]	0	294	1278	4577	7088	10,784
Hydrogen Fuel Production—as above, but in PJ/a	[PJ/a]	0	1059	4601	16,478	25,517	38,822
Synthetic Fuel Production— Electricity demand	[TWh$_{electric/a}$]	0	0	82	364	1118	1533
Synthetic Fuel Production— as above, but in PJ/a	[PJ/a]	0	0	296	1310	4023	5517

stranded assets by repurposing the existing fossil-fuel infrastructure, such as pipelines.

Figure 2 shows a possible structure for the decarbonized *Energy* and *Utility* sectors. The (primary) energy industry will focus on utility-scale power generation and the production of hydrogen and synthetic fuels for the supply of energy and chemical feedstock. Gas utilities will focus on the transport of hydrogen and fuels and offer decentralized hydrogen production and storage services to the power sector. Power utilities will concentrate on the power grid, the management of the electricity system, and the integration of decentralized renewable power generation and storage systems, including those from 'prosumers'.

Table 25 Global utilities sector—electricity and gas distribution under the OECM 1.5 °C scenario

Sub-sector	Units	2019	2025	2030	2035	2040	2050
Power							
Total public power generation (incl. CHP, excluding auto producers, losses)	[TWh/a]	25,817	29,139	36,660	54,689	64,650	76,130
Compared with 2019	[%]		13%	42%	112%	150%	195%
Coal: public power generation (incl. CHP, excluding auto producers)	[TWh/a]	8338	5134	1879	497	193	0
Compared with 2019	[%]		−38%	−77%	−94%	−98%	−100%
Lignite: public power generation (incl. CHP, excluding auto producers)	[TWh/a]	1871	390	287	292	84	0
Compared with 2019	[%]		−79%	−85%	−84%	−96%	−100%
Gas: public power generation (incl. CHP, excluding auto producers)	[TWh/a]	6127	5611	5003	3977	2597	0
Compared with 2019	[%]		−8%	−18%	−35%	−58%	−100%
Nuclear: power generation	[TWh/a]	2764	2113	1521	923	281	0
Renewables: public power generation (incl. CHP, excluding auto producers)	[TWh/a]	6716	15,892	27,970	49,000	61,495	76,130
Compared with 2019	[%]		137%	316%	630%	816%	1033%
Electricity carbon intensity	[gCO$_2$/kWh]	509	291	135	52	24	0
Electricity intensity: variation compared with 2019	[%]		−43%	−73%	−90%	−95%	−100%
Gas							
Gas: transport & distribution	[BCM/year]	3693	3558	3178	2609	1796	238
	[PJ/a]	129,888	125,132	111,785	91,766	63,182	8371
Compared with 2019	[%]	0%	−4%	−14%	−29%	−51%	−94%
Synthetic & hydrogen fuels	[PJ/a]	0	720	3563	13,009	22,651	34,945
Total energy transport & distribution (gas, synthetic fuels, & hydrogen)	[PJ/a]	129,888	125,851	115,348	104,776	85,833	43,316

Climate Sensitivity Analysis—All Greenhouse Gases and Aerosols

The IPCC Assessment Report 6 (AR6), published in August 2021, contains five scenarios, each of which represents a different emissions pathway. These scenarios are called the *Shared Socioeconomic Pathway* (*SSP*) scenarios. The most optimistic

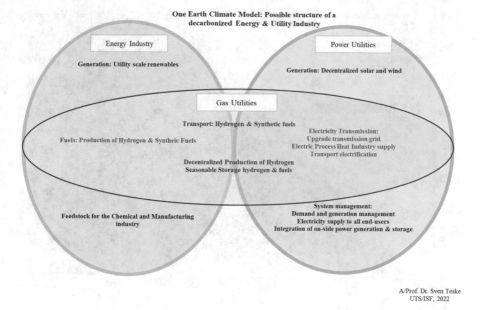

Fig. 2 One Earth Climate Model: Possible structure of a decarbonized Energy and Utilities industries

scenario, in which global CO_2 emissions are cut to net zero around 2050, is the SSP1-1.9 scenario. The number at the end (1.9) stands for the approximate end-of-century radiative forcing, a measure of how hard human activities are pushing the climate system away from its pre-industrial equilibrium. The most pessimistic is SSP5-8.5.

Climate Resource[1] has added CO_2 emissions that fall under other fossil fuel and industrial activities, such as fugitive emissions, cement production, and waste disposal and management, from the SSP1-1.9 scenario, with energy-related CO_2 emissions data of the OECM 1.5 °C pathway.

Here, we provide the global mean probabilistic temperature projections, including their medians and 5–95% ranges, for the OECM scenarios analysed (Fig. 3). These probabilistic ranges are sourced from the underlying 600 ensemble members, which are calibrated against the IPCC AR6 WG1 findings.

---

[1]This section is based on the analysis of *Climate Resource* under contract to the University of Technology Sydney (UTS) as part of the Net-Zero Sectorial Industry Pathways Project (UTS/ISF 2021). The study is an update of the previous One Earth Climate Model (OECM) publication (Teske et al. 2019). However, the Generalized Quantile Walk (GQW) methodology used (Meinshausen & Dooley 2019) has been developed further. The energy and industrial CO_2 emissions pathways are based on the OECM 1.5 °C energy scenario described in previous chapters, whereas the non-CO_2 GHG emission time series have been described with the advanced GQW methodology.

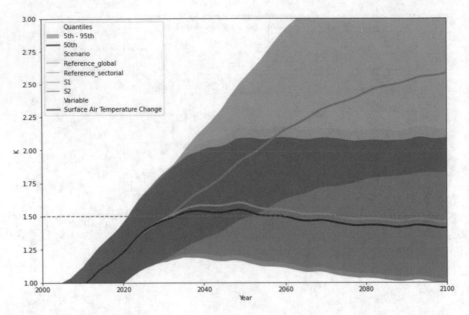

Fig. 3 Probabilistic global mean surface air temperature (GSAT) projections relative to 1850–1900

Similar to the SSP1-1.9 scenario in IPCC AR6 WG1, the OECM 1.5 °C pathways slightly overshoot the 1.5 °C pathway in their median values during the middle of the century, before dropping back to below 1.5 °C warming towards the end of the century. The likelihood that the OECM 1.5 °C scenario will stay below 1.5 °C throughout the century, despite strong mitigation actions, does not exceed 67%. Figure 3 shows the probabilistic global mean surface air temperature (GSAT) projections relative to 1850–1900 for the scenarios analysed.

OECM 1.5 °C Pathway for the Global Energy Supply

The supply side of this 1.5 °C energy scenario pathway builds upon modelling undertaken in an interdisciplinary project led by the University of Technology Sydney (UTS). The project modelled sectorial and regional decarbonization pathways to achieve the Paris climate goals—to maintain global warming well below 2 °C and to 'pursue efforts' to limit it to 1.5 °C. That project produced the One Earth Climate Model (OECM), a detailed bottom-up examination of the potential to decarbonize the energy sector. The results of this on-going research were published in 2019 (Teske et al., 2019). For the present analysis, the 1.5 °C supply scenario has been updated to match the detailed bottom-up analysis for the industry and service sectors, as well as the buildings and transport sectors.

Global Final Electricity Demand

Figure 4 shows the development of the final electricity demand by sector between 2019 and 2050. The significant increase in the demand is due to the electrification of heat, for both space and process heating, and to a lesser extent for the manufacture of hydrogen and synthetic fuels. The overall global final demand in 2050 will be 2.5 times higher than in the base year, 2019. In 2050, the production of fuels alone will consume the same amount of electricity as the total global electricity demand in 1991. Therefore, the demand shares will change completely, and 47% of all electricity (Fig. 5) will be for heating and fuels that are mainly used in the industry and service sectors. Electricity for space heating—predominantly from heat pumps—will also be required for residential buildings.

Global power plant capacities will quadruple between 2019 and 2050, as shown in Fig. 6. Capacity will increase more than actual power generation because the capacity factors for solar photovoltaic and wind power are lower than those for fuel-based power generation. By 2030, solar photovoltaic and wind will make up 70% of the generation capacity, compared with 15% in 2019, and will clearly dominate by 2050, with 78% of the total global generation capacity.

However, fossil-fuel-based power generation must be decommissioned, and the global total capacity will not increase over current levels but will remain within the greenhouse gas (GHG) emissions limits. By 2025, global capacities of 63 GW from hard coal power plants and 55 GW from brown coal power plants must go offline. All coal power plants in OECD countries must cease electricity generation by 2030, and the last coal plants must finish operation globally by 2045 to remain within the carbon budget for power generation required to limit the global mean temperature increase to +1.5 °C. Specific CO_2 emission per kilowatt-hour will decrease from 509 g of CO_2 in 2019 to 136 g by 2030, and 24 g in 2040, to be entirely CO_2 free by 2050.

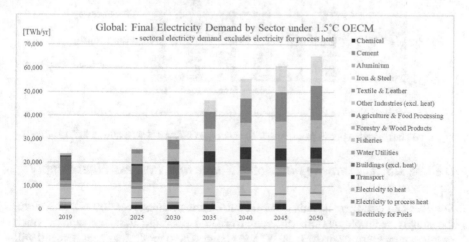

Fig. 4 Electricity demand by sector under the OECM 1.5 °C pathway in 2019–2050

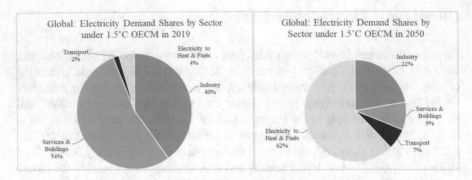

Fig. 5 Electricity demand shares by sector under the OECM 1.5 °C pathway in 2019 and 2050

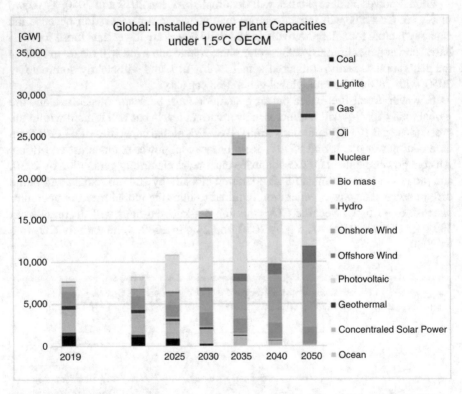

Fig. 6 Global installed power plant capacities under the OECM 1.5 °C pathway in 2019–2050

OECM 1.5 °C Pathway for Global Space and Process Heat Supply

Services and buildings usually do not require temperatures over 100 °C. Therefore, the supply technologies are different from those of the industry sector, which requires temperatures up to 1000 °C and above. The overall final heat demand will increase globally under the OECM 1.5 °C pathway, but the demand shares will

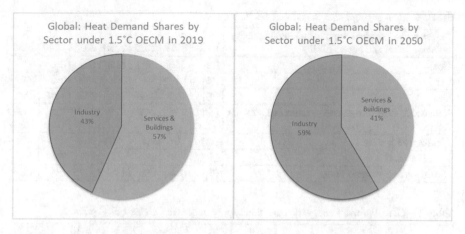

Fig. 7 Electricity demand shares by sector under the OECM 1.5 °C pathway in 2019 and 2050

change significantly. With energy efficiency measures for buildings, the overall space heating demand will decrease globally, even with increased floor space. However, the industrial process heat demand is projected to increase because energy efficiency measures will not compensate for the increasing production arising from the expected increase in global GDP to 2050. In 2019, the industry sector consumed 43% of the global heat demand and the service and buildings sector the remaining 57%. By 2050, these shares will be exchanged, and the industry sector will consume close to 60% of the global heat demand (Fig. 7).

Table 26 shows the assumed trajectory for the generation of industry process heat between 2019 and 2050. In 2019, gas and coal dominated global heat production. Renewables only contributed 9%—mainly biomass, and electricity had a minor share of 1%. District heat—mainly from gas-fired heating plants—contributed the remaining 7% of the process heat supply, whereas hydrogen and synthetic fuels contributed no measurable proportion. The global OECM 1.5 °C pathway phases out coal and oil for process heat generation between 2035 and 2040, and gas is phased out as the last fossil fuel by 2050. The most important process heat supply technologies are electric heat systems, such as heat pumps, direct electric resistance heating, and arc furnace ovens for process heat; the share will increase to 22% by 2030 and to 60% by 2050. Bio-energy will remain an important source of heat, accounting for 25% in 2050—2.5 times more than in 2019. Synthetic fuels and hydrogen are projected to grow to 8% of the total industry heat supply by 2050.

Global Primary Energy Demand—OECM 1.5 °C Pathway

The global primary energy demand under the OECM 1.5 °C pathway is shown in Table 27. Primary energy includes all losses and defines the total energy content of a specific energy source. In 2019, coal and oil made the largest contributions to the

Table 26 Heat supply under the OECM 1.5 °C pathway

Industry Process Heat Supply, including industry combined heat and power (CHP)	Units	2019	2025	2030	2035	2040	2050
Coal	[%]	33%	18%	11%	6%	0%	0%
Oil	[%]	14%	5%	3%	1%	0%	0%
Gas	[%]	36%	38%	25%	22%	17%	0%
Renewable Heat (bio-energy, geothermal, & solar thermal)	[%]	9%	24%	32%	27%	21%	25%
Electricity for Heat	[%]	1%	8%	22%	36%	49%	60%
Heat (District)	[%]	7%	6%	6%	7%	7%	7%
Hydrogen & Synthetic Fuels	[%]	0%	0%	1%	2%	6%	8%

Table 27 Global primary energy demand and supply under the OECM 1.5 °C pathway

		2019	2025	2030	2035	2040	2050
Total (including non-energy-use)	[PJ/a]	564,549	536,105	513,324	487,632	470,211	461,442
Fossil (excluding non-energy use)	*[PJ/a]*	*418,757*	*330,140*	*235,409*	*136,281*	*72,225*	*0*
Hard coal	[PJ/a]	138,615	80,288	33,904	13,228	3000	0
Lignite	[PJ/a]	20,955	5724	3276	3062	695	0
Natural gas	[PJ/a]	121,586	117,698	103,982	83,904	55,084	0
Crude oil	[PJ/a]	137,600	126,431	94,247	36,087	13,446	0
Nuclear	[PJ/a]	30,156	24,148	17,194	10,303	3082	0
Renewables	*[PJ/a]*	*76,332*	*144,057*	*221,713*	*302,449*	*355,774*	*420,974*
Hydro	[PJ/a]	15,534	15,601	17,614	19,576	20,963	23,029
Wind	[PJ/a]	4694	14,626	26,724	44,372	57,899	80,601
Solar	[PJ/a]	3433	30,123	68,563	134,363	164,964	190,239
Biomass	[PJ/a]	52,300	79,302	100,710	90,488	92,198	94,061
Geothermal	[PJ/a]	366	4113	7495	11,911	17,255	28,438
Ocean energy	[PJ/a]	4	293	607	1740	2496	4606
Total Renewable Energy Share, including electricity & synfuel imports	[PJ/a]	76,329	144,057	221,713	302,449	355,774	420,974
Renewable Energy Share	[%]	15%	30%	49%	69%	83%	100%

global energy supply, followed by natural gas, whereas renewable energies contributed only 15%. The table also provides the projected trajectories for supplies for non-energy uses, e.g., oil for the petrochemical industry. The OECM does not phase-out fossil fuels for non-energy use, because their direct replacement with biomass is not always possible. A detailed analysis of the feedstock supply for non-energy uses was beyond the scope of this research.

Global CO₂ Budget

The remaining carbon budget for each of the following sectors has been defined based on the bottom-up demand analysis of the 12 main industry and service sectors, as documented in Chaps. 5, 6, 7, and 8. Each of those industry and service sectors must complete the transition to fully decarbonized operation within the carbon budget provided. It is very important that the carbon budget shows the cumulative emissions up to 2050, and not the annual emissions. A rapid reduction in annual emissions is therefore vital.

The shares of the cumulative carbon budget required to achieve the 1.5 °C net-zero target are shown in Fig. 8. The total energy-related CO_2 for the aluminium industry between 2020 and 2050 is calculated to be 6.1 Gt, 1.5% of the total budget. For the steel industry, the remaining budget is 19.1 Gt of CO_2 (4.8%), whereas the chemicals industry has the highest carbon budget of 24.8 $GtCO_2$ or 6.2% of the total carbon budget. All other remaining industries can emit 27.1 $GtCO_2$ (6.8%), and all other energy-related activities, such as for buildings, transport, and residential uses, have a combined remaining emissions allowance of 323 $GtCO_2$, or 80.7% of the budget.

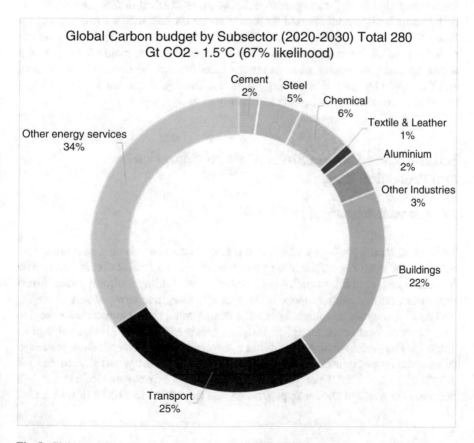

Fig. 8 Global carbon budget by sub-sector under 1.5 °C OECM pathway in 2020–2050

Scopes 1, 2, and 3—Global Summary

A global assessment of *Scopes 1*, *2*, and *3* for the whole *Industry* sector is a new research area, and changes had to be made to the method of determining those emissions, which was originally developed by the World Resource Institute (WRI), as documented in Chap. 4.

The OECM methodology differs from the original concept primarily insofar as the interactions between industries and/or other services are kept separate. A primary class is defined for the primary energy industry, a secondary class for the supply utilities, and an end-use class for all the economic activities that consume energy from the primary- or secondary-class companies, to avoid double counting. All the emissions by defined industry categories (e.g., those defined by GICS) are also separated, streamlining the accounting and reporting systems. The volume of data required is reduced and reporting is considerably simplified with the OECM methodology.

Figure 9 shows the global energy-related *Scope 1*, *2*, and *3* CO_2 emissions in 2030 as a Sanky flow chart. The primary energy emissions are on the left and the end-use-related emissions are on the right. The carbon budgets remain constant, from production to end-use, apart from losses and statistical differences. A simplified description is that all *Scope 1* emissions are on the left, with the primary energy industry as the main emitter, and all *Scope 3* emissions are on the right, with the consumers of all forms of energy and for all purposes as the main emitters. In the secondary energy industry, utilities are the link between the demand of end-users and the supply by the primary energy industry. The figure also shows the complex interconnections between demand and supply.

Nature-Based Carbon Sinks: Carbon Conservation and Protection Zones

Ecosystem Restoration Pathways

The OECM model presents a 1.5 °C-compatible scenario combining ecosystem restoration with deep decarbonization pathways, called the RESTORE scenario. The five ecosystem restoration pathways involve forests and agricultural lands: forest restoration, reforestation, reduced harvest, agroforestry, and silvopasture.

The median gross cumulative potential of additional CO_2 removal under the five ecosystem restoration pathways is 100 Gt of carbon (C) until 2100, as shown in Table 28. The peak annual sequestration rate for all ecosystem restoration pathways (forest restoration, reforestation, reduced harvest, agroforestry, and silvopasture) is 2.5 Gt of carbon (GtC) per year, although this rate is only maintained for 1–2 decades. The average annual sequestration rate from 2020 to 2100 will be 1.2 GtC

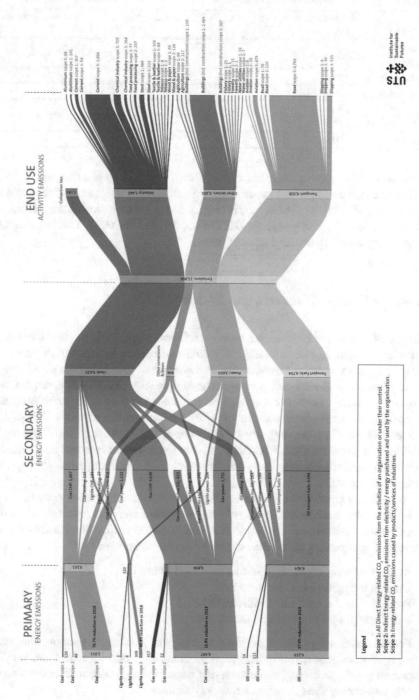

Fig. 9 Global Scope 1, 2, and 3 energy-related CO_2 emissions in 2030 under the OECM 1.5 °C pathway

Table 28 Summary statistics for the cumulative uptake of CO_2 in all pathways

Pathway	Cumulative uptake ($GtCO_2$ in 2020–2100) (global average)	Land area (million ha)
Forest restoration	19	449
Reforestation	35.57	211
Reduced harvest	27.25	685
Agroforestry	9.68	276
Silvopasture	17.42	307
All pathways	99.88	

per year. This is approximately 9% higher than the carbon uptake that would occur if the same land management pathways were modelled in a dynamic global vegetation model (DGVM). The difference is largely due to the inclusion of the soil carbon response to land-use changes in the DGVM (Littleton et al. 2021). This removal will be offset by on-going net land-use emissions.

Creating Carbon Conservation Zones (CCZ)

The role of nature and ecosystem services as climate solutions is gaining increasing attention. As well as their climate mitigation and carbon sequestration potential, ecosystem approaches have co-benefits that contribute to sustainable development goals in terms of livelihoods, productivity, biodiversity conservation, health, and ecosystem services. However, it is important to note that even with ambitious land-use restoration, carbon removal can still only compensate for, at most, 15% of current emissions. The vast majority of emissive activities must cease if we are to achieve an approximately 1.5 °C target, and all the available removal strategies are required to achieve net-negative emissions pathways and reduce the atmospheric concentrations of CO_2.

Feasible approaches to CDR using land-based mitigation options must be predicated on a 'responsible development' framework that includes broader social and environmental objectives. Carbon conservation zones, which implement different ecosystem approaches, must address these broader objectives:

- Respecting indigenous rights and knowledge of land
- Understanding financial implications
- Protecting and conserving biodiversity
- Influencing supply chains and investment portfolios

Forests and forest products are important parts of a number of supply chains for food, consumer goods, transport, etc., and companies and investors can play an important role in protecting and conserving nature through corporate commitments and by influencing their downstream supply chains.

Conclusion—High-Level Summary

To comply with the Paris Climate Agreement and limit the global mean temperature rise to +1.5 °C, rapid decarbonization of the energy sector with currently available technologies is necessary, and is possible.

However, to achieve the transformation to a fully renewable energy supply, all available efficiency potentials must be combined to reduce the total demand. To reach Net Zero by 2050, the complete phase-out of fossil fuels for all combustion processes is essential.

For the Industry sector, the transition from fossil-fuel-based process heat to renewable energy or electrical systems is the single most important measure. The further reduction of non-energy-related process emissions—mainly from cement and steel manufacture—by altering or optimizing manufacturing processes is also essential. The remaining process emissions might be compensated by natural carbon sinks, so the Industry sector must actively support the Service sector in terms of soil regeneration and reforestation measures.

For the Service sector, especially agriculture and forestry, reducing GHG emissions must clearly involve reducing the greenhouse gas (GHG) emissions arising from land-use changes. Increasing yield efficiency to avoid the further expansion of agricultural land at the expense of forests and other important ecosystems is key. However, feeding the growing world population without increasing the area committed to agriculture will require more than just an increase in technical efficiency. Moreover, there seems to be no alternative to reducing the consumption of meat and dairy products.

The Forestry sector is the single most important sector for the implementation of nature-based carbon sinks. Deforestation must cease immediately. Reforestation with native trees and plants that are typical of specific regions and climate zones must replace the forest areas that have been lost since 1990.

To reduce the demand of the Transport sector, a shift from resource-intensive air and road transport to more-efficient and electrified means of transport is required, together with an overall reduction in transport activity, especially in high-income countries. Phasing-out the production of combustion engines for passenger cars by 2030 and introducing synthetic fuels for long-distance freight transport are essential elements for the future transportation sector. Even with this ambitious goal, the full decarbonization of the road transport sector will not be achieved before 2050, because cars are used, on average, for 15–20 years. There is also significant potential for efficiency gains in shipping and aviation. However, due to the foreseeable further growth in traffic volume and the lack of alternatives, the large-scale use of synthetic fuels from renewable electricity will also be necessary for these modes of transport. Since not all regions will be able to produce this with domestic resources at reasonable costs, a global trade of these new energy sources must be established.

The decarbonization of the Buildings sector will require a significant reduction in the energy demand for climatization—heating and cooling—per square metre. A key result of our research is that the global energy demand for buildings can be halved with currently available technologies. The utilization of this efficiency potential will require high renovation rates and changed building codes for new constructions. The widespread use of heat pumps and heat grids are important elements on the supply side. In some areas, however, the supply of renewable gases can substitute today's natural gas consumption with a long-term perspective, especially where there is an industrial gas demand. The conversion of today's gas networks and the local/regional availability of resources for the production of green gases play a decisive role here.

Significant electrification across all sectors before 2030—especially for heating, process heat, and to replace combustion engines in the Transport sector—is the decisive and most urgent step. Increased electrification will require sector coupling, demand-side management, and multiple forms of storage (for heat and power), including hydrogen and synthetic fuels. Accelerating the implementation of renewable heat technologies is equally important because half the global energy supply must derive from thermal processes by 2050.

The transition of the global energy sector will only be possible with significant policy changes and reforms in the energy market.

The complete restructuring of the Energy and Utilities sectors is required. The primary Energy sector—the oil, gas, and coal industry—must wind-down all fossil -fuel extraction and mining projects and move towards utility-scale renewable energy projects, such as offshore wind and the production of hydrogen and synthetic fuels.

Power utilities will play a key role in providing the rapidly increasing electricity demand, generated from renewable power. The nexus of the global energy transition will be the power grid. Replacing oil and gas with electricity means that power grids must transport most energy, instead of oil and gas pipelines.

Therefore, the expansion of power grid capacities is one of the most important and also most overlooked measures required. In addition, converting existing gas pipelines and using them for the long-range transport of hydrogen and synthetic methane can significantly reduce the infrastructural demands on the power system and increase efficiency.

According to the scenario, global transmission and distribution grids must transport at least three times more electricity by 2050 than in 2020. The upgrades and expansion of power grids must start immediately because infrastructure projects, such as new power lines, can take 10 years or more to implement. Conversions of existing gas pipelines will be possible first where industrial users need large quantities of hydrogen for decarbonized processes.

Limiting the global mean temperature rise to +1.5 °C cannot be achieved by the decarbonization of the energy sector alone. As stated earlier, it will also require significant changes in land use, including the rapid phase-out of deforestation and significant reforestation. These measures are not alternative options to the decarbonization of the energy sector but must be implemented in parallel. If governments fail to act and mitigation is delayed, we face the serious risk of exceeding the carbon budget. In the One Earth Climate Model (OECM) 1.5 °C pathway, the land-use sequestration pathways complement very ambitious energy-mitigation pathways, and should therefore be regarded as necessary to reduce the CO2 concentrations that have arisen from the overly high emissions in the past, and not as compensatory measures that can be extended indefinitely into the future.

Policy Recommendations

The OECM is an integrated assessment tool for the development of science-based targets for all major global industries in a granularity. It includes the key performance indicators (KPIs) required to make informed investment decisions that will credibly align with the global net-zero objective in the short, medium, and long terms. The key finding of our work on the OECM 1.5 °C cross-sectorial pathway is that it is still possible to remain within the 1.5 °C limit if governments, industries, and the financial sector act immediately. The technology required to decarbonize the energy supply with renewable energy is available, market ready, and in most cases, already cost competitive. The energy efficiency measures needed to reduce the energy demand have also been understood for years and can be introduced without delay. Finally, the finance industry—for instance, the Net-Zero Asset Owner Alliance—is committed to implementing carbon targets for its investment portfolios. However, policies are required to ensure that all measures are implemented in the rather short time frame required.

Implementing Short-Term Targets for 2025 and 2030

To implement the documented short-term targets for 2025 and 2030, the following actions are required:

Government Policies

1. Immediate cessation of public and private investment in new oil, coal, and gas projects.
2. Implementation of carbon pricing with a reliable minimum CO_2 price, consistent with the underlying OECM emissions caps.
3. All OECD countries must phase-out coal by 2030.
4. The automobile industry must phase-out internal combustion engines for passenger cars by 2030.
5. Legally binding efficiency standards must be instituted for all electrical applications, vehicles, and buildings.
6. Renewable energy targets must be based on IPCC-carbon-budget-based 1.5 °C scenarios or detailed country-specific master plans.
7. Mandatory transparent forward-looking and historic disclosure of the most relevant KPIs: energy intensity, share of renewable energy supply, energy demand, carbon emissions, and carbon intensities per production unit.

8. A global phase-out of all fossil-fuel subsidies by 2025.
9. Pursuing a nationally and internationally to globally integrated and coordinated policy with the aim of creating investment security and incentives for the necessary transformation processes.
10. Conducting a comprehensive scientific analysis of feasible national pathways and formulate corresponding NDCs for 2025/2030 and beyond.
11. Establishing global governance of the transformation of energy systems, including monitoring of the necessary political, social, economic, environmental, and legal requirements.

Actions Needed by Industry and Financial Institutions

Industry

1. Setting and implementing a climate strategy consistent with 1.5 °C no- or low-overshoot sector models.
2. Immediate cessation of investments in new oil, coal, and gas projects.
3. Utilities must rapidly up-scale renewable electricity to provide logistical support for reducing *Scope 2* emissions for all industries and services. This is a huge market opportunity for utilities.
4. Development of efficient technologies to implement electric mobility.
5. Mandatory transparent forward-looking and historic disclosure of the most relevant KPIs, such as carbon emissions, energy demand, and carbon intensities per production unit.

Financial Institutions

1. Setting and implementing decarbonization targets for investment, lending, and underwriting portfolios that are consistent with the 1.5 °C no- or low-overshoot sector models
2. Cessation of investment in new oil, coal, and gas projects
3. Ensured coal phase-out in OECD countries by 2030, and in all regions between 2030 and 2045
4. Scaled climate solution investments, especially in emerging economies
5. Disclosure of:

   - climate mitigation strategies
   - short- and mid-term target setting
   - target achievements
   - progress of climate solution investments
   - engagement outcomes

Contents

List of Figures

List of Tables

Part I
Introduction

Chapter 1
Introduction

Sven Teske

Abstract This is a brief introduction to the status of the international climate negotiations of the United Nations Framework Convention on Climate Change (UNFCCC) and its latest scientific publications, the status of global greenhouse gas emissions, and the impact of the pandemic on energy-related CO_2 emissions. The research focus of this book is presented, and how the second part of the book relates to our first book *Achieving the Paris Climate Agreement Goals* is explained.

The background to the creation of the book is given. The parameters upon which the authors focused when documenting the assumptions used for all calculations are explained. The results and their derivation are presented.

Keywords United Nations Framework Convention on Climate Change (UNFCCC) · Net-zero targets · Achieving the Paris Climate Agreements

The climate and energy debate continues to be high on the political agenda at intergovernmental summits. However, since the publication of our first edition *Achieving the Paris Climate Agreement Goals* in February 2019, the situation has changed dramatically. The COVID-19 pandemic dominates almost every conversation, both in the private sphere and in international political discussions. For the first time since the beginning of the United Nations Framework Convention on Climate Change (UNFCCC) with Climate Conference COP1 in 1995 in Berlin, Germany, a conference was cancelled. COP26 was meant to be in November 2020 but had to be pushed back by 12 months in response to the pandemic.

S. Teske (✉)
University of Technology Sydney – Institute for Sustainable Futures (UTS-ISF),
Sydney, NSW, Australia
e-mail: sven.teske@uts.edu.au

S. Teske (ed.), *Achieving the Paris Climate Agreement Goals*,
https://doi.org/10.1007/978-3-030-99177-7_1

As a consequence of travel restrictions and lockdowns in almost all countries worldwide, the oil demand decreased by nine billion barrels per day compared with 2020 (BP, 2021, p. 23). Industry production dropped because workers could not come to work, restaurants had to close, and public life almost came to a halt in many countries. Global energy-related CO_2 emissions declined by 5.8% in 2020, equal to about 2 Gt (IEA, 2021)—an unprecedented event. Even the global financial crisis of 2009 did not have such a profound impact on global emissions. At the time of writing this book—December 2021—the pandemic persists. However, global CO_2 emissions have bounced back and increased by 4.8% in 2021—to almost the same level as before the pandemic.

Extreme weather events (extreme rainfall and floods, cyclones, and bushfires) have increased in frequency. Australia experienced the worst bushfire season on record between September 2019 and March 2020—known as the *Black Summer* (Cook et al., 2021). In June 2020, the Arctic region of Siberia experienced a heat wave with temperatures up to 38 °C and wildfires covering almost one million hectares. The World Meteorological Organization (WMO) recognised this as a new Arctic temperature record (WMO, 2021).

Time is running out. In August 2021, the Sixth Assessment Report (AR6) of the United Nations Intergovernmental Panel on Climate Change (IPCC) was published. The First Assessment Report was launched in 1990 and underlined the importance of climate change as a challenge with global consequences that required international co-operation (IPCC, 2021). Thirty years later, the IPCC states unequivocally that the world is already in the middle of climate change. UN Secretary-General António Guterres said the Working Group's report was nothing less than 'a code red for humanity. The alarm bells are deafening, and the evidence is irrefutable' (UN, 2021).

On the positive side, the international finance industry is increasingly engaged in the international and national climate debate. Initiatives such as the UN-convened Net-Zero Asset Owner Alliance (NZAOA, 2021) and the Glasgow Financial Alliance for Net Zero (GFANZ, 2021) represent leading financial institutions committed to achieving the goals of the Paris Climate Agreement and transitioning their investment portfolios to achieve net-zero greenhouse gas (GHG) emissions by 2050.

Our first book laid out global and regional 100% renewable energy scenarios with non-energy GHG pathways for +1.5 °C or +2 °C warming scenarios and compared them with a reference case. Those scenarios were calculated under the leadership of the Institute for Sustainable Futures (ISF) at the University of Technology Sydney (UTS) in close co-operation with the German Aerospace Center (DLR) and the University of Melbourne, Australia. The model used for that project became known as the *OneEarth Climate Model* (OECM) in 2020 during the numerous debates that followed the book launch in February 2019.

The second book focuses on sectorial pathways and provides key performance indicators (KPIs) for industry sectors to limit the global temperature increase to 1.5 °C.

The OECM is an integrated energy assessment model to be used for developing science-based net-zero targets for all major industries in a granularity and with the

KPIs needed to make short-, mid-, and long-term investment decisions. The 1.5 °C emission pathways developed by UTS are no or low overshoot scenarios (SSP 1), as defined by the IPCC. This means that a carbon budget overshoot is avoided and that the CO_2 already released is not assumed to be 'removed' by unproven technologies still under development, such as carbon capture and storage. The OECM does take 'technical' negative emissions into account, but only natural carbon sinks, such as forests, mangroves and seaweed, which will compensate for the process emissions that are currently unavoidable, such as those from cement production.

A number of climate modelling organisations, including the Energy Transitions Commission, the Potsdam Institute for Climate Impact Research, the Science-Based Targets Initiative, the Carbon Risk Real Estate Monitor (CRREM), and the World Wide Fund for Nature (WWF), were invited to peer-review the OECM-derived net-zero pathways between mid-2020 and mid-2021.

The book documents all the steps in the scenario development and provides a detailed analysis of the main assumptions and scenario narratives. The results of the OECM 1.5 °C pathways for 12 industry and service sectors include the total remaining carbon budget and *Scope 1, 2,* and *3* emissions for each sector.

References

BP. (2021). *Statistical review of world energy 2021.* https://www.bp.com/content/dam/bp/business-sites/en/global/corporate/pdfs/energy-economics/statistical-review/bp-stats-review-2021-full-report.pdf

Cook, G., Dowdy, A., Knauer, J., et al. (2021). Australia's Black Summer of fire was not normal—and we can prove it. In: *CSIROscope.* https://blog.csiro.au/bushfires-linked-climate-change/. Accessed on 20th Dec 2021.

GFANZ. (2021). Bringing together the financial sector to accelerate the transition to a net-zero economy. In: *Glasgow Finance Alliance Net Zero.* https://www.gfanzero.com/membership/. Accessed 20 Dec 2021.

IEA. (2021). *Global energy review 2021. Assessing the effects of economic recoveries on global energy demand and CO_2 emissions in 2021.* Paris 2021. https://www.iea.org/reports/global-energy-review-2021

IPCC. (2021). The reports. In: *History of the IPCC.* https://www.ipcc.ch/about/history/. Accessed on 13th Dec 2021.

NZAOA. (2021). *UN-convened Net-Zero Asset Owner Alliance—Institutional investors transitioning their portfolios to net zero GHG emissions by 2050.* https://www.unepfi.org/net-zero-alliance/. Accessed on 3rd Oct 2021.

UN. (2021). *IPCC Report: 'Code red' for human driven global heating', warns UN chief.* UN News, Clim. Environ.

WMO. (2021). *WMO recognizes new Arctic temperature record of 38 °C.* Press Release. https://public.wmo.int/en/media/press-release/wmo-recognizes-new-arctic-temperature-record-of-38%E2%81%B0c

Part II
State of Research

Chapter 2
Science-Based Industry Greenhouse Gas (GHG) Targets: Defining the Challenge

Sven Teske and Thomas Pregger

Abstract Background information is given on the Paris Climate Agreement and the role of nationally determined contributions and net-zero pledges. An overview of historical energy-related CO_2 emissions since 1750 and how they relate to economic development, measured in gross domestic product (GDP), is provided, together with the cumulative energy-related CO_2 emissions by region. The future energy demand if historical trends in energy efficiency and carbon intensity continue until 2050 is projected. The term 'science-based target setting' is defined, and how it relates to the carbon budget published in the Sixth Assessment Report of the IPCC is discussed. The energy-related CO_2 emission pathway required to achieve the 1.5 °C target is outlined.

Keywords Science-based GHG targets · GHG development · GDP · Population · Nationally determined contributions (NDCs)

S. Teske (✉)
University of Technology Sydney – Institute for Sustainable Futures (UTS-ISF), Sydney, NSW, Australia
e-mail: sven.teske@uts.edu.au

T. Pregger
German Aerospace Center (DLR), Institute for of Networked Energy Systems (VE), Department of Energy Systems Analysis, Stuttgart, Germany
e-mail: thomas.pregger@dlr.de

Paris Agreement—Article 2

1. *This Agreement, in enhancing the implementation of the Convention, including its objective, aims to strengthen the global response to the threat of climate change, in the context of sustainable development and efforts to eradicate poverty, including by:*

   (a) *Holding the increase in the global average temperature to well below 2 °C above pre-industrial levels and pursuing efforts to limit the temperature increase to 1.5 °C above pre-industrial levels, recognising that this would significantly reduce the risks and impacts of climate change;*

   (b) *Increasing the ability to adapt to the adverse impacts of climate change and foster climate resilience and low greenhouse gas emissions development, in a manner that does not threaten food production; and*

   (c) *Making finance flows consistent with a pathway towards low greenhouse gas emissions and climate-resilient development.*

2. *This Agreement will be implemented to reflect equity and the principle of common but differentiated responsibilities and respective capabilities, in the light of different national circumstances.*

To understand the challenges involved in implementing the Paris Climate Agreement, it is helpful to look at historic trends in the world's population, its economic growth, its increasing energy demand, and—as a result of the energy sources chosen—the trajectory of energy-related CO_2 emissions. CO_2 concentrations are increasing in the global atmosphere, causing global warming (IPCC, 2021). The Paris Climate Agreement Goal is to limit this temperature increase to 1.5 °C above pre-industrial levels. In 2021, a new scientific report defined the remaining global carbon budget.

2.1 The Sixth Assessment Report of the IPCC: Climate Change Is Here

The Intergovernmental Panel on Climate Change (IPCC) is the United Nations (UN) body that assesses the science related to climate change. In August 2021, it launched the Working Group I contribution to its Sixth Assessment Report (AR6) *Climate Change 2021: The Physical Science Basis*. The IPCC concluded that the emission of GHGs from human activities is responsible for approximately 1.1 °C of warming that has occurred since 1850–1900. Based on the improved observational datasets that are used to assess historical warming and the progress in scientific understanding of the climate system's response to anthropogenic GHG emissions, the IPCC expects that the increase in the global temperature will reach or exceed 1.5 °C (IPCC, 2021) (Table 2.1).

Table 2.1 Assumed population and GDP developments by region in 2020–2050

		Units	2019	2025	2030	2040	2050
OECD North America	Population	[Million]	499	524	543	575	599
	GDP	[Billion $]	24,255	27,650	30,513	37,562	45,788
Latin America	Population	[Million]	526	552	571	599	616
	GDP	[Billion $]	7415	8807	10,141	13,761	18,675
OECD Europe	Population	[Million]	579	587	592	598	598
	GDP	[Billion $]	23,433	26,076	28,269	32,807	36,963
Africa	Population	[Million]	1321	1522	1704	2100	2528
	GDP	[Billion $]	6865	9247	11,376	17,498	26,403
Middle East	Population	[Million]	250	276	295	331	363
	GDP	[Billion $]	6120	7230	8857	12,112	17,587
Eurasia	Population	[Million]	346	347	346	343	339
	GDP	[Billion $]	6685	7919	9081	11,853	15,025
Non-OECD Asia	Population	[Million]	1189	1269	1329	1428	1499
	GDP	[Billion $]	11,101	14,577	17,794	25,876	34,234
India	Population	[Million]	1368	1452	1513	1605	1659
	GDP	[Billion $]	10,816	17,084	22,652	37,966	54,074
China	Population	[Million]	1427	1447	1450	1426	1374
	GDP	[Billion $]	26,889	37,997	47,427	64,986	84,825
OECD Pacific	Population	[Million]	208	208	208	204	198
	GDP	[Billion $]	8761	9644	10,407	11,842	13,081
Global	Population	[Million]	7713	8185	8551	9210	9772
	GDP	[Billion $]	132,339	166,230	196,516	266,263	346,656

The IPCC also identified the global carbon budget required to avoid exceeding 1.5 °C. Between 2020 and 2050, the global cumulative CO_2 emissions must not surpass 400 $GtCO_2$ if we are achieving this target with 67% likelihood. This likelihood decreases to 50% if total emissions reach 500 $GtCO_2$ (Table 2.2) between 2020 and 2050 (IPCC, 2021).

The IPCC media statement was unusually clear and unambiguous for a high-level scientific organization:

> Stabilising the climate will require strong, rapid, and sustained reductions in greenhouse gas emissions, and reaching net zero CO_2 emissions. Limiting other greenhouse gases and air pollutants, especially methane, could have benefits both for health and the climate...
> *IPCC media release, Geneva 9 August 2021*

2.2 The OneEarth Climate Model: The Context

The UN-convened *Net-Zero Asset Owner Alliance* is a *Program for Responsible Investment* and a United Nations Environment Programme Finance Initiative (UNEP FI)-supported initiative. The members of the Alliance have committed to transitioning their investment portfolios to net-zero GHG emissions by 2050, consistent with

Table 2.2 Estimates of remaining carbon budgets and their uncertainties—IPCC AR6, WG1, Technical Summary

Global surface temperature change: 2010–2019	Global surface temperature change: 1850–1900[a(1)]	Estimated remaining carbon budgets—starting from 1 January 2020 and subject to variations and uncertainties quantified in the columns on the right				
°C	°C	Percentiles of TCRE[b(2)]				
		GtCO$_2$				
		17th	33rd	50th	67th	83rd
0.43	1.5	900	650	500	400	300
0.53	1.6	1200	850	650	550	400
0.63	1.7	1450	1050	850	700	550
0.73	1.8	1750	1250	1000	850	650
0.83	1.9	2000	1450	1200	1000	800
0.93	2	2300	1700	1350	1150	900

Source: IPCC AR6, WG1, Technical Summary, Table TS.3, page 150

[a](1) Human-induced global surface temperature increases in 1850–1900 and 2010–2019 are assessed to be 0.8–1.3 °C, with a best estimate of 1.07 °C. Combined with a central estimate of the transient climate response to cumulative carbon emissions (TCRE) of 1.65 °C EgC-1, this uncertainty in isolation results in a potential variation in the remaining carbon budgets of ±550 GtCO$_2$. However, this is not independent of the assessed uncertainty of TCRE and is thus not fully additional

[b](2) TCRE: transient climate response to cumulative emissions of CO$_2$, assessed to probably be 1.0–2.3 °C EgC-1, with a normal distribution, from which the percentiles are taken

a maximum global temperature rise of 1.5 °C above pre-industrial levels. This requires intermediate targets to be established for 5-year intervals and regular reporting on progress.

The Alliance commissioned the Institute for Sustainable Futures (ISF) at the University of Technology Sydney (UTS) to utilize its pre-existing OneEarth Climate Model (Teske et al., 2019) to derive 1.5 °C decarbonization pathways for key high-emitting sectors, on a global level, to achieve net-zero emissions by 2050, and to inform the development of sector-based targets for decarbonization. This book presents the results of that research, undertaken between late 2019 and December 2021. We hope it will clarify investor expectations for decarbonization strategies for the sectors in which they invest.

2.2.1 Development of GHG Emissions: A Look Back

The global economy must decarbonize the energy system entirely within the next 30 years—in one generation. In historical terms, this means breaking the connection between population growth, steady economic development fuelled by fossil energy, and the increase in CO$_2$ emissions of the past 120 years and reversing those trends within the next 5 years. Between 2025 and 2030, global energy-related CO$_2$ must peak and start to decline to zero by 2050.

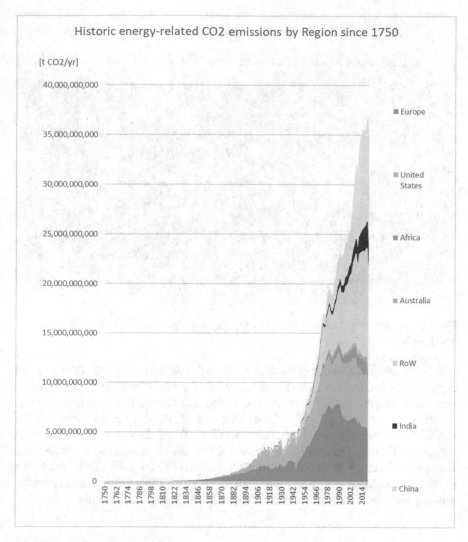

Fig. 2.1 Development of global energy-related CO_2 emissions by region in 1750–2020 (Ritchie & Roser, 2020)

Figure 2.1 shows the development of the annual energy-related CO_2 emissions between 1750 and 2020 based on data from the *Global Carbon Project* of the *Integrated Carbon Observation System* (Global Carbon Project, 2021). Global annual CO_2 emissions rose from 2 Gt in 1900 to 4 Gt 1935, to 6 Gt in 1950, and to 12 $GtCO_2$ in 1966. By 1996, global emissions had reached 24 $GtCO_2$—and only 10 years later, emissions increased by another 10 Gt. Since 2012, the increase in emissions has at least slowed, and in 2020, emissions were around 35 $GtCO_2$.

A closer look into regional emissions shows that Europe was responsible for 43% of all historic CO_2 emissions between 1750 and 1990, followed by the USA

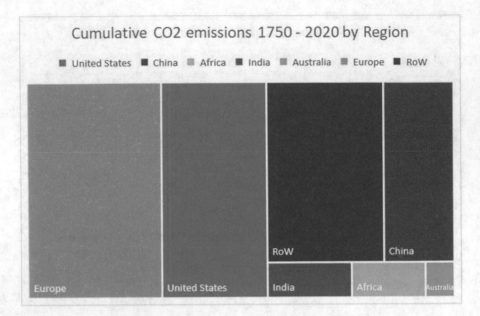

Fig. 2.2 Cumulative global energy-related CO_2 emissions by region in 1750–2020 (Ritchie & Roser, 2020)

with 31%, whereas China emitted 5%, Africa 2%, and India only 1%. However, the regional distribution changed dramatically after 1990, with China's double-digit economic growth over the past decades: China and Europe both contributed 21%, followed by the USA with 19%, India 5%, and Africa 4%. Figure 2.2 shows the cumulative CO_2 emissions by region between 1750 and 2020, based on data from Integrated Carbon Observation Systems (ICOS 2021). According to these data, Europe emitted 31% of all cumulative CO_2, followed by the USA (25%) and China (14%). The remaining 30% was distributed across all other regions and countries outside those three main economic hubs (the USA, Europe, and China).

2.2.2 Global Economic Development: A Look Back

On average, global economic development has steadily increased. Based on the World Bank data (World Bank, 2020), the global median GDP growth between 1970 and 2015 was 3.5%, although with significant regional differences. In 1966, the total global output of the world economy increased by over US$20 trillion and then doubled within 20 years to $40 trillion by 1986. Thirty years later—in 2006—this value surpassed $80 trillion (Fig. 2.3). In 2020, the global GDP reached $132 trillion. For this analysis, we follow the World Bank projection—which was also used for the World Energy Outlook 2017 and was the basis for the first OECM book published in 2019 (Teske et al., 2019) (Table 2.2).

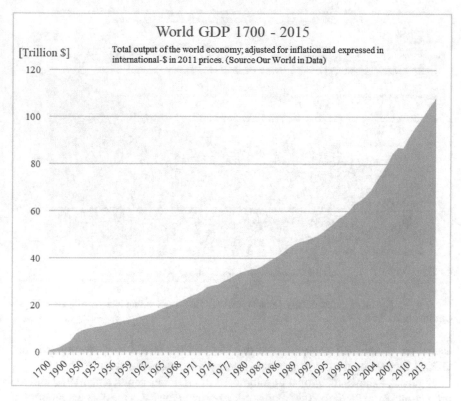

Fig. 2.3 Global GDP development in 1700–2015 (Ritchie & Roser, 2020)

2.2.3   Socio-economic Assumptions for the OECM 1.5 °C Scenario

The assumed development of regional populations is based on the projections of the United Nations Department of Economic and Social Affairs (UN DESA, 2019), whereas the regional GDP developments are based on World Bank projections. The global values for population and GDP are identical throughout the entire analysis, across all sectors. Regional values are used for the buildings and transport sectors, whereas for all other sectors, the resulting (summed) global values are used.

2.2.4   Outlining the Task: Trend Reversals Until 2025

The first step in the development of sectorial 1.5 °C pathways is to decide on the basic drivers of the future energy demand: population growth and economic development. To ensure that the OECM is transparent and comparable with other

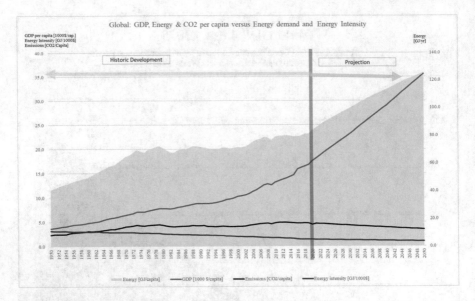

Fig. 2.4 Projection of global energy demand under the assumption that historic efficiency trends continue until 2050

scenarios, established projections of the UN and the World Bank were used. The OECM focuses on the development of energy-relevant parameters.

Figure 2.4 shows the global development of GDP per capita since 1950 and the projections from 2020 to 2050. Economic energy intensity is the average amount of energy units required for each dollar of economic value. In 1950, the energy intensity was around 11 GJ per US$1000 GDP, on a global average. This value includes electricity and fuel demands, e.g. for heating and transport. Energy intensity decreased over time, which indicated the successful implementation of efficiency measures. Different economic sectors have very different energy intensities. A highly industrialized country with large manufacturing capacities, e.g. for steel production, has a significantly higher energy intensity than a service-based economy that is focused on tourism, for example. Therefore, a low energy intensity is not necessarily a sign of a very efficient economy, but could indicate an economy that is largely based on agriculture. However, on a global average, energy intensity is an important parameter reflecting advances in efficiency.

Between 1950 and 2020, the global energy intensity decreased by 1.2% annually, leading to an energy intensity of 4.8 GJ per US$1000 GDP—about half the value in 1950. The projection of the energy demand shown in Fig. 2.4 was calculated under the assumption that the energy intensity will continue to decrease at 1% per year, while GDP continues to grow by 3.5%, on average, between 2020 and 2050.

The third relevant parameter is the average energy demand per capita, which is simply the overall primary energy demand divided by the population. The per capita energy demand doubled from 40.5 GJ per year in 1950 to around 80 GJ per year in

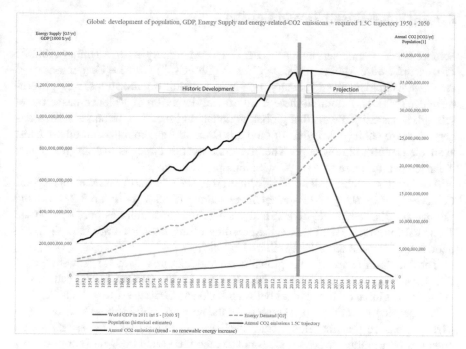

Fig. 2.5 Global development of key parameters

2020. With the assumed increase in economic energy intensity and the overall economic development, the per capita energy use should increase again by over 50% to 125 GJ per year in 2050. Finally, the emission intensity, or the average amount of energy-related CO_2 emissions per capita, results from the energy demand and the energy source selected. If coal is used to supply the entire energy demand, GHG emissions will steadily increase, whereas a supply of renewable energy will lead to a decarbonized economy.

Figure 2.5 shows the historic development of the global population, GDP, energy demand, and the resulting annual CO_2 emissions between 1950 and 2020 on the left side and the projected trend development until 2050 on the right side. Based on the projected population and economic growth until 2050 and under an assumed annual decline of 1% in both energy and emission intensities, the global energy demand will double, whereas CO_2 emissions will remain at around current levels.

The OECM does not question the development of the population or the global economy projected by international organizations, but focuses on technical measures to increase energy efficiencies and decarbonize the energy supply by a transition to renewable energies to achieve the 1.5 °C decarbonization trajectory (marked with the red line). This will require a bottom-up assessment of the energy demand combined with an alternative energy supply concept for power, heating, and transport, which are documented in the following chapters of this book.

2.3   Science-Based Target Setting

Science-based target setting has been discussed widely at the *United Nations Framework Convention on Climate Change* (UNFCCC) Climate Conferences and among stakeholders from industry, non-governmental organizations, and government departments. Although there is no official definition of 'science-based target setting', it basically means that global, regional, and sectorial carbon emission targets are set to achieve the goals of the Paris Climate Agreement based on the latest available scientific knowledge. Therefore, the overall target is to limit the global mean temperature rise to +1.5 °C with high probability.

The latest available scientific information is IPCC's Sixth Assessment Report *Climate Change 2021: The Physical Science Basis* (Sect. 2.1). Table 2.2 shows the estimates of the remaining carbon budgets and their uncertainties published in the Technical Summary (IPCC, 2021). According to the IPCC definition, 67% likelihood is 'good', whereas 50% likelihood is 'fair'.

The OECM aims to limit the global mean temperature rise to 1.5 °C with 'good' likelihood. Therefore, the 'science-based target' for the OECM 1.5 °C pathway in terms of the global carbon budget between 2020 and 2050 is set to 400 Gt CO_2.

The development of sectorial targets for the needs of specific countries or industries will ensure that the global sum of all energy-related CO_2 emissions for all countries or all industry sectors does not exceed the global budget. Therefore, any approach undertaken in isolation, such as for only a single industry sector, will involve the risk that one industry sector will claim a higher CO_2 budget and push the responsibility to reduce CO_2 emissions onto other sectors.

2.3.1   Science-Based Targets for the Finance Industry

Investment decisions for the decarbonization of investment portfolios are underpinned by highly complex considerations. In November 2020, the European Central Bank published a 'Guide on climate-related and environmental risks', which maps out a detailed process for 'climate stress tests' for investment portfolios. For the global finance industry to implement the Paris Climate Agreement, decarbonization targets and benchmarks for industry sectors are required.

The estimation of carbon budgets for specific industry sectors requires a holistic approach, and the interconnection of all sectors and regions must be considered. To estimate the carbon budget for a single industry sector in an isolated 'silo approach' based on current emission shares will inevitably lead to inaccurate results because this approach does not consider the possible technical developments in that sector or its interactions with other industry sectors. Therefore, the total of all sub-concepts for certain industries will exceed the actual CO_2 emitted, and/or the responsibilities for CO_2 reduction will be shifted to other areas.

2.3.2   Nationally Determined Contributions (NDCs)

The Paris Climate Agreement was adopted by 196 countries and regions (e.g. the European Union) in 2015 and came into force on 4 November 2016. It is a legally binding international treaty on climate change. Each signatory country must submit *nationally determined contributions* (NDCs). An NDC is basically a plan for a country that outlines specific measures that will be implemented to reduce GHG emissions. This usually includes an energy scenario but can also include targets for emissions related to land-use changes, such as in forestry and agriculture.

NDCs play a central role in the Paris Climate Agreement and are defined in Article 4, paragraphs 2, 3, and 4:

Paris Climate Agreement, Article 4 (UNFCCC, 2015)

§ 2. *Each Party shall prepare, communicate and maintain successive nationally determined contributions that it intends to achieve. Parties shall pursue domestic mitigation measures, with the aim of achieving the objectives of such contributions.*

§ 3. *Each Party's successive nationally determined contribution will represent a progression beyond the Party's then current nationally determined contribution and reflect its highest possible ambition, reflecting its common but differentiated responsibilities and respective capabilities, in the light of different national circumstances.*

§ 4. *Developed country Parties should continue taking the lead by undertaking economy-wide absolute emission reduction targets. Developing country Parties should continue enhancing their mitigation efforts, and are encouraged to move over time towards economy-wide emission reduction or limitation targets in the light of different national circumstances.*

Nationally determined contributions must be submitted every 5 years. The first submission was in 2020, so the subsequent NDCs are required in 2025 and 2030. All NDCs are publicly available and collected at the 'NDC registry'. At the time of writing (December 2021), the modalities and procedures of the NDC registry were still under negotiation, and an interim NDC registry was in place.

All submitted NDCs are regularly analysed, and their targets are summarized in order to maintain an overview of the projected GHG emissions over the next 5-year period and to assess whether global emissions are on track to meet the 1.5 °C target. The estimated emissions are reported in CO_2 equivalents, in order to include all GHGs, not only energy-related CO_2 emissions.

There is a not-entirely-fixed template for NDCs, and each country structures its NDC differently. However, all NDCs are expected to cover the following five sectors:

1. Energy: Energy demand and supply scenarios and political measures for the country or region.
2. Industrial Processes and Product Use (IPPU): Projection of industry-related emissions, such as from the cement and steel industries and the planned development of feedstock for the chemical industry.

3. Agriculture: Emissions especially from land-use changes, including agriculture, forestry, and other land-use (AFOLU) emissions (see Chap. 11).
4. Land Use, Land-Use Change, and Forestry (LULUCF): This sector mainly focuses on activities that increase the removal of GHGs from the atmosphere or reduce emissions by halting the loss of carbon stocks.
5. Waste: Aims to reduce emissions, such as methane (CH_4), from water treatment plants, but also emission from food waste and other disposal substances.

NDCs should include not only CO_2 emissions but also CH_4, nitrous oxide (N_2O), and gases and aerosols that fall under the Montreal Protocol (see Chap. 11).

2.3.3 Net-Zero Pledges

To support the NDC process, the UNFCCC started the 'Race to Zero' campaign, with the aim of obtaining 'net-zero pledges' in the run-up to and during COP26 in November 2021. The target group for these pledges were industry sectors and/or industry companies, finance sectors, and/or finance institutions, but also countries, which would submit pledges in addition to NDCs. The campaign received significant positive feedback. An analysis by the International Energy Agency (IEA) in November 2021 concluded that all the pledges announced by 3 November 2021 will—under the assumption that they will be implemented by 2050—reduce annual global CO_2 emissions from around 35 Gt currently to just over 20 Gt. Although this is already a notable reduction, it will not limit the global temperature increase to 1.5 °C, but instead to around 2.0 °C (Birol, 2021).

According to the UNFCCC Race to Zero website (UNFCCC, 2021), the *net-zero pledge* consists of four steps (see Box 2.1):

Box 2.1: Race to Zero Criteria
1. Pledge: Pledge at the head-of-organization level to reach (net) zero in the 2040s or sooner, or by mid-century at the latest, in line with global efforts to limit warming to 1.5 °C.
2. Plan: In advance of COP26, explain what steps will be taken towards achieving net zero, especially in the short to medium term. Set an interim target to achieve in the next decade, which reflects a fair share of the 50% global reduction in CO_2 by 2030 identified in the IPCC Special Report on Global Warming of 1.s5 °C.
3. Proceed: Take immediate action towards achieving net zero, consistent with delivering interim targets specified.
4. Publish: Commit to report progress at least annually, including via, to the extent possible, platforms that feed into the UNFCCC Global Climate Action Portal.

The OneEarth Climate Model aims to support the development of NDCs and Net-Zero Pledges. The following chapters document the detailed bottom-up assessment of the energy demand, the energy supply concept, and the changes in land-use required to achieve the Paris Agreement goals.

References

Birol, F. (2021). *COP26 climate pledges could help limit global warming to 1.8 °C, but implementing them will be the key*; https://www.iea.org/commentaries/cop26-climate-pledges-could-help-limit-global-warming-to-1-8-c-but-implementing-them-will-be-the-key

Global Carbon Project. (2021). *Supplemental data from Global Carbon Budget 2021 (version 1.0) [Data set]*. In: Global Carbon Project. https://www.icos-cp.eu/science-and-impact/global-carbon-budget/2021. Accessed on 22nd Dec 2021.

IPCC. (2021). *Climate change 2021: The physical science basis. Contribution of working group I to the sixth assessment report of the intergovernmental panel on climate change*. Cambridge University Press.

Ritchie, H., & Roser, M. (2020). *CO₂ and greenhouse gas emissions—Our world in data*. OurWorldInData.org. https://ourworldindata.org/co2-and-other-greenhouse-gas-emissions. Accessed on 22nd Dec 2021.

Teske, S., Pregger, T., Naegler, T., et al. (2019). Energy scenario results. In S. Teske (Ed.), *Achieving the Paris climate agreement goals. Global and regional 100% renewable energy scenarios with non-energy GHG pathways for +1.5°C and +2°C*. Springer Open. https://link.springer.com/book/10.1007/978-3-030-05843-2

UN DESA. (2019). *World population prospects 2019*. United Nations Dep. Econ. Soc. Aff. https://population.un.org/wpp/Download/Standard/Population/. Accessed on 25th Oct 2021.

UNFCCC. (2015). *Paris Agreement*. UNFCCC. https://unfccc.int/process-and-meetings/the-paris-agreement/the-paris-agreement

UNFCCC. (2021). *Race to zero*. United nations climate change. https://unfccc.int/climate-action/race-to-zero-campaign. Accessed on 1st Oct 2021.

World Bank. (2020). *GDP, PPP (constant 2017 international $) | Data*. World Bank. https://data.worldbank.org/indicator/NY.GDP.MKTP.PP.KD. Accessed on 22nd Dec 2021.

Part III
Methodology

Chapter 3
Methodology

Sven Teske, Jaysson Guerrero Orbe, Jihane Assaf, Souran Chatterjee, Benedek Kiss, and Diana Ürge-Vorsatz

Abstract The OneEarth Climate Model (OECM), its background, and program architecture are described. How the OECM is broken down into two independent modules to calculate demand and supply is explored. The basic program logic of the MATLAB-based bottom-up demand module, with high technical resolution, is described for various sectors, including the input and output parameters. The description includes numerous figures and tables for both demand and supply modules. The sub-sectors used for the OECM 1.5 °C pathway are listed, including outputs and the areas of use.

The second part of the chapter documents the high-efficiency building (HEB) model of the Central European University, which was used for the global and regional bottom-up analyses of the building sector. Its methodology, including the programme architecture, the workflow, and the equations used, is provided.

Keywords Methodology · OneEarth Climate Model (OECM) · MATLAB · High-efficiency building (HEB) model

The Paris Climate Agreement (UNFCCC, 2015) 'notes that … emission reduction efforts will be required … to hold the increase in the global average temperature to below 2 °C above pre-industrial levels…'. The Intergovernmental Panel on Climate Change (IPCC) further quantified the carbon budget to achieve this target in its Sixth Assessment Report of the Working Group (IPCC, 2021). According to the IPCC, a global carbon budget of 400 $GtCO_2$ is required to limit the temperature rise to 1.5 °C, with 67% likelihood, by 2050.

S. Teske (✉) · J. G. Orbe · J. Assaf
University of Technology Sydney – Institute for Sustainable Futures (UTS-ISF),
Sydney, NSW, Australia
e-mail: sven.teske@uts.edu.au

S. Chatterjee · B. Kiss · D. Ürge-Vorsatz
Central European University, Department of Environmental Sciences and Policy,
Budapest, Hungary

To implement these targets, energy and climate mitigation pathways are required. Numerous computer models for the analysis and development of energy and emission pathways have been developed over the last few decades. Many different calculation methods have been established, which mainly differ in the principal task of the model and the level of detail in the GHG emissions and/or energy systems calculated. The various methods of climate-economy modelling use different ways to describe the economy- and climate-relevant parameters as parts of a highly interconnected process (Nikas et al., 2019). In this context, the economy includes all aspects of the energy system and the policy framework, whereas the climate module reflects various GHG emissions from energy-related and non-energy-related processes, such as land use.

A comprehensive review of energy models, focusing on the usability of those models for decision-making, found 'that a better understanding of user needs and closer co-operation between modellers and users is imperative to truly improve models and unlock their full potential to support the transition towards climate neutrality …' (Süsser et al., 2022).

3.1   The OneEarth Climate Model

The UN-convened Net-Zero Asset Owner Alliance (NZAOA) is an international group of institutional investors committed to transitioning their investment portfolios to net-zero emissions by 2050 (NZAOA, 2021). Detailed industry sector-based energy scenarios are required to implement those net-zero commitments. On the basis of the OneEarth Climate Model (OECM; Teske et al. 2019a, b), the Institute for Sustainable Futures, University of Technology Sydney (UTS/ISF), in close co-operation with institutional investors, has developed an integrated energy assessment model for industry-specific 1.5 °C pathways, with high technical resolution, for the finance sector. In this article, we describe the detailed methodology and the architecture of the energy model in the 2021 edition of the advanced OneEarth Climate Model (OECM 2.0).

3.1.1   The OneEarth Climate Model Architecture

The OneEarth Climate Model has been developed on the basis of established computer models. The energy system analysis tool consisted of three independent modules:

1. Energy system model (EM): a mathematical accounting system for the energy sector (Simon et al., 2018)
2. Transport scenario model TRAEM (transport energy model) with high technical resolution (Pagenkopf et al., 2019)

3. Power system analysis model [R]E 24/7, which simulates the electricity system on an hourly basis and at geographic resolution to assess the requirements for infrastructure, such as grid connections between different regions and electricity storage types, depending on the demand profile and power generation characteristics of the system (Teske, 2015)

The advanced OneEarth Climate Model, OECM 2.0, merges the energy system model (EM), the transport energy model (TRAEM), and the power system model [R]E 24/7 into one MATLAB-based energy system module. The Global Industry Classification Standard (GICS) was used to define sub-areas of the economy. The global finance industry must increasingly undertake mandatory climate change stress tests for GICS-classified industry sectors in order to develop energy and emission benchmarks to implement the Paris climate protection agreement. This requires very high technical resolution for the calculation and projection of future energy demands and the supply of electricity, (process) heat, and fuels that are necessary for the steel and chemical industries. An energy model with high technical resolution must be able to calculate the energy demand based on either projections of the sector-specific gross domestic product (GDP) or market forecasts of material flows, such as the demand for steel, aluminium, or cement in tonnes per year.

To decarbonize the energy supply, fossil fuels must be phased out and replaced by a renewable energy supply. However, the supply of high-temperature process heat for various production processes cannot yet be fully electrical, and a simple fuel switch from oil, gas, or coal to biomass is also impossible, given the limited availability of sustainable bioenergy (Seidenberger et al., 2010; Farjana et al., 2018). To develop a detailed sector-specific solution, the temperature level required must be considered when developing an energy scenario. An energy model with such high technical resolution can provide detailed results for various industry sectors but requires a highly complex and data-intensive model architecture. Separate modules for the calculation of different sectors of the energy system are not practicable for such high technical resolution because high electrification rates lead to increased sector coupling, and the interactions between sectors cannot be captured if the energy model uses separate modules.

Furthermore, the geographic distribution of the energy demand and supply must be accommodated to calculate the import and export of energy, especially for energy-intensive industries. Finally, the simulation of 100% renewable energy systems requires high time resolution to accommodate the high proportions of variable solar and wind energy.

The MATLAB model has an object-oriented structure and two modules—to calculate demand and supply—that can be operated independently of each other. Therefore, an energy demand analysis independent of the specific supply options or the development of a supply concept based on demand from an external source is possible.

3.1.2 The OECM Demand Module

The demand module uses a bottom-up approach to calculate the energy demand for a process (e.g. steel production) or a consumer (e.g. a household) in a region (e.g. a city, island, or country) over a period of time. One of the most important elements in this approach is the strict separation of the original need (e.g. to get from home to work), how this need can be satisfied (e.g. with a tram), and the kind of energy required to provide this service (in this case, electricity). This basic logic is the foundation for the energy demand calculations across all sectors: *buildings*, *transport*, *services*, and *industry*. Furthermore, the energy services required are defined: electricity, heat (broken down into four heat levels: <100 °C, 100–500 °C, 500–1000 °C, and > 1000 °C), and fuels for processes that cannot (yet) be electrified. Synthetic fuels, such as hydrogen, are part of both the demand module, because electricity is required to produce it, and the supply module.

The energy requirements are assigned to specific locations. This modular structure allows regions to be defined and, if necessary, the supply from other areas to be calculated.

Demand and generation modules are independent and can be used individually or sequentially. Energy demands can be calculated either as synthetic load profiles, which are then summed to annual energy demands, or as annual consumption only, without hourly resolution. Whether or not hourly resolution is selected depends to a large extent on the availability of data. Load profiles, such as those for the chemical industry, are difficult to obtain and are sometimes even confidential.

3.1.2.1 Input Parameters

As in basic energy models, the main drivers of the energy demand are the development of the population and of economic activity, measured in GDP. Figure 3.1 shows the basic methodology of the OECM demand module. Tier 1 inputs are population and GDP by region and sector. 'Population' defines the number of individual energy services, which determines the energy required per capita, and 'economic activity' (in GDP) defines the number of services and/or products manufactured and

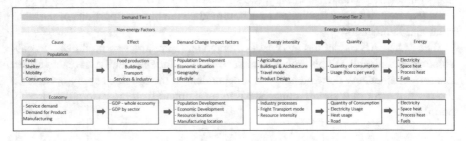

Fig. 3.1 Tier 1 and tier 2 input parameters for the assessment of energy demand

sold. Tier 1 demand parameters are determined by the effect that a specific service requires. For population, the demand parameters are defined by the need for food, shelter (buildings), and mobility and—depending on the economic situation and/or lifestyle of the population—the demand for goods and services.

Economic activity (measured in GDP) is a secondary input and is directly and indirectly dependent upon the size of the population. However, a large population does not automatically lead to high economic activity. Both population and projected GDP are inputs from external sources, such as the United Nations or the World Bank. Tier 1 input parameters themselves are strictly non-technical. The need to produce food can be satisfied without electricity or (fossil) fuels, just as a service can be provided with physical strength.

Tier 2 demand parameters are energy-relevant factors and describe technical applications, their energy intensities, and the extent to which the application is used. For example, if lighting is required, the technical application 'light bulb' is chosen to satisfy the demand.

In this example, the energy intensity is the capacity of a light bulb, e.g. 100 W. The use of the application (e.g. for 5 h per day) defines the daily demand (5 h × 100 W = 500 Wh per day). The quantity of consumption per year is 365 days at 500 Wh per day = 1825 Wh or 182.5 kWh per year. This very basic and simple principle is used for every application in each of the main sectors: *residential + buildings*, *industry*, and *transport*. These sectors are broken down into multiple sub-sectors, such as aviation, navigation, rail, and road for *transport*, and further into applications, such as vehicle types. The modular programming allows the addition of as many sub-sectors and applications as required.

3.1.2.2 Structure of the Demand Module

Each of the three sectors, *residential and buildings*, *industry*, and *transport*, has standardized sub-structures and applications. The residential sector R (first layer) has a list of household types (second layer), and each household type has a standard set of services (third layer), such as 'lighting', 'cooling', and 'entertainment'. Finally, the applications for each of the services are defined (fourth layer), such as refrigerator or freezer for 'cooling'. The energy intensity of each application can be altered to reflect the status quo in a certain region and/or to reflect improvements in energy efficiency. An illustrative example of the layers of the residential sector is shown in Fig. 3.2.

Figure 3.3 shows an example of the model structure of the *industry* sector. In the second layer are different industries—the OECM uses the GICS classification system for industry sub-sectors. The quantity of energy for each of the sub-sectors is driven by either GDP or the projected quantity of product, such as the tonnes of steel produced per year. The market shares of specific manufacturing processes are defined, and each process has a specific energy intensity for electricity, (process) heat, and/or fuels.

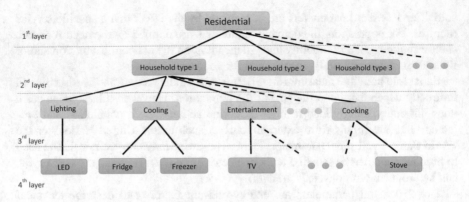

Fig. 3.2 Residential sector sub-structures

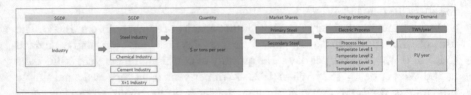

Fig. 3.3 Calculation of the *industry* energy demand

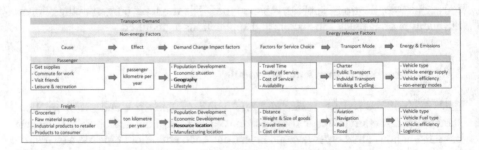

Fig. 3.4 Calculation of *transport* energy demand

Figure 3.4 shows the structure for the *transport* sector. Again, the demand is driven by 'non-energy' factors, such as passenger-kilometres and freight-kilometres, and energy-related factors, such as the transport mode and the energy intensity of the different vehicle options.

3.1.2.3 Demand Module Architecture in MATLAB

The demand module is implemented in MATLAB, a widely used programming language for mathematics and science computing. MATLAB allows the integration of a range of tools and databases and has the flexibility to add and develop new functions. Specifically, the model has been developed using an object-oriented programming approach, allowing extensibility and modularity.

Figure 3.5 shows the demand module developed in MATLAB. The demand module encompasses eight classes: (1) demand class, (2) household class, (3) household application class, (4) sub-sector, (5) industry class, (6) industry application class, (7) transport modes, and (8) vehicles class (Fig. 3.6).

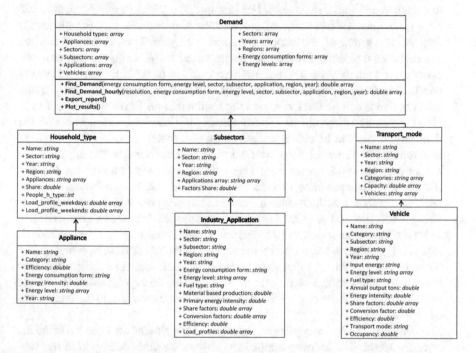

Fig. 3.5 A unified modelling language (UML) diagram of the demand module in MATLAB, showing its classes, attributes, methods, and associations

Household_type
+ Name: *string* = "Rural – Phase 1"
+ Sector: *string* = "Residential"
+ Year: *string* = "2020"
+ Region: *string* = "Global"
+ Appliances: *string array* = matrix(35,1)
+ Share: *double* = 0.2
+ People_h_type: = 5
+ Load_profile_weekdays: *double array* = matrix(35,120)
+ Load_profile_weekends: *double array* = matrix(35,48)

Fig. 3.6 An example of a *household* type object, showing the assigned attributes

- *Demand class*: This is the main class, which describes the *residential, industry*, and *transport* sectors, which are defined by household type, sub-sector, and transport mode classes, respectively. The attributes that define this class include a range of years, energy consumption forms, energy levels, list of sectors, household types, sub-sectors, applications, appliances, and vehicles. The demand class also has two main types of methods: (i) calculation demand methods and (ii) printing results methods. The calculation methods use equations and algorithms to calculate and find the demand. For example, the 'Find Demand' method can be used to find a wide range of calculations and outputs, e.g. the electricity demand of a group of households for a specified year. The calculation method can calculate the demand for single or aggregated sectors, sub-sectors, or applications, for a single year or a range of years, unique or multiple forms of energy consumption, and single or various types of vehicle categories. The printing result methods can be used to export the results into an Excel spreadsheet or to plot the results using the MATLAB interface. Therefore, the outputs of the demand module can be either predefined graphs, tables, or data for a standardized report. See Table 3.1 for a brief description of each method in this class.
- *Household and appliance classes*: These classes are used to define the *residential* sector. The appliance objects are embedded within the household-type objects. Attributes include names, sectors, and regions, which are defined as string inputs (i.e. text or character inputs) or numerical inputs, which are defined as int (i.e. integers) or double (i.e. numeric variables holding numbers with decimal points). Attributes can also include arrays of strings or double values. Array variables are helpful in input time series data, such as load profiles. Because households and appliances have their own classes, this architecture is flexible and allows the addition of households with different attributes and different types of appliances.
- *Sub-sector and industry application classes*: These classes are used to define the *industry* sector. The industry application objects are embedded within the sub-sector objects. As shown in Fig. 3.7, these classes have their own lists of attributes. Therefore, the module developed can accommodate different types of

Table 3.1 Methods within the demand class

Type of method	Method	Description
Calculation	Find Demand() and Find_Demand_ hourly()	These methods calculate the annual or hourly aggregated energy demand for the specified region and energy form (i.e. power, heat, or hydrogen). The calculations can be aggregated by sector, sub-sector, transport mode, or any other object class
Printing results	Export_report()	This method exports the specified results to external Excel spreadsheets and can be used to print results on predefined report tables
Printing results	Plot_results()	This method can be used to plot results using the MATLAB interface

Industry_Application
+ Name: *string* = "Primary steel production"
+ Sector: *string* = "Industrial"
+ Subsector: *string* = "Iron & Steel"
+ Region: *string* = "Global"
+ Year: *string* = "2020"
+ *Energy consumption form: string = "Electricity & Heat"*
+ *Energy level: string array = matrix(2,5)*
+ Material based production: *double* = 1.178×10^3
+ Primary energy intensity: *double* = 1.184×10^3
+ Share factors: *double array* = matrix(1,3)
+ Conversion factors: *double array* = matrix(1,3)
+ Efficiency: *double* = 0.98
+ Load_profiles: *double array* = matrix(1,168)

Fig. 3.7 An example of an *industry* application object, showing the assigned attributes

sub-sectors (e.g. steel, cement, etc.) and incorporate various types of applications under each sub-sector.

- *Transport modes and vehicle classes*: These classes are used to define the *transport* sector. The vehicle objects are embedded within the transport mode objects. Therefore, multiple types of transport modes can be defined, such as aviation and navigation, as well as various types of vehicles, such as planes and cruise ships.

Figures 3.6 and 3.7 show the high-level class definitions for *residential* and *industry* sub-sector objects, respectively. The blue-marked text indicates the defined value for each attribute. For example, one household object with five residents is defined by the name 'Rural–Phase 1' and has a list of 35 appliance objects, defined with a string array. It is assigned a share factor for 2020 of 0.2, which means that 20% of the households in that specific region and year are defined by this type of household and its attributes. Furthermore, 24 h load profiles are defined for each application for every day, with numerical arrays. For example, weekend load profiles have a size of 35 rows and 48 columns, representing 35 applications and 24 time slots for each weekend day.

The object-oriented architecture allows all these input attributes to be updated or modified easily. These attributes can also be read from a predefined Excel spreadsheet. This facilitates a data input process that follows the array structure, such as the load profile.

Figure 3.7 shows an example of an industrial application object that belongs to the sub-sector *iron and steel*. In this case, the energy consumption form is defined as electricity and heat, which means that it considers the electrical and heat demand. The 'share factors' represent the portions of the demand assigned to electricity and heat. The energy-level array also allows the predefined network to which the application is connected to be defined, as well as the temperature levels. In this particular case, the demand is defined based on the total annual primary energy intensity and the material-based production, which are 1184 GJ/tonnes and 1178 Mt, respectively, for the specified region and year. The input and output units must be predefined when the MATLAB modules are initialized. Other attributes that can be assigned are conversion factors, such as from primary energy to the final energy via an efficiency factor.

Additional attributes and methods can be defined for each class if required and the data are available. Therefore, the demand module class can be extended by defining new classes, attributes, and methods.

3.1.3   The OECM Supply Module

The supply module consists of three main elements: supply technologies, storage technologies, and the infrastructure for the power supply (capacities of power lines). For the generation of electricity and heat, the programme considers all the technologies of the energy market, from both renewable and non-renewable sources. In addition to the generation of pure electricity and heat, the entire range of combined heat and power systems is included.

Storage technologies include batteries and the use of hydrogen from electrolysers. The calculation of heat storage is possible, but has not yet been used in the OECM scenarios.

A dispatch strategy is defined for electricity and heat generation that reflects market and policy factors. Whether electricity from photovoltaics and onshore and offshore wind turbines have priority dispatch ahead of fossil-fuel power plants and how storage systems are used can be determined. Each technology has a specific conversion efficiency.

Heat generation technologies are also defined by the temperature levels they can provide. For example, residential solar collectors can only supply low-temperature heat and will therefore not be considered for high-temperature process heat (Table 3.2).

The regional energy demand—as defined in the previous section—can be met by neighbouring regions, with importation from or, in the case of oversupply, exportation to them. The extent to which electricity can be imported or exported from one region to another is defined by the capacity of regional interconnections, which represent the available power line capacities.

3.1.3.1   The OECM Dispatch Module

The methodology of the dispatch module of the MATLAB-based OECM is based on the previous version of the model (Teske et al. 2019a). The key inputs are related to the supply technologies, storage types, dispatch strategy, and the

Table 3.2 Example of generation and storage technologies

Generation			Storage		
Power plants	Combined heat and power plants	Heating plants	Electrical	Thermal	Hydrogen
Hard coal	Hard coal	Coal	Lithium battery	Water tank	Tank
Lignite	Lignite	Lignite	Pumped hydro	Molten salt	
Gas	Gas	Gas			
Oil	Oil	Oil			
Diesel	Biomass	Biomass			
Biomass	Geothermal	Solar collectors			
Hydro	Hydrogen	Geothermal			
Wind		Hydrogen			
Photovoltaic					
Concentrated solar (CSP)					
Geothermal					
Solar thermal					
Ocean energy					
Hydrogen					

Table 3.3 Input parameters for the dispatch model

Input parameter		
$L_{Cluster}$	Load cluster	[MW]
$L_{Interconnection}$	Maximum power-line capacity (import/export)	[MW]
$L_{Initial}$		[MW]
$Cap_{Var.RE}$	Installed capacity for *variable renewables*	[MW]
$Meteo_{Norm}$	Meteorological data for solar and wind	[MW/MW$_{INST}$]
$L_{Post\_Var.RE}$	Load after *variable renewable* supply	[MW]
$Cap_{Storage}$	*Storage* capacity	[MW]
$CapFact_{Max\_Storage}$	Maximum capacity factor for storage technologies	[h/yr]
$L_{Post\_Storage}$	Load after *storage* supply	[MW]
$Cap_{Dispatch}$	Capacity of *dispatch power plants*	[MW]
$CapFact_{Max\_Dispatch}$	Maximum capacity factor for *dispatch power plants*	[h/a]
$L_{Post\_Dispatch}$	Load after *dispatch power plant* supply	[MW]
$Cap_{Interconnection}$	*Interconnection* capacity	[MW]

interconnections among regions for possible power exchange (Table 3.3). Different supply technologies can be selected, each with its technical characteristics, including its efficiency, available installed capacity, fuel type, and regional meteorological data (solar radiation or wind speed). Meteorological data define the capacity factors of solar and wind energy generators as their levels of availability at 1-h resolution for an entire year (Table 3.4).

The supply technologies can be either dispatchable (e.g. gas power plants) or non-dispatchable (e.g. solar photovoltaic without storage). The model allows the

Table 3.4 Output parameters for the dispatch model

Output parameter		
$L_{Initial}$	Initial load (cluster)	[MW]
$L_{Post\_Var.RE}$	Load after *variable renewable* supply	[MW]
$S_{EXECC\_VAR.RE}$	Access supply *renewables*	[MW]
$L_{Post\_Storage}$	Load after *storage* supply	[MW]
$S_{Storage}$	Storage requirement/curtailment	[MW]
$CapFact_{Actual\_Storage}$	Utilization factor for storage	[h/a]
$L_{Post\_Dispatch}$	Load after *dispatch power plant* supply	[MW]
$S_{Dispatch}$	Dispatch requirement	[MW]
$CapFact_{Actual\_Dispatch}$	Utilization factor for *dispatch power plants*	[h/a]
$L_{Post\_Interconnection}$	Load after *interconnection* supply	[MW]
$S_{Interconnection}$	Interconnection requirement	[MW]
$CapFact_{Actual\_Interconnection}$:	Utilization factor for *interconnection*	[h/a]

Table 3.5 Technology groups for the selection of dispatch order

Technology options	Input: assumed order marked with (1) to (4)
1. Variable renewables	Variable renewables (1)
2. Storage	Dispatch generation (3)
3. Dispatch generation	Storage (2)
4. Interconnector	Interconnector (4)

Table 3.6 Technology options—variable renewable energy

Variable renewable power technology options	Input: assumed order of generation priority marked with (1) to (5)
1. Photovoltaic—rooftop	Photovoltaic—utility scale (2)
2. Photovoltaic—utility scale	Photovoltaic—rooftop (1)
3. Wind—onshore	Wind—offshore (4)
4. Wind—offshore	Wind—onshore (3)
5. CSP (dispatchable)	CSP (5)

order in which the supply technologies and storage functions are utilized to be adjusted to satisfy the demand. However, storage and interconnections cannot be selected as the first elements of supply (Table 3.5).

Tables 3.6, 3.7, and 3.8 provide an overview of the possible supply technologies and examples of different dispatch scenarios. Although concentrated solar power (CSP) plants with storage are dispatchable to some extent—depending on the storage size and the available solar radiation—they are part of the renewable variable group in the MATLAB model. Although the model allows the dispatch order to be changed, the 100% renewable energy analysis always follows the same dispatch logic. The model identifies excess renewable production, which is defined as any potential wind and solar photovoltaic generation greater than the actual hourly demand in MW during a specific hour. To avoid curtailment, the surplus renewable electricity must be

Table 3.7 Technology options—dispatch generation

Dispatch generation technology options	Input: assumed order of generation priority marked with (1) to (13)
1. Bioenergy	Hydropower (3)
2. Geothermal	Bioenergy (1)
3. Hydropower	CoGen bioenergy (7)
4. Ocean	Geothermal (2)
5. Oil	CoGen geothermal (8)
6. Gas	Ocean (4)
7. CoGen bioenergy	Gas (6)
8. CoGen geothermal	CoGen gas (9)
9. CoGen gas	Coal (11)
10. CoGen coal	CoGen coal (10)
11. Coal	Brown coal (12)
12. Brown coal	Oil (5)
13. Nuclear	Nuclear (13)

Table 3.8 Technology options—storage technologies

Storage technology option	Input: assumed priority order for storage technologies marked with (1) to (3)
1. Battery	Hydro pump (2)
2. Hydro pump	Battery (1)
3. Hydrogen	Hydrogen (3)

stored with some form of electric storage technology or exported to a different cluster or region. Within the model, the excess renewable production accumulates through the dispatch order. If storage is present, it will charge the storage within the limits of the input capacity. If no storage is present, this potential excess renewable production is reported as 'potential curtailment' (pre-storage) (Table 3.9).

Limitations: It is important to note that calculating the possible interconnection capacities for transmission grids between subregions does not replace technical grid simulations. Grid services, such as the inductive power supply, frequency control, and stability, should be analysed, although this is beyond the scope of the OECM analysis. The results of [R]E 24/7 provide a first rough estimate of whether increased use of storage or increased interconnection capacities or a mix of both will reduce systems costs.

3.1.3.2 Regional Interconnections

Interconnection capacities are set as a function of the total generation capacity within a cluster. Interconnections between defined regions are the only ones considered, and all intra-regional interconnections or line constraints are excluded. Therefore, a region is considered a 'copper plate'—and a transmission system

Table 3.9 Dispatch module—inputs, intermediate outputs, and outputs

Inputs, intermediate outputs, outputs		
Inputs	Maximum capacity for interconnections among regions	[MW]
Inputs	Initial load (cluster or region)	[MW]
Inputs	Technical specifications of supply technologies and storage strategies	
Inputs	Meteorological data	
Intermediate output	Dispatch order of technologies	
Intermediate output	Load after *variable renewable* supply	[MW]
Intermediate output	Load after *storage* supply	[MW]
Intermediate output	Load after *dispatch power plant* supply	[MW]
Intermediate output	Load after *interconnection* supply	[MW]
Output	Deficit and curtailment	[MWh]
Output	*Renewable penetration*	[MWh]

where electricity can flow unconstrained from any generation site to any demand site is found in most energy modelling tools (Avrin, 2016). This simplification is required to achieve a short calculation time while maintaining high technical and time resolution. The algorithm devised for the function of the interconnectors is based on the following information for each region:

- Unmet load in the region
- Excess generation in other regions
- Interconnection capacity between the undersupplied region and each of the other regions
- Priority of the closest region(s) in exporting power to the undersupplied region

The excess generation capacity and unmet load are calculated by running the model without the interconnections to determine the excess or shortfall in generation when the load within the region is met. These excesses and shortfalls are calculated at the point in the dispatch cascade at which the interconnectors provide or consume power, for example, after the variable renewables and dispatchable generators and before the storage technologies.

The interconnection capacity between regions is defined based on a percentage of the maximum regional load. The capacity is defined in a matrix, both to and from each region to every other region. A priority order for each region to every other region is given based on proximity, so that if a region has an unmet load, it will be served sequentially with the excess generation of loads in other regions in their defined order of proximity.

For every hour and every region in each cluster (a cluster is a group of regions), the possible interconnections required for the importation or exportation of energy to balance the load are calculated. Each region is considered in turn, and the algorithm attempts to meet the unmet load with excess generation by other regions, keeping track of the residual excess loads and the interconnector capacities. Each region's internal load is met first, before its generation resources are considered for other interconnected regions.

For regions sending generation capacity to other regions, the interconnector element behaves as an increase in load, whereas for regions accepting power from neighbouring regions, the interconnector element behaves as an additional generator, from the model's perspective.

Once the total inflow and outflow of the interconnectors are calculated, the hourly values for the total supply in each region are updated, together with any residual deficit in supply or any curtailed (= forced to shut down) electricity generator that does not have priority dispatch.

Similar to the supply technologies, different storage technologies (electrical, thermal, or hydrogen) can be defined and selected, together with their technical characteristics, such as their round-trip efficiency, new or installed capacities in each year of the modelled period, lifetime, maximum depth of discharge, maximum energy out in a time step, and costs. When the total energy delivered by the supply technologies in a region does not meet the demand, energy is discharged from storage (if the storage technology has energy available), following the constraints of the storage operation (maximum energy out per time step, maximum depth of discharge, maximum depth of charge, state of charge) and the order of operation for the defined storage technologies. In the case of a demand deficit after storage, electricity from other regions will be imported. When there is surplus energy generation, the surplus will charge any storage appliances (if available), also according to the same constraints of energy storage operation and sequential order.

3.1.3.3 Supply Module Architecture in MATLAB

Analogous to the demand module, inputs can be made directly into the supply module via MATLAB or a standardized Excel sheet. The supply module in MATLAB is also based on an object-oriented structure, in which classes and the objects belonging to those classes are built based on attributes and methods.

Figure 3.8 shows the UML class diagram for the supply module developed in MATLAB. Specifically, the supply module has three main classes:

1. *Supply class*: This is the main class and it is built on the supply and storage technology objects. Attributes that describe the supply class include years, region, energy supply form, fuel, and generation and storage technologies. The supply class has two main types of methods: (i) calculation supply methods and (ii) printing result methods. The calculation methods implement equations and algorithms to calculate the dispatch and fuel consumption. Table 3.10 presents a brief description of each method.
2. *Supply technology class*: This class is used to define supply technologies. Attributes include name, type, efficiency, year, region, and energy supply form and are defined as text inputs. Additional attributes are defined as numerical inputs, such as lifetime, cost, and capacity factors. The structure adopted allows

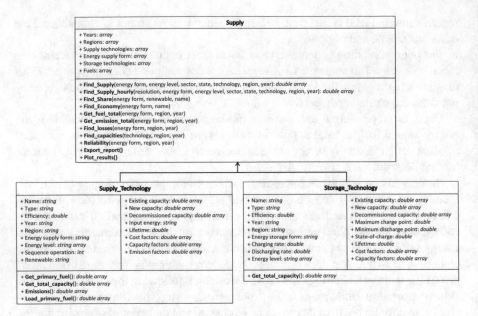

Fig. 3.8 A UML diagram of the supply module in MATLAB, showing its classes, attributes, methods, and associations

the addition of new attributes if required. This class has methods that are used by the main supply class to calculate the primary fuel, emissions, or installed capacity of a specific technology.

3. *Storage technology class*: This class is used to define storage technologies. The attributes include name, type, efficiency, year, region, and energy storage form and are defined as text inputs. Other numerical attributes include charging and discharging rates, capacity, cost factors, and state of charge.

Figures 3.9 and 3.10 show the high-level class definitions for supply technologies and storage objects, respectively. The text in blue indicates the defined value for each attribute. For example, the supply technology object in Fig. 3.9 has the name 'coal power plant', its input energy is defined as hard coal, and the object is associated with the electricity energy form. The attributes in Fig. 3.9 consider the year 2020 and a global scenario. For example, the existing capacity is defined as 989.5 GW and the decommissioned capacity is 23 GW. The lifetime of this object is 35 years.

An example of a storage object is shown in Fig. 3.10. The attributes of this object include text inputs, such as its name 'battery lithium' and its type 'electrical'. This object has numerical attributes such as the efficiency (equal to 0.95 for this object) and the charging and discharging rates (fixed at 5 kW). Note that the units for each attribute are defined when the module is initialized in MATLAB.

Table 3.10 Methods within the supply class

Type of method	Method	Description
Calculation	Find_Supply() and Find_Supply_hourly()	These methods calculate the annual or hourly aggregated energy supply for the specified region and the energy form (i.e. power, heat, or hydrogen). The calculations can be made by individual or group supply technology type or storage type. These methods can also be used to calculate the emissions and primary fuel associated with each supply technology
Calculation	Find_Share()	This method calculates the share factor results for predefined supply scenarios; for example, the share factors of power generated from renewable energy sources and non-renewable sources. Another example is the portion of the transport sector that requires electricity or hydrogen
Calculation	Find_Economy()	This method calculates the costs associated with supply technologies
Calculation	Get_fuel_total()	This method calculates the total fuel or total primary fuel required for demand and supply
Calculation	Get_emission_total()	This method calculates the total emissions, considering all the demand sectors and supply technologies
Calculation	Find_losses()	This method calculates the losses for a specified energy form. For example, it can be used to calculate the electricity losses or heat losses arising from transport and distribution
Calculation	Find_capacities()	This method calculates the installed capacity for a specified technology when decommissioning or new capacity parameters have been defined
Calculation	Reliability()	This method calculates the total energy deficit and curtailment based on the total demand and generation, for the specified energy form
Printing results	Export_report()	This method exports the specified results to external Excel spreadsheets and can be used to print results on predefined report tables
Printing results	Plot_results()	This method can be used to plot results using the MATLAB interface

Supply_Technology	
+ Name: *string* = "Coal power plant"	+ Existing capacity: *double* = 989.5
+ Type: *string* = "Coal"	+ New capacity: *double* = 0
+ Efficiency: *double* = 0.37	+ Decommissioned capacity: *double* = 23
+ Year: *string* = "2020"	+ Input energy: *string* = "Hard coal"
+ Region: *string* = "Global"	+ Lifetime: *double* = 35
+ Energy supply form: *string* = "Power"	+ Cost factors: *double array* = matrix(2,1)
+ Energy level: *string array* = matrix(1,3)	+ Capacity factors: *double* = 0.57
+ Sequence operation: *int* = NA	+ Emission factors: *double* = 93
+ Renewable: *string* = "N"	

Fig. 3.9 An example of a supply technology object, showing the assigned attributes

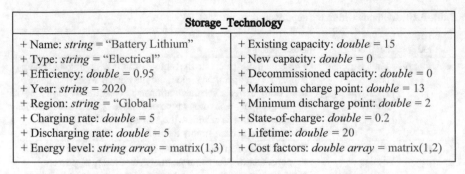

Storage_Technology	
+ Name: *string* = "Battery Lithium"	+ Existing capacity: *double* = 15
+ Type: *string* = "Electrical"	+ New capacity: *double* = 0
+ Efficiency: *double* = 0.95	+ Decommissioned capacity: *double* = 0
+ Year: *string* = 2020	+ Maximum charge point: *double* = 13
+ Region: *string* = "Global"	+ Minimum discharge point: *double* = 2
+ Charging rate: *double* = 5	+ State-of-charge: *double* = 0.2
+ Discharging rate: *double* = 5	+ Lifetime: *double* = 20
+ Energy level: *string array* = matrix(1,3)	+ Cost factors: *double array* = matrix(1,2)

Fig. 3.10 An example of a storage technology object, showing the assigned attributes

The supply module architecture developed is flexible to accommodate different types of supply and storage technologies. Additional attributes or methods can be easily added to the model.

3.1.4 Databases and Model Calibration

The OECM model uses several databases for energy statistics, energy intensities, technology market shares, and other market or socio-economic parameters. The calculation of the energy balance for the base year is based on the International Energy Agency (IEA) Advanced World Energy Balances (IEA, 2020, 2021).

The energy statistics for a calculated country and/or region are uploaded via an interface module. The data for each year from 2005 onwards until the last year for which data are available are used to calibrate the model. This process is based on the energy system model (EM), developed by the German Aerospace Center DLR, and is implemented in the energy simulation platform Mesap/PlaNet (Schlenzig, 1999; Seven2one, 2012). The market shares are calculated based on the IEA statistics and a technical database for energy intensities for various appliances and applications across all sectors. These data are input and the calibration processes performed with a standardized Excel tool. The calibration method is briefly outlined below using the *transport* sector.

To calibrate the model, the transport demand of the past decade is recalculated on the basis of the available energy statistics. The IEA's Advanced World Energy Balances provides the total final energy demand by transport mode—aviation, navigation, rail, or road—by country, by region, or globally. However, it provides no further specification of the energy use within each of the transport modes. Therefore, further division into passenger and freight transport is calculated using percentage shares. These proportions are determined with a literature search, together with the average energy intensity for each of the transport modes for passenger and freight vehicles.

The annual transport demand in passenger-kilometres per year (pkm/year) or tonne-kilometres per year (tkm/year) is calculated as the annual energy demand

Table 3.11 Calibration for calculating the transport demand

Calculation concept	Process	Until 2019	Units	Comment
Transport demand				
Aviation, navigation, rail, and road—*past to present*				
Annual demand	Data	Database	[PJ/yr]	Data: IEA Advanced World Energy Balances
Passenger share	Input	Literature	[%]	Shares of the total energy demand from the literature
Freight share	Input	Literature	[%]	Shares of the total energy demand from the literature
Average energy intensity—passenger transport	Data	Literature	[MJ/pkm]	Literature review—based on current supply mix
Average energy intensity—freight transport	Data	Literature	[MJ/tkm]	Literature review—based on current supply mix
Passenger-kilometres	Calculation	= Annual demand/ energy intensity	[pkm]	Checked against OECD statistics
Tonne-kilometres	Calculation		[tkm]	Checked against OECD statistics
Annual growth/ reduction—passenger-kilometres	Calculation	= Annual demand in the previous year/ annual demand in the calculated year	[%/yr]	Calculated to understand the trend between 2005 and 2020
Annual growth/ reduction—tonne-kilometres	Calculation		[%/yr]	
Population—indicator of passenger transport development	Data	Database	[Million]	Data: UN
GDP per capita—indicator of passenger and freight transport development	Data	Database	[$GDP/ capita]	Data: World Bank
GDP—indicator of freight transport development	Data	Database	[$GDP]	Data: World Bank

divided by the average energy intensity by mode. These results are then compared with the Organisation for Economic Co-operation and Development (OECD) transport statistics, which provide both parameters——pkm/year and tkm/year. Calibrating the model on the basis of historical data ensures that the basis of the scenario projections for the coming years and decades is correctly mapped and ensures that the changes are calculated most realistically (Tables 3.11 and 3.12).

Table 3.12 Projection of transport demand based on the changing demand in kilometres

Process	2020–2050	Units	Comment
Aviation, navigation, rail, road—projections			
Calculation	= (pkm in previous year) × (increase/reduction in % per year)	[pkm]	Starting point: base year 2019
Calculation	= (tkm in previous year) × (increase/reduction in % per year)	[tkm]	Starting point: base year 2019
Input	Input in %/year	[%/yr]	Assumption
Input	Input in %/year	[%/yr]	Assumption
Calculation	Input in %/year	[million]	Assumption based on UN projection
Calculation	= $GDP/population	[$GDP/capita]	
Calculation	INPUT in %/year	[$GDP]	Assumption based on the World Bank projection
Result	Time series 2020–2050: pkm per year and region	[pkm/yr]	Input for energy demand calculation
Result	Time series 2020–2050: tkm per year and region	[tkm/yr]	Input for energy demand calculation

For the forward projection of the transport demand, the calculation method is reversed: the transport demand for each transport mode is calculated on the basis of the annual change, as a percentage. The calculated total annual pkm and tkm are the inputs for the energy demand calculation.

This methodology for calibration and projection is used across all sectors.

The developed MATLAB tool can access online data and databases through available *application programming interfaces* (APIs). For example, the API for the World Bank Indicators provides access to nearly 16,000 time series indicators, including population estimates and projections (World Bank, 2021). Likewise, the OECD provides access to datasets through an API. This allows a developer to easily call the API and access data using the code lines in MATLAB.

3.1.5   Sectors and Sub-sectors

The OECM was designed to calculate energy pathways for geographic regions, as documented in Chap. 2. The OECM was developed further to meet the requirements of the financial industry and to design energy and emission pathways for clearly defined industry sectors (sectorial pathways). The finance industry uses different classification systems to describe sub-areas of certain branches of industry. The most important system is the Global Industry Classification Standard (GICS; MSCI, 2020). However, the GICS sub-sectors do not match the IEA statistical breakdown of the energy demands of certain industries. Table 3.13 shows examples of the finance sector calculated with the OECM model, the GICS codes, and the statistical

Table 3.13 Examples of industry sub-sectors based on the Global Industry Classification Standard (GICS)

Financial sector	GICS	IEA statistical categories	Sector definition
Agriculture	3010 Food and staple retailing	**Farming**	Food and tobacco production, excluding the energy demand for agriculture, as defined under the IEA energy statistic *other sectors*. Additional statistics from industry partners are required because the IEA statistics only provide the accumulated energy demand for agriculture and forestry
	3020 Food, beverages, and tobacco	Food production and supply	
Forestry	1510 Materials	**Agricultural and forestry**	Energy demand for all wood and wood products, including pulp and paper and printing. Also includes all energy demands for agricultural services not included in food and tobacco production
	1510 50 Paper and forest products	Paper and forest products	
	1510 5010 Forest products		
	1510 5020 Paper products		
Chemicals	1510 Materials	**Chemical industry**	Energy demand for all chemical, petrochemical, glass, and ceramic products
	1510 10 Chemicals	Chemical products	
		Petrochemical products	
		Glass and ceramics	
Aluminium	1510 40 Metals and mining	**Aluminium**	Energy demand for the production of primary and secondary aluminium, as well as bauxite mining
	1510 4010 Aluminium		
Textiles and leather	2520 Consumer durables and apparel	**Textiles and leather industry**	This sector covers the energy demand for the textiles and leather industry
	2520 30 Textiles, apparel, and luxury goods		

information used. Although the OECM model allows all the GICS code sub-sectors to be calculated, the availability of statistics is the factor limiting the resolution of the sectorial pathways. For example, the statistical data for the textile and leather industry are stored in the IEA database, but the database does not separate the two industries further.

3.1.6 Cost Calculation

The costs linked to the energy supply in each year of the modelled period include the investment costs related to 'new capacities' for technologies and storage (including replacement or decommissioning, based on the assumed technical lifetime = vintaging), operation and maintenance (O&M) costs as a percentage of the total installed capacities, and fuel costs. Other inputs for each technology and storage type include the capital cost per unit ($/kW), O&M costs as a percentage of the capital cost, and unit fuel cost ($/GJ).

Therefore, for each technology or storage type:

- It is assumed that the change in 'installed capacity' between each of the years modelled is linear and a linear interpolation between these is considered.
- The 'installed capacities' and 'new capacities' are interrelated (one depends upon the other) in each of the modelled and interpolated years, based on the cumulative capacities in the calculated year and the assumed technology lifetime.
- The capital costs per unit and the fuel costs in each of the modelled years are also interpolated linearly between the modelled years. Therefore, if a scenario is calculated in 5-year steps, e.g. the development from 2025 and 2030, the years 2026 to 2029 are calculated as a linear interpolation.
- Replacement capacities, if required, are also included in each year as part of the investment costs.
- The O&M costs in each of the interpolated years are calculated based on the interpolated installed capacities and the annual O&M input costs (as a percentage of the capital cost).
- Annual fuel costs for non-renewable technologies are calculated based on their output energy (running time) and interpolated fuel costs.
- The resulting 'specific costs' ($/kWh) are also calculated from the interpolated energy supplied in each year.

The total specific costs ($/kWh) of a scenario, as practically distributed over the interpolated years, allow the incurred costs for a scenario to be determined. *Limitations*: The economic model does not consider the change in the value of money over time. Each year of the modelled period is regarded as if it were the present year, with the multiple costs incurred. Future additions to the model could include the net present costs and the contemporary value of money.

3.1.7 OECM 2.0 Output and Area of Use

The added value of OECM 2.0 is its high resolution of the sector-specific parameters for both demand and supply, which are required as key performance indicators (KPIs) by the finance industry. Table 3.14 provides an overview of the main parameters and the areas of their use, with a focus on the needs of institutional investors.

Commodities and GDP are the main drivers of the energy demand for industries. The projection of, for example, the global steel demand in tonnes per year over the next decades is discussed with the industry and/or client. The OECM 2.0 can calculate either a single specific sector only or a whole set of sectors. For the development of global scenarios, various industry projections are combined to estimate both the total energy supply required and the potential energy-related emissions. Therefore, a global carbon budget can be broken down into carbon budgets of specific industries.

Energy intensities are both input data for the base year and a KPI for future projections. The effect of a targeted reduction in the energy intensity in a given year and the resulting energy demand and carbon emissions can be calculated, for example, for the steel industry.

All sector demands are supplied by the same energy supply structure in terms of electricity, process heat (for each level), and total final energy. Finally, specific emissions, such as CO_2 per tonne of steel or per cubic metre of wastewater treatment, are calculated and can be used to set industry targets.

All input and output OECM data are available as MATLAB-based tables or graphs or as standard Excel-based reports.

3.1.8 Further Research Demand

Industry-specific energy intensities and energy demands are not available for a variety of industries. In particular, the energy intensities for sub-sectors of the chemical industry are either totally unavailable or confidential. A database of energy intensities is required to develop more detailed scenarios. Although energy intensities can be estimated based on the available data, the input parameters are usually derived from various sources, which may not follow the same methodology. Energy intensities based on GDP, for example, are calculated with either nominal GDP, real GDP, or purchasing power parity GDP. Furthermore, energy intensities can be provided as final energy or primary energy. In some cases, this information is not available at all. A database of industry-specific energy demands and energy intensities, with a consistent methodology, is required to improve the accuracy of calculations in future research.

To capture the complexity of regional and global building demand projection, both in terms of data availability and high technical resolution, the high-efficiency building (HEB) model was used to develop four bottom-up demand scenarios. The HEB was developed by the Central European University (CEU) of Budapest under

Table 3.14 Energy-related key performance indicators (KPIs) for net-zero target setting, calculated with OECM 2.0

Sector	Parameter	Units	Base year 2019	Projections 2025, 2030, 2035, 2040, 2045, 2050
Commodities				
Water utilities	Water withdrawal	[Billion m³/yr]	Input	Calculated projection with annual growth rates discussed with client
Chemical industry	Economic development	[$GDP/yr]	Input	
Steel industry	Product-based market projection	[Tonnes steel/yr]	Input	
Energy intensities				
Water utilities	Wastewater treatment	[kWh/m³]	Input	Technical target (KPI) Calculated with annual progress ratio based on technical assessment
Chemical industry	Industry-specific energy intensity	[MJ/$GDP]	Input	
Steel industry	Energy intensity	[MJ/tonne steel]	Input	
Energy demand				
Water utilities	Final energy demand	[PJ/yr]	Input	Output—industry-specific scenario(s)
Chemical industry	Electricity demand	[TWh/yr]	Input	
Steel industry	Process heat demand by temperature level	[PJ/yr]	Input	
	Total final energy demand	[PJ/yr]		
Energy supply				
Water utilities	Electricity generation by technology	[TWh/yr]	Input	Output—based on scenario developed Supply for all (sub-)sectors
Chemical industry	(Process) heat by technology	[PJ/yr]	Input	
Steel industry	Fuel supply by fuel type	[PJ/yr]	Input	
	Total final energy supply by fuel type	[PJ/yr]	Input	
Energy-related emissions				
	Electricity—specific CO_2 emissions	[gCO_2/kWh]	Calculated	Output—KPI for utilities
	Electricity—total CO_2 emissions	[$t\ CO_2$/yr]	Calculated	Output—KPI for utilities
	(Process) heat—specific CO_2 emissions	[gCO_2/kWh]	Calculated	Output—KPI for industry
	(Process) heat—total CO_2 emissions	[tCO_2/yr]	Calculated	Output—KPI for industry

(continued)

Table 3.14 (continued)

Sector	Parameter	Units	Base year 2019	Projections 2025, 2030, 2035, 2040, 2045, 2050
Product-specific emission				
Water utilities	Emissions intensity	[kgCO$_2$/m^3]	Calculated	KPI—water utilities
	Total energy-related CO$_2$ emissions	[tCO$_2$]	Calculated	KPI—water utilities
Chemical industry	Emissions intensity	[kgCO$_2$/$GDP]	Calculated	KPI—chemical industry
	Total energy-related CO$_2$ emissions	[tCO$_2$]	Calculated	KPI—chemical industry
Steel industry	Emissions intensity	[kgCO$_2$/t steel]	Calculated	KPI—steel industry
	Total energy-related CO$_2$ emissions	[tCO$_2$]	Calculated	KPI—steel industry

the scientific leadership of Prof. Dr. Diana Uerge-Vorsaz. The following section documents the methodology of the HEB based on the paper by Chatterjee, S.; Kiss, B.; and Ürge-Vorsatz, D. (2021). The results are documented in Sects. 7.1 and 7.2.

3.2 The High-Efficiency Building (HEB) Model

Modelling the energy demand for buildings is a complex task because the building sector-related energy demand depends on several factors, such as spatial resolution, temporal resolution, building physics, and the different technologies of building construction (Prieto et al., 2019; Chatterjee & Ürge-Vorstaz, 2020). The majority of demand models do not incorporate these factors and therefore provide insights into the future energy demand scenarios of the building sector that can be far from realistic (Prieto et al., 2019; Chatterjee & Ürge-Vorstaz, 2020). Therefore, in this study, we use the HEB model to understand the future energy demand potentials for building in key regions across the globe.

The HEB model was originally developed in 2012 to calculate the energy demand and CO$_2$ emissions of the residential and tertiary building sectors until 2050 under three different scenarios (Ürge-Vorsatz & Tirado Herrero, 2012). Since then, the model has been developed and updated several times. With the latest update, the model calculates the energy demand under four scenarios until 2060 based on the most recent data for macroeconomic indicators and technological development. This model is novel in its methodology compared with earlier global energy analyses and reflects an emerging paradigm—the performance-oriented approach to the energy analysis of buildings. Unlike component-oriented methods, a systemic perspective is taken: the performance of whole systems (e.g. whole buildings) is studied, and these performance values are used as the input in the scenarios. The model calculates the overall energy performance levels of buildings, regardless of the measures applied to achieve them. It also captures the diversity of solutions required in each region by including region-specific assumptions about advanced and suboptimal technology mixes. The elaborated model uses a bottom-up approach, because it includes rather detailed technological information for one sector of the economy. However, it also exploits certain macroeconomic (GDP) and socio-demographic data (population, urbanization rate, floor area per capita, etc.). The key output of the HEB model is floor area projections for different types of residential and tertiary buildings in different regions and their member states, the total energy consumption of residential and tertiary buildings, the energy consumption for heating and cooling, the energy consumption for hot water energy, the total CO_2 emissions, the CO_2 emissions for heating and cooling, and the CO_2 emissions for hot water energy.

3.2.1 The High-Efficiency Building Model Methodology

The HEB model conducts a scenario analysis for the entire building sector, in which the building sector is distinguished by location (rural, urban, and slum), building type (single-family, multifamily, commercial, and public buildings, with subcategories), and building vintage (existing, new, advanced new, retrofitted, and advanced retrofitted). This detailed classification of buildings is undertaken for 11 regions (Ürge-Vorsatz & Tirado Herrero, 2012), extended with country-specific results for the EU-27 countries, China, India, and the USA. Furthermore, within each region, different climate zones are considered to capture the differences in building energy uses and renewable energy generation caused by variations in climate. The climate zones are calculated based on four key climatic factors—heating degree days (HDD), cooling degree days (CDD), relative humidity (RH) in the warmest month, and average temperature in the warmest month (T). These parameters are processed using the GIS5 tool—spatial analysis—and performed with the ArcGIS software. The detailed classification categories are summarized in Table 3.2.

The purpose of the detailed classification of building categories and scenario assessments is to explore the consequences of certain policy directions or decisions that inform policy-making (Table 3.15).

The key input data used in the HEB are region-specific forecasts of GDP, population, rate of urbanization, and the proportion of the population living in urban slums. The time resolution of the model is yearly, so that socio-economic input data can be easily obtained from various credible sources, such as the databases of the World Bank, United Nations Development Programme (UNDP), EUROSTAT, and the OECD. Besides these socio-economic parameters, many others are included, and in the case of data absence, assumptions are made in the HEB model to calculate the final energy demand. Figure 3.11 shows the main workflow of the HEB model.

The HEB model includes several calculation steps, from considering the input data to obtaining the final output. Each of these calculation steps is discussed in the sections below.

Table 3.15 Building classification scheme of HEB

Classification scope	Categories	Subscript notation
Regions	11 key geographic regions +30 focus countries	r
Climate zones	17 different climate zones	c
Urbanization	Urban/rural areas	u
Building category	Residential/commercial and public/slums	b
Building type	Single-family houses (SF)/multifamily houses (MF) (residential sector) Educational/hotel and restaurant/hospital/retail/office/others (commercial and public sector)	t
Building vintage	Existing/new/advanced new/retrofitted/advanced retrofitted	v

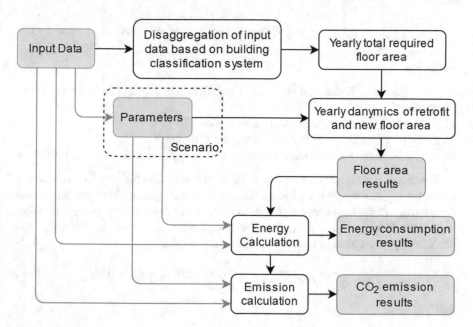

Fig. 3.11 The main workflow of the HEB. Input data and parameters can be modified by the user (green). Main outputs are the floor areas of different building vintage types and the energy consumption and CO_2 emissions of the stock (blue)

3.2.2 Disaggregation

In the first step of the calculation, after all the socio-economic input data are obtained, the input is disaggregated into the detailed building classification scheme, and the total floor area required to satisfy the year-specific population and GDP needs (the year is denoted with Y in subscript) is determined. The core concept for calculating the floor area differs for residential and commercial buildings:

- For residential buildings, the total occupied floor area correlates with the population, and thus, population forecasts are used to determine the floor areas of buildings in each region.
- For commercial and public buildings, the floor area correlates more strongly with GDP, so GDP forecasts are used as a proxy to determine the total floor space areas of commercial and public buildings.

The region-specific population data—as the input for the calculation—is further disaggregated into urban and rural populations based on the urbanization rate and into the different climate zones based on GIS data:

$$P_{r,c,u,Y} = P_{r,Y} \times U_{r,Y} \times Sc_{r,c} \quad \text{if } u = \text{urban} \tag{3.1}$$

and

$$P_{r,c,u,Y} = P_{r,Y} \times (1 - U_{r,Y}) \times Sc_{r,c} \quad \text{if } r = \text{rural} \tag{3.2}$$

where

$P_{r,c,u,Y}$ [capita]   is the total urban/rural population of region r and climate zone c in year Y
$P_{r,Y}$ [capita]   is the total population of region r in year Y
$U_{r,Y}$ [−]   is the urbanization rate of region r in year Y
$Sc_{r,c}$ [%]   is the share of the population within region r living in climate zone c

The urban population is then further disaggregated into the population living in slums (in regions where a significant number of people do not have access to standard living conditions) and the population living in conventional residential buildings. The latter group is split into the populations living in single-family and multifamily houses based on region-specific fixed values:

$$P_{r,c,u,b,Y} = P_{r,c,u,Y} \times Ss_{r,Y} \quad \text{where } u = \text{urban and } b = \text{slum} \tag{3.3}$$

and

$$P_{r,c,u,b,Y} = P_{r,c,u,Y} \times (1 - Ss_{r,Y}) \quad \text{where } u = \text{urban and } b = \text{residential} \tag{3.4}$$

then

$$P_{r,c,u,b,t,Y} = P_{r,c,u,b,Y} \times Ssf_{r} \quad \text{where } u = \text{urban, } b = \text{residential and } t = \text{SF} \tag{3.5}$$

and

$$P_{r,c,u,b,t,Y} = P_{r,c,u,Y} \times (1 - Ssf_{r}) \quad \text{where } u = \text{urban, } b = \text{residential and } t = \text{MF} \tag{3.6}$$

where

$P_{r,c,u,b,Y}$ [capita] is the total urban/rural population of region r, climate zone c, and building category b in year Y.

$P_{r,c,u,Y}$ [capita] is the total urban/rural population of region r and climate zone c in year Y.

$P_{r,c,u,b,t,Y}$ [capita] is the total urban/rural population of region r, climate zone c, building category b, and building type t in year Y.

$Ss_{r,Y}$ [%] is the share of the urban population living in slums in region r and year Y.

Ssf_r [%] is the share of the urban population living in single-family houses in region r.

The population living in rural areas is assumed to live in single-family houses.

The disaggregation of GDP follows the same pattern, except that the share of GDP that can be associated with rural commercial or public buildings is fixed within the modelling period:

$$\text{GDP}_{r,c,u,Y} = \text{GDP}_{r,Y} \times \left(1 - U_{r,Y}\right) \times \text{Sc}_{r,c} \quad \text{if } u = \text{urban} \tag{3.7}$$

and

$$\text{GDP}_{r,c,u,Y} = \text{GDP}_{r,Y} \times U_{r,Y} \times \text{Sc}_{r,c} \quad \text{if } u = \text{rural} \tag{3.8}$$

where

$\text{GDP}_{r,c,u,Y}$ [USD] is the total GDP that can be associated with urban/rural commercial or public buildings in region r and climate zone c in year Y.

$\text{GDP}_{r,Y}$ [USD] is the total GDP of region r in year Y.

$U_{r,Y}$ [−] is the urbanization rate of region r in year Y.

$\text{Sc}_{r,c}$ [%] is the share of climate zone c within region r.

The share of different commercial building types is also determined with fixed ratios based on data from the literature:

$$\text{GDP}_{r,c,u,t,Y} = \text{GDP}_{r,c,u,Y} \times \text{Scp}_t \tag{3.9}$$

where

$\text{GDP}_{r,c,u,t,Y}$ [USD] is the total GDP that can be associated with urban/rural commercial or public buildings of type t in region r and climate zone c in year Y.

$\text{GDP}_{r,c,u,Y}$ [USD] is the total GDP that can be associated with urban/rural commercial or public buildings in region r and climate zone c in year Y.

Scp_t [%] is the share of commercial and public buildings of type t in the commercial and public building stock.

3.2.3 Determining the Total Floor Area

Different equations are used for the calculation of the floor area of residential buildings and non-residential buildings. The floor area of residential buildings can be calculated with the following equation, using specific floor area values (the floor area that is occupied by one person):

$$\text{TFA}_{r,c,u,b,t,Y} = P_{r,c,u,b,t,Y} \times \text{SFAc}_{r,u,b,t,Y} \quad \text{where } b = \text{residential / slum} \quad (3.10)$$

where

$\text{TFA}_{r,c,u,b,t,Y}$ $[m^2]$ is the total urban/rural floor area of building category b and building type t in region r and climate zone c in year Y.
$P_{r,c,u,b,t,Y}$ [capita] is the total urban/rural population of region r, climate zone c, building category b, and building type t in year Y.
$\text{SFAc}_{r,u,b,t,Y}$ $[m^2/\text{capita}]$ is the specific floor area of building category b and building type t in region r in year Y.

Similarly, the floor area of commercial and public buildings is calculated using specific floor area values (the floor area that is required to produce one unit of GDP):

$$\text{TFA}_{r,c,u,b,t,Y} = \text{GDP}_{r,c,u,t,Y} \times \text{SFAg}_{r,b,Y} \quad \text{if } b = C \& P \quad (3.11)$$

where

$\text{TFA}_{r,c,u,b,t,Y}$ $[m^2]$ is the total urban/rural floor area of commercial or public buildings of building type t in region r and climate zone c in year Y.
$\text{GDP}_{r,c,u,t,Y}$ [USD] is the total GDP that can be associated with urban/rural commercial or public buildings of type t in region r and climate zone c in year Y.
$\text{SFAg}_{r,b,Y}$ $[m^2/\text{USD}]$ is the specific floor area of commercial or public buildings in region r in year Y.

Specific floor area values are determined from statistical data for each region. To take socio-economic development into account, the floor area per capita and the floor area per GDP are modelled as values that change yearly, reaching the average for OECD countries by the end of the modelling period in developing regions.

3.2.4 Yearly Dynamics of Floor Area Changes

The yearly dynamics of this floor area model transition the existing building stock into the future state determined by the scenarios. This includes the retrofitting or demolition of existing buildings, as well as the introduction of new buildings to the stock. In some cases, the floor area is left abandoned, which might result from a reduction in the population (e.g. in developed regions) or an increased rate of urbanization due to which buildings located in rural areas are abandoned after a certain time. It is important to capture this phenomenon, because abandoned buildings do not contribute to energy consumption or the emissions of the building stock. This yearly dynamic of the vintage types of buildings is presented in Fig. 3.12.

The demolished floor area is calculated with region-specific demolition rates. After the demolished floor area is subtracted from the existing total, the remaining existing floor area is classified into different building vintages. Similarly, the retrofitted floor area is calculated by applying the yearly changing region-specific retrofitting rate to the total existing building stock. The retrofitted floor area is further classified into two types: advanced retrofitted floor area and normal retrofitted floor area. For each of the regions, the shares of retrofitted and advanced retrofitted floor area differ, and the shares of advance retrofitted, advance new, and retrofitted floor areas also vary under different scenarios. The floor area from new constructions is classified into two building vintages: new and advanced new. Like the retrofitted floor area, the share of advanced new floor area also varies under different scenarios.

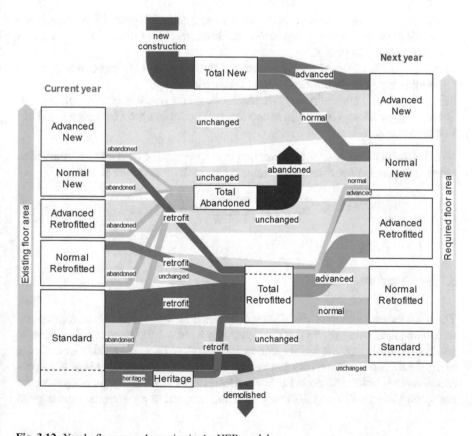

Fig. 3.12 Yearly floor area dynamics in the HEB model

3.2.5   Calculating the Energy Consumption of Buildings

The energy consumption for heating and cooling depends on the floor area. Therefore, in the HEB model, energy consumption is calculated after the year-specific floor area is calculated. The key input required to calculate the energy consumption for heating and cooling is the average consumption data for heating and cooling, which are usually obtained from data reported in the literature, for each of the regions, climate zones, and building types, because different building vintages have different consumption requirements. Therefore, different vintage types are modelled by assuming different energy intensities (denoted with subscript v). The values also depend on the scenario (denoted with subscript s). Energy intensity is multiplied by the corresponding floor area to determine the energy consumption for heating and cooling the stock:

$$HCE_{r,c,u,b,t,Y,v,s} = TFA_{r,c,u,b,t,Y} \cdot EUhc_{r,c,u,b,t,v,s} \tag{3.12}$$

where

$HCE_{r,c,u,b,t,Y,v,s}$ [kWh/year]   is the total energy demand for heating and cooling of buildings with vintage type v in scenario s, building type t in region r, and climate zone c in year Y.

$TFA_{r,c,u,b,t,Y}$ [m^2]   is the total urban/rural floor area of building category b and building type t in region r and climate zone c in year Y.

$EIhc_{r,c,u,b,t,v,s}$ [kWh/m^2/year]   is the heating and cooling energy intensity of buildings of vintage type v in scenario s and building type t in region r and climate zone c.

After the detailed energy consumption is calculated, the data can be summed to arrive at the region-specific, yearly aggregated results for a given scenario:

$$\text{Total Energy}_{r,Y,s} = \sum_c \sum_u \sum_b \sum_t \sum_v \text{Total Energy}_{r,c,u,b,t,Y,v,s} \tag{3.13}$$

3.2.6   Implementation

The most recent version of the HEB model was developed in the Python programming language, using the PyData ecosystem to handle large datasets. This ecosystem ensures quite large flexibility among the modelling parameters, and the diversity of input data and its granularity can be properly handled. This model is not an open-access model, but the Central European University has received funding from the European Union's Horizon 2020 research and innovation programme (under grant agreement no. 837089) in the Sentinel[1] project, to develop HEB further. In this project, the HEB model will be made an open-source model that users can use without cost.

References

Avrin, A-P. (2016). *Implications of electricity transmission for integrating variable renewable energies – Young scientists summer program – IIASA*. International Institute for Applied Systems Analysis. https://iiasa.ac.at/web/scientificUpdate/2015/cb/Avrin-Anne-Perrine3.html. Accessed 20 Dec 2021.

Chatterjee, S, Ürge-Vorstaz, D. (2020) *D3.1: Observed trends and modelling paradigms. Topic: LC-SC3-CC-2-2018 of the horizon 2020 work program: Modelling in support to the transition to a low-carbon energy system in Europe. Building a low-carbon, climate resilient future: Secure, clean and efficient energy*. SENTINEL. https://sentinel.energy/wp-content/uploads/2021/02/D-3.1-837089-EC.pdf

Chatterjee, S., Kiss, B., & Ürge-Vorsatz, D. (2021). *How far can building energy efficiency bring us towards climate neutrality?* Central European University.

Farjana, S. H., Huda, N., Mahmud, M. A. P., & Saidur, R. (2018). Solar process heat in industrial systems – A global review. *Renewable and Sustainable Energy Reviews, 82*, 2270–2286. https://doi.org/10.1016/j.rser.2017.08.065

IEA. (2020). *World energy balances 2020*. IEA. https://www.iea.org/data-and-statistics?country=WORLD&fuel=Energytransitionindicators&indicator=TFCShareBySector. Accessed 6 Apr 2021.

IEA. (2021). *IEA world energy statistics and balances*. IEA. https://doi.org/10.1787/enestats-data-en. https://www.oecd-ilibrary.org/energy/data/iea-world-energy-statistics-and-balances_enestats-data-en. Accessed 10 Oct 2021.

IPCC. (2021). Summary for policymakers. In V. P. Masson Delmotte, A. Zhai, S. L. Pirani, et al. (Eds.), *Climate change 2021: The physical science basis. Contribution of Working Group I to the Sixth Assessment Report of the Intergovernmental Panel on Climate Change*. Cambridge University Press.

MSCI. (2020). *Global industry classification standard (GICS®) methodology*. Guid. Princ. Methodol. GICS. https://www.msci.com/documents/1296102/11185224/GICS+Methodology+2020.pdf/9caadd09-790d-3d60-455b-2a1ed5d1e48c?t=15784059356. Accessed 31 Jan 2021.

Nikas, A., Doukas, H., & Papandreou, A. (2019). A detailed overview and consistent classification of climate-economy models. In *Understanding risks and uncertainties in energy and climate policy: Multidisciplinary methods and tools for a low carbon society* (pp. 1–54). https://doi.org/10.1007/978-3-030-03152-7_1

NZAOA. (2021). *UN-convened Net-Zero Asset Owner Alliance – Institutional investors transitioning their portfolios to net zero GHG emissions by 2050*. https://www.unepfi.org/net-zero-alliance/. Accessed 3 Oct 2021.

Pagenkopf, J., van den Adel, B., Deniz, Ö., & Schmid, S. (2019). Transport transition concepts. In S. Teske (Ed.), *Achieving the Paris climate agreement goals: Global and regional 100% renewable energy scenarios with non-energy GHG pathways for +1.5 C and +2 C* (pp. 131–159). Springer International Publishing. https://doi.org/10.1007/978-3-030-05843-2_6

---

[1] https://sentinel.energy/

Prieto, M. P., de Uribarri, P. M. Á., & Tardioli, G. (2019). Applying modeling and optimization tools to existing city quarters. *Urban Energy Syst Low-Carbon Cities*, 333–414. https://doi.org/10.1016/B978-0-12-811553-4.00010-X

Schlenzig, C. (1999). Energy planning and environmental management with the information and decision support system MESAP. *International Journal of Global Energy Issues, 12*(1/2/3/4/5/6), 81–91. https://ideas.repec.org/a/ids/ijgeni/v12y1999i1-2-3-4-5-6p81-91.html

Seidenberger, T., Thrän, D., Offermann, R., et al. (2010). Global biomass potentials. Investigation and assessment of data, remote sensing in biomass potential research, and country-specific energy crop potentials. *Epic [r]evolution – A Sustain World Energy Outlook, 3*, 166–168.

Seven2one. (2012). Mesap/PlaNet Software Framework. *Mesap4, Release, 4*(14), 1.9.

Simon, S., Naegler, T., & Gils, H. C. (2018). Transformation towards a renewable energy system in Brazil and Mexico—Technological and structural options for Latin America. *Energies, 11*, 907. https://doi.org/10.3390/EN11040907

Süsser, D., Gaschnig, H., Ceglarz, A., Stavrakas, V., Flamos, A., & Lilliestam, J. (2022). Better suited or just more complex? On the fit between user needs and modeller-driven improvements of energy system models. *Energy, 239*(Part B), 121909., ISSN 0360-5442. https://doi.org/10.1016/j.energy.2021.121909

Teske, S. (2015). Thesis, bridging the gap between energy and grid models. In *Developing an integrated infrastructural planning model for 100% renewable energy systems in order to optimize the interaction of flexible power generation, smart grids and storage technologies, chapter 2*. University Flensburg.

Teske, S., Pregger, T., Simon, S., et al. (2019a). Methodology. In S. Teske (Ed.), *Achieving the Paris climate agreement goals: Global and regional 100% renewable energy scenarios with non-energy GHG pathways for +1.5°C and +2°C* (pp. 25–78). Springer International Publishing.

Teske, S., Pregger, T., Naegler, T., Simon, S., Pagenkopf, J., van den Adel, B., & Deniz, Ö. (2019b). Energy scenario results. In *Achieving the Paris climate agreement goals: Global and regional 100% renewable energy scenarios with non-energy GHG pathways for +1.5 C and +2 C* (pp. 175–401). https://doi.org/10.1007/978-3-030-05843-2_8

UNFCCC. (2015). *United Nations Climate Change, website, Report of the Conference of the Parties on its twenty-first session*, December 2015, at https://unfccc.int/sites/default/files/resource/docs/2015/cop21/eng/10a01.pdf

Ürge-Vorsatz, D., & Tirado Herrero, S. (2012). Building synergies between climate change mitigation and energy poverty alleviation. *Energy Policy, 49*, 83–90. https://doi.org/10.1016/J.ENPOL.2011.11.093

World Bank. (2021). *World Bank Indicators API*. Available at https://datahelpdesk.worldbank.org/knowledgebase/articles/889386-developer-information-overview; and OECD API, available at https://data.oecd.org/api/sdmx-json-documentation/

Chapter 4
Classification Systems for Setting Net-Zero Targets for Industries

Sven Teske, Kriti Nagrath, and Sarah Niklas

Abstract The structure of the Global Industry Classification Standard (GICS) system and how it is used in the OECM are described, as well as how the statistical data of the International Energy Agency (IEA) were merged with the GICS structure. The development of the pathways for the industry and service sectors, based on the GICS and IEA data, is explained, together with the parameters that are important for the financial industry. In this context, the definitions of *Scope 1*, *2*, and *3* emissions newly developed for the OECM are explained, as well as how the systemic error of double counting in the original procedure can now be avoided.

Keywords Global Industry Classification Standard (GICS) · IEA Statistic · *Scope 1 · 2 · and 3* emissions

Investment decisions, such as the decarbonization targets for the finance industry (see also Chap. 2), are highly complex processes. In November 2020, the European Central Bank published a *guide on climate-related and environmental risks*, which maps a detailed process for undertaking 'climate stress tests' for investment portfolios. To achieve the Paris Climate Agreement goals in the global finance industry, decarbonization targets and benchmarks for individual industry sectors are required. This opens up a whole new research area for energy modelling because although decarbonization pathways have been developed for countries, regions, and communities, few have been developed for industry sectors. The OneEarth Climate Model (OECM) is an integrated assessment model for climate and energy pathways that focuses on 1.5 °C scenarios (Teske et al., 2019) and has been further improved to meet this need. To

S. Teske (✉) · K. Nagrath · S. Niklas
University of Technology Sydney – Institute for Sustainable Futures (UTS-ISF),
Sydney, NSW, Australia
e-mail: sven.teske@uts.edu.au

develop energy scenarios for industry sectors classified under the Global Industry Classification Standard (GICS), the technological resolution of the OECM required significant improvement. Furthermore, all demand and supply calculations had to be broken down into industry sectors before the individual pathways could be developed.

4.1 Role of the Global Industry Classification Standard (GICS) in Achieving Net-Zero Targets

The GICS was developed by the American investment research firm Morgan Stanley Capital International (MSCI) and Standard & Poor's (S & P), a finance data and credit rating company, in 1999. According to MSCI, the GICS was designed to define specific industry classifications for reporting, comparison, and investment transaction processes (MSCI, 2020). The GICS has 4 classification levels and includes 11 sectors, 24 industry groups, 69 industries, and 158 sub-industries. The 11 GICS sectors are energy, materials, industrials, consumer discretionary, consumer staples, health care, financials, information technology, communication services, utilities, and real estate (Table 4.1).

This section provides an overview of the 1.5 °C sectorial pathways and the associated GICS sectors. The individual end-use sectors are subdivided into four major sections:

1. Industry (Chap. 5)
2. Service (Chap. 6)
3. Buildings (Chap. 7)
4. Transport (Chap. 8)

The focus of each of these sections is documented in dedicated chapters (see above) that focus exclusively on current and future market developments and their energy-related aspects. The non-energy-related greenhouse gas (GHG) emissions are described in a separate section (Chaps. 11 and 14).

Table 4.1 GICS: 11 main industries

10	Energy
15	Materials
20	Industrials
25	Consumer discretionary
30	Consumer staples
35	Health care
40	Financials
45	Information technology
50	Communication services
55	Utilities
60	Real estate

The primary energy sector, fossil-fuel-producing companies, and the secondary energy industries, energy-distributing utilities, make up their own two GICS groups.

4.1.1   OECM 1.5 °C Industry Pathways and the Associated GICS Sectors

Table 4.2 provides an overview of the OECM 1.5 °C industry pathways. The majority are in the *materials* sector (1510) and the related sub-sectors *chemicals* (1510 10), *construction materials/cement* (1510 20), *aluminium* (15104010), and *steel* (15104050). Textiles and leather, which are classified as *consumer durables and apparel* (2520) in the subgroup *textiles* (2520 3030), are included because textiles and leather production are part of the *industry* sector in the International Energy Agency (IEA) World Energy Balances. To maintain consistency in the data sources across all the sectors analysed and to integrate the supply side with the OECM, this sub-sector cannot always follow the GICS categorization.

Table 4.2   OECM 1.5 °C industry pathways and the associated GICS sectors

1510 Materials	
1510 10	*Chemicals*
	1510 1010 Commodity chemicals
	1510 1020 Diversified chemicals
	1510 1030 Fertilizers and agricultural chemicals
	1510 1040 Industrial gases
	1510 1050 Specialty chemicals
1510 20	*Construction materials*
	1510 2010 Construction materials (including cement)
1510 30	*Containers and packaging*
	1510 3010 Metal and glass containers
	1510 3020 Paper Packaging
1510 40	*Metals and mining*
	15104010 Aluminium
	1510 4020 Diversified metals and mining
	1510 4025 Copper
	1510 4030 Gold
	1510 4040 Precious metals and minerals
	1510 4045 Silver
	15104050 Steel
1510 50	*Paper and forest products*
	1510 5010 Forest products
	1510 5020 Paper products
25	*Consumer discretionary*
	2520 3030 Textiles

4.1.2 OECM 1.5 °C Service and Energy Pathways and the Associated GICS Sectors

The four service sectors analysed are distributed across four GICS sectors. Agriculture and food processing is part of *consumer staples* (30), forestry and wood products are part of the *materials* group (1510 50), the fisheries industry is only represented by its actual product (fish), and the fishing fleet is not part of the GICS classification (Table 4.3). Finally, water utilities are part of the wider *utilities* group (55). The OECM 1.5 °C pathways for the primary energy supply are all included in the *energy* group (10), whereas the secondary energy supply is part of the *utilities* group (55) (Table 4.4).

4.1.3 1.5 °C Pathways for Buildings and Transport and the Associated GICS Sectors

The OECM pathways for buildings are all included in GICS sector 60—*real estate* (Table 4.5). However, it is unclear to what extent the actual electricity demand, especially of residential buildings, can be considered as part of an economic activity and therefore as the responsibility of the real estate industry itself. Whereas the energy demand for climatization (heating and cooling) is directly related to the building envelope and architecture and is therefore the responsibility of the real estate industry, the electricity demand for appliances is not related and is the responsibility of private households.

Table 4.3 1.5 °C OECM service pathways and the associated GICS sectors

30 Consumer staples	
3010	Food and staple retailing
	30101010 Drug retail
	30101020 Food distributors
	30101030 Food retail
	30101040 Hypermarkets and supercentres
3020	Food, beverages, and tobacco
	30201010 Brewers
	30201020 Distillers and vintners
	30201030 Soft drinks
302020	Food products
	30202010 Agricultural products
	30202030 Packaged foods and meats (including fish)
302030	Tobacco
	30203010 Tobacco

Table 4.4   OECM 1.5 °C energy pathways and the associated GICS sectors

10 Energy	
1010 10	Energy Equipment & Services
	10101010 Oil & Gas Drilling
	10101020 Oil & Gas Equipment & Services
1010 20	Oil, Gas & Consumable Fuels
	10102010 Integrated Oil & Gas
	10102020 Oil & Gas Exploration & Production
	10102030 Oil & Gas Refining & Marketing
	10102040 Oil & Gas Storage & Transportation
	10102050 Coal & Consumable Fuels
55 Utilities	
5510	Electric Utilities
5520	Gas Utilities
5530	Multi Utilities
5540	Water Utilities
5550	Independent Power and Renewable Electricity Producers

Table 4.5   OECM 1.5 °C energy pathways and the associated GICS sectors

60 Real estate	
6010 Real estate	
601010	Equity real estate investment trusts (REITs)
	60101010 Diversified REITs
	60101020 Industrial REITs
	60101030 Hotel and resort REITs
	60101040 Office REITs
	60101050 Healthcare REITs
	60101060 Residential REITs
	60101070 Retail REITs
	60101080 Specialized REITs
601020	Real estate management and development
	60102010 Diversified real estate activities
	60102020 Real estate operating companies
	60102030 Real estate development
	60102040 Real estate services

Finally, the transport sector is part of the *industrials* group (20) and is repre-sented as a subgroup under *transportation* (20) (Table 4.6).

Table 4.6 OECM 1.5 °C energy pathways and the associated GICS sectors

20 Industrials	
2030 Transportation	
203010	Air freight and logistics
	20301010 Air freight and logistics
203020	Airlines
	20302010 Airlines
203030	Marine
	20303010 Marine
203040	Road and rail
	20304010 Railroads
	20304020 Trucking
203050	Transportation infrastructure
	20305010 Airport services
	20305020 Highways and rail tracks
	20305030 Marine ports and services

4.2 Adaptation of Energy Statistical Databases to the GICS Industry System

The OECM uses the IEA World Energy Statistics and Balances (IEA, 2021b) as one of the main input sources for energy demand and supply data for the base year and the historical time series for model calibration, as described in Chap. 3. To develop energy scenarios that are based on the GICS classification, the IEA final energy demand sectors used for statistical data must be adapted to GICS sectors. This section provides an overview of the two different categorization systems and how they differ.

4.2.1 The Industry Sector

The IEA database documentation (IEA, 2020) provides detailed information about various statistical parameters. Table 4.7 shows the IEA *industry* sector and how it is broken down into four main sub-sectors:

1. Mining and quarrying
2. Construction
3. Machinery
4. Manufacturing

The *manufacturing* sector consists of 11 industries, among the largest of which are iron and steel, chemical and petrochemical industries, and non-metallic minerals, which includes the cement industry. The IEA (IEA DB, 2020) identifies the

Table 4.7 IEA World Energy Balances—definition of the *industry* sector (IEA, 2021a; ISIC, 2008)

IEA statistic—industry		
Mining and quarrying		[ISIC Rev. 4 Divisions 07 and 08 and Group 099] Mining (excluding fuels) and quarrying
		[Div. 7] Mining of metal ores
		[Div. 8] Other mining and quarrying
		[Div. 099] Support activities for other mining and quarrying
Construction		[ISIC Rev. 4 Divisions 41–43]
		[Div. 41] Construction of buildings
		[Div. 42] Civil engineering
		[Div. 43] Specialized construction activities
Machinery		[ISIC Rev. 4 Divisions 25–28] Fabricated metal products, machinery, and equipment other than transport equipment
		[Div. 25] Manufacture of fabricated metal products, except machinery and equipment
		[Div. 26] Manufacture of computer, electronic, and optical products
		[Div. 27] Manufacture of electrical equipment
		[Div. 28] Manufacture of machinery and equipment
Manufacturing	'Manufacturing' refers to the sum of the following industrial sub-sectors:	
	Iron and steel	[ISIC Rev. 4 Group 241 and Class 2431]
	Chemicals and petrochemicals	[ISIC Rev. 4 Divisions 20 and 21] Excluding petrochemical feedstock
		[Div. 20] Manufacture of chemicals and chemical products
		[Div. 21] Manufacture of pharmaceuticals, medicinal chemicals, and botanical products
	Non-ferrous metals	[ISIC Rev. 4 Group 242 and Class 2432] Basic industries
		[Div. 2420] Manufacture of basic precious and other non-ferrous metals: gold, silver, platinum, zinc, aluminium
		[Div. 2432] Casting of non-ferrous metals
	Non-metallic minerals	[ISIC Rev. 4 Division 23] Including glass, ceramic, cement, etc.
		[Div. 23] Manufacture of other non-metallic mineral products: glass and glass products, ceramic products, tiles and baked clay products, and cement and plaster, from raw materials to finished articles
	Transport equipment	[ISIC Rev. 4 Divisions 29 and 30]
		[Div. 29] Manufacture of bodies (coachwork) for motor vehicles, manufacture of trailers and semi-trailers
		[Div. 30] Manufacture of other transport equipment including ship and boat building, manufacture of locomotives, aircraft, and spacecraft
	Food and tobacco	[ISIC Rev. 4 Divisions 10–12]

(continued)

Table 4.7 (continued)

IEA statistic—industry		
		[Div. 10] Manufacture of food products
		[Div. 11] Manufacture of beverages
		[Div. 12] Manufacture of tobacco products
	Paper, pulp, and print	[ISIC Rev. 4 Divisions 17 and 18]
		[Div. 17] Manufacture of paper and paper products
		[Div. 18] Printing and reproduction of recorded media
	Wood and wood products	[ISIC Rev. 4 Division 16] Wood and wood products other than pulp and paper
		[Div. 16] Manufacture of wood and of products of wood and cork, except furniture, manufacture of articles of straw and plaiting materials
	Textile and leather	[ISIC Rev. 4 Divisions 13–15]
		[Div. 13] Manufacture of textiles
		[Div. 14] Manufacture of wearing apparel
		[Div. 15] Manufacture of leather and related products
	Industries not specified elsewhere	[ISIC Rev. 4 Divisions 22, 31, and 32] Any manufacturing industry not included above

subgroups of all economic sectors based on the International Standard Industrial Classification of All Economic Activities (ISIC) of the United Nations (ISIC, 2008).

The *iron and steel* sector, for example, includes all activities listed under ISIC divisions 241 and 2431. The ISIC lists under Division 241 *manufacture of basic iron and steel* and under 2431 *casting of iron and steel: This class includes the casting of iron and steel,* i.e. *the activities of iron and steel foundries. This class includes:*

* *Casting of semi-finished iron products*
* *Casting of grey iron castings*
* *Casting of spheroidal graphite iron castings*
* *Casting of malleable cast iron products*
* *Casting of semi-finished steel products*
* *Casting of steel castings*
* *Manufacture of tubes, pipes and hollow profiles, and tube or pipe fittings of cast iron*
* *Manufacture of seamless tubes and pipes of steel by centrifugal casting*
* *Manufacture of tube or pipe fittings of cast steel*

However, the IEA statistics do not provide a further breakdown of the energy demand for the specific economic activities listed under the ISIC divisions but lump them together. In terms of iron and steel, only one value is provided, and no further details are available. To match the IEA sector with the GICS sectors, the *industry* and *service* sectors of the IEA have been grouped according to GICS classes. The iron and steel industry is part of the GICS industry sector 15 *materials* (Table 4.2), subclass 151,040 *metals and mining*, and the sub-industry 15104050 *steel*. This group includes iron ore, as identified in the documentation. However, the same group (15 *materials*) also lists the *aluminium* industry (15194010)—a separate IEA statistical sector. Although the industry sectors of the IEA and the GICS systems correspond to a large extent, the *service* sector has significant differences.

4.2.2   The Service Sector

The IEA statistics do not have a *service* sector category as such. Under *other sectors*, the energy demand is broken down into four subgroups: (1) residential, (2) commercial and public services, (3) agriculture and forestry, and (4) fisheries.

Detailed data for *water utilities*, for example, are not available, and are part of *commercial and public services*, as highlighted in Table 4.8. When sector-specific data are not available, the energy demand has been estimated from the energy intensities based on GDP ([MJ/$GDP] or commodities, such as energy demand per cubic meter of water withdrawn [MJ/billion m^3 water]). Furthermore, the service sectors *agriculture and food* and *forestry & wood products* (Chap. 6) are partly from IEA's *other sectors* and partly from the *industry* section. Therefore, the current and future energy demand for agriculture and forestry has been derived bottom-up from energy intensities and calibrated with statistical data from the IEA for the years 2005–2019.

4.2.3   The Buildings Sector

The 1.5 °C OECM pathway for buildings (Chap. 7) consists of three sub-sectors: residential and commercial buildings and construction. The IEA statistics for the buildings sector is comprised of 'residential' and 'commercial and public services', excluding water utilities. There are also economic activities, such as Div. 38 ('waste collection, treatment and disposal activities; materials recovery'), that are outside the OECM scenario breakdown. Therefore, the *buildings* sector has been calculated separately with a bottom-up approach, from the floor space and energy intensities per square meter, to project the current and future energy demands. The energy data for *construction*, which is part of the *industry* group, are taken from the IEA statistics.

4.2.4   The Transport Sector

Statistical data for the transport sector in the IEA database best match the GICS classification '2030 Transportation', and the development of the OECM 1.5 °C pathways for aviation, shipping, and road transport is based directly on the IEA statistics. Table 4.9 describes the IEA data series for transport.

Table 4.8 IEA World Energy Balances—definition of *other sectors*

Other sectors	
Residential	Includes consumption by households, excluding fuels used for transport. Includes households with employed persons [ISIC Rev. 4 Divisions 97 and 98], which is a small part of total residential consumption
	[Div. 96] Other personal service activities
	[Div. 99] Activities of extraterritorial organizations and bodies
Commercial and public services	[ISIC Rev. 4 Divisions 33, 36–39, 45–47, 52, 53, 55–56, 58–66, 68–75, 77–82, 84 (excluding Class 8422), 85–88, 90–96, and 99]
	[Div. 33] Repair and installation of machinery and equipment
GICS: 55 *utilities*—**5510 40** *water utilities*	**[Div. 36] Water collection, treatment, and supply**
	[Div. 37] Sewage
	[Div. 38] Waste collection, treatment, and disposal activities, materials recovery
	[Div. 39] Remediation activities and other waste management services
	[Div. 45] Wholesale and retail trade and repair of motor vehicles and motorcycles
	[Div. 46] Wholesale trade, except of motor vehicles and motorcycles
	[Div. 47] Retail trade, except of motor vehicles and motorcycles
	[Div. 52] Warehousing and support activities for transportation
	[Div. 53] Postal and courier activities
	[Div. 55] Accommodation
	[Div. 56] Food and beverage service activities
	[Div. 58] Publishing activities
	[Div. 59] Motion picture, video, and television programme production, sound recording and music publishing activities
	[Div. 60] Programming and broadcasting activities
	[Div. 61] Telecommunications
	[Div. 62] Computer programming, consultancy, and related activities
	[Div. 63] Information service activities
	[Div. 64] Financial service activities, except insurance and pension funding
	[Div. 65] Insurance, reinsurance, and pension funding, except compulsory social security
	[Div. 66] Activities auxiliary to financial service and insurance activities
	[Div. 68] Real estate activities
	[Div. 69] Legal and accounting activities
	[Div. 70] Activities of head offices, management consultancy activities
	[Div. 71] Architectural and engineering activities, technical testing and analysis
	[Div. 72] Scientific research and development
	[Div. 73] Advertising and market research
	[Div. 74] Other professional, scientific, and technical activities
	[Div. 75] Veterinary activities
	[Div. 77] Rental and leasing activities
	[Div. 78] Employment activities
	[Div. 79] Travel agency, tour operator, reservation service, and related activities

(continued)

Table 4.8 (continued)

Other sectors	
	[Div. 80] Security and investigation activities
	[Div. 81] Services to buildings and landscape activities
	[Div. 82] Office administrative, office support, and other business support activities
	[Div. 84] Public administration and defence, compulsory social security
	[Div. 85] Education
	[Div. 88] Social work activities, without accommodation
	[Div. 90] Creative, arts, and entertainment activities
	[Div. 91] Libraries, archives, museums, and other cultural activities
	[Div. 92] Gambling and betting activities
	[Div. 93] Sports activities and amusement and recreation activities
	[Div. 94] Activities of membership organizations
	[Div. 95] Repair of computers and personal and household goods
	[Div. 96] Other personal service activities
	[Div. 99] Activities of extraterritorial organizations and bodies
Agriculture/forestry	Agriculture, hunting, and forestry—excluding agricultural highway use, power, or heating [ISIC Rev. 4 Divisions 01 and 02]
	[Div. 01] Crop and animal production, hunting, and related service activities
	[Div. 02] Forestry and logging
Fishing	Includes fuels used for inland, coastal, and deep-sea fishing. Fishing covers fuels delivered to ships of all flags that have refuelled in the country (including international fishing) and the energy used in the fishing industry [ISIC Rev. 4 Division 03]
	[Div. 3] Fishing and aquaculture

Table 4.9 IEA World Energy Balances—definition of the transport sector

Transport	
World aviation bunkers	Covers fuels delivered to aircraft of all countries that are engaged in international aviation (*international aviation bunkers*) for the world total aviation bunker demand
Domestic aviation	Aviation fuels to aircraft for domestic aviation—commercial, private, agricultural use
Road	Fuels used in road vehicles and for agricultural and industrial highway use. Excludes military consumption and the motor gasoline used in stationary engines and the diesel oil used in tractors that are not for highway use
Rail	Rail traffic, including industrial railways, and rail transport laid in public roads as part of urban or suburban transport systems (trams, metros, etc.)
Pipeline transport	Energy used in the support and operation of pipelines transporting gases, liquids, slurries, and other commodities, including the energy used for pump stations and the maintenance of pipelines
World marine bunkers	Fuels delivered to ships of all flags not engaged in international navigation (*international marine bunkers*) for the whole world marine bunker demand
Domestic navigation	Fuels delivered to vessels of all flags not engaged in international navigation
Transport not specified elsewhere	Transport not specified elsewhere

The reported differences between IEA and GICS categorization systems lead to some inconsistencies, and discrepancies between the available statistical energy data and the actual energy demands for specific economic activities are unavoidable. The advantage of the high technical resolution of the OECM is also a disadvantage because it requires a significant amount of data, which are sometimes unavailable. Therefore, the energy demand projections may vary from those in other sectorial analyses.

4.2.5   From Sectorial Energy Scenarios for Industry Sectors to Emissions

The finance industry requires sectorial energy scenarios for the industry and service sectors to set sector-specific decarbonization targets. Increasingly, investment decisions of international and national banks, insurance companies, and investor groups are driven by key performance indicators (KPIs) not only for profitability but also with regard to the embedded GHG emissions of a company. For asset managers, it has become increasingly important to have access to detailed information about GHG emissions, e.g. whether or not a steel manufacturer is on a decarbonization trajectory. The emissions must be further divided according to the responsibility for those emissions. This is done by calculating the so-called *Scopes 1*, *2*, and *3*.

4.3   Methodologies for Calculating *Scopes 1, 2*, and *3*

4.3.1   Calculation of Scopes 1, 2, *and* 3

Reporting corporate GHG emissions is important, and the focus is no longer only on direct energy-related CO_2 emissions but includes other GHGs emitted by industries. These increasingly include the indirect emissions that occur in supply chains (Hertwich & Wood, 2018). The Greenhouse Gas Protocol, a global corporate GHG accounting and reporting standard (WRI & WBCSD, 2021), distinguishes between three 'scopes':

- *Scope 1*—Emissions are direct emissions from owned or controlled sources.
- *Scope 2*—Emissions are indirect emissions from the generation of purchased energy.
- *Scope 3*—Emissions are all indirect emissions (not included in *Scope 2*) that occur in the value chain of the reporting company, including both upstream and downstream emissions.

The United States Environmental Protection Agency (US EPA) defines *Scope 3* emissions as 'the result of activities from assets not owned or controlled by the reporting organization, but that the organization indirectly impacts in its value chain. They include upstream and downstream of the organization's activities' (EPA, 2021). According to the EPA, *Scope 3* emissions include all sources of emissions not within an organization's *Scope 1* and *2* boundaries, and *Scope 3* emissions of one organization are *Scope 1* and *2* emissions of another organization. *Scope 3* emissions, also referred to as 'value chain emissions', often represent the majority of an organization's total GHG emissions (EPA, 2021).

Whereas the methodologies for calculating *Scope 1* and *Scope 2* emissions are undisputed, the method of calculating *Scope 3* emissions is an area of ongoing discussion and development (Baker, 2020; Liebreich, 2021; Lombard Odier, 2021). The main issues discussed are data availability, reporting challenges, and the risk of double counting. MSCI, for example, avoids double counting by using a 'deduplication multiplier of approximately 0.205' (Baker, 2020). This implies that the allocation of emissions based on actual data is not possible. Accounting methodologies for *Scope 3* emissions have been developed for entity-level accounting and reporting (WRI & WBCSD, 2013).

By contrast, the OECM model focuses on the development of 1.5 °C net-zero pathways for industry sectors classified under the GICS (MSCI, 2020), for countries or regions or at the global level. Emission-calculating methodologies for entity-level *Scope 3* require bottom-up entity-level data to arrive at exact figures. Therefore, data availability and accounting systems for whole industry sectors on a regional or global level present significant challenges.

Therefore, *Scope 3* calculation methodology must be simplified for country-, region-, and global-level calculations and to avoid double counting. In the Greenhouse Gas Protocol, *Scope 3* emissions are categorized into 15 categories, shown in Table 4.10.

To include all the upstream and downstream categories shown in Table 4.10 for an entire industry sector is not possible because, firstly, complete data are not available, for example, how many kilometres employees commute—and, secondly, it is impossible to avoid double counting, for example, when calculating *Scope 3* for the car industry.

The OECM methodology is based on the *Technical Guidance for Calculating Scope 3 Emissions* of the World Resource Institute (WRI & WBCSD, 2013) but is

Table 4.10 Upstream and downstream *Scope 3* emission categories (WRI & WBCSD, 2013; Baker, 2020)

Upstream		Downstream	
U1	Business travel	D1	Use of solid products
U2	Purchased goods and services	D2	Downstream transportation and distribution
U3	Waste generated in operations	D3	End-of-life treatment of solid products
U4	Fuel- and energy-related activities	D4	Investments
U5	Employee commuting	D5	Downstream leased assets
U6	Upstream transportation and distribution	D6	Processing of solid products
U7	Capital goods	D7	Franchises
U8	Upstream leased assets		

simplified to reflect the higher levels of industry- and country-specific pathways. The OECM defines the three emission scopes as follows:

Scope 1—All direct emissions from the activities of an organization or under its control, including fuel combustion on site (such as gas boilers), fleet vehicles, and air-conditioning leaks.
   Limitations of the OECM Scope 1 analysis: Only economic activities covered under the sector-specific GICS classification that are counted for the sector are included. All energy demands reported by the IEA *Advanced World Energy Balances* (IEA, 2021a) for the specific sector are included.
Scope 2—Indirect emissions from electricity purchased and used by the organization. Emissions are created during the production of energy that is eventually used by the organization.
   Limitations of the OECM Scope 2 analysis: Because data availability is poor, the calculation of emissions focuses on the electricity demand and 'own consumption', e.g. that reported by the IEA, 2021b for power generation.
Scope 3—GHG emissions caused by the analysed industry that are limited to sector-specific activities and/or products classified by the GICS.
   Limitations of the OECM Scope 3 analysis: Only sector-specific emissions are included. Traveling, commuting, and all other transport-related emissions are reported under *transport*. The lease of buildings is reported under *buildings*. All other financial activities, such as *capital goods*, are excluded because no data are available for the GICS industry sectors and would lead to double counting. The OECM is limited to energy-related CO_2 and energy-related methane (CH_4) emissions. All other GHG gases are calculated outside the OECM by Meinshausen and Dooley (2019).

The main difference between the OECM and the World Resources Institute (WRI) concept is that the interactions between industries and other services are kept separate. The OECM reports only emissions directly related to the economic activities classified by GICS. Furthermore, the industries are broken down into three categories: primary class, secondary class, and end-use activity class.

Table 4.11 shows a schematic representation of the OECM *Scope 1, 2*, and *3* calculation methods according to GICS class, which are used to avoid double counting. The sum of *Scopes 1, 2*, and *3* for each of the three categories is equal to the actual emissions. Example: The total annual global energy-related CO_2 emissions are 35 Gt in a given year.

- The sum of *Scopes 1, 2*, and *3* for the primary class is 35 $GtCO_2$.
- The sum of *Scopes 1, 2*, and *3* for the secondary class is 35 $GtCO_2$.
- The sum of *Scopes 1, 2*, and *3* for end-use activities is 35 $GtCO_2$.

Double counting can be avoided by defining a primary class for the primary energy industry, a secondary class for the supply utilities, and an end-use class for all the economic activities that use the energy from the primary- and secondary-class companies. The separation of all emissions by the defined industry categories—such as GICS—also streamlines the accounting and reporting systems. The

Table 4.11 Schematic representation of OECM *Scopes 1, 2,* and *3* according to GICS classes, to avoid double counting

	Primary class				Secondary class			End-use activity class		
	GICS 10 *energy*				**GICS 55** *utilities*			**All other industries and services**		
	Scope 1	Scope 2	Scope 3		Scope 1	Scope 2	Scope 3	Scope 1	Scope 2	Scope 3
CO_2				CO_2						
CH_2 AFOLU				CH_2 AFOLU						
CH_4				CH_4						
N_2O				N_2O						
				CFCs						
Total GHG	Sum of *Scopes 1, 2,* and *3* equals total emissions			Total GHG	Sum of *Scopes 1, 2,* and *3* equals total emissions			Sum of *Scopes 1, 2,* and *3* equals total emissions		

volume of data required is reduced, and reporting is considerably simplified under the OECM methodology.

For a specific industry sector to achieve the global targets of a 1.5 °C temperature increase and net-zero emissions by 2050 under the Paris Agreement requires that all its business activities are with other sectors that are also committed to a 1.5 °C and net-zero emission targets.

The results of the OECM *Scope 3* analysis are documented in Chap. 13.

References

Baker, B. (2020) *Scope 3 carbon emissions: Seeing the full picture.* MSCI. https://www.msci.com/www/blog-posts/scope-3-carbon-emissions-seeing/02092372761

EPA. (2021). *Scope 3 inventory guidance.* https://www.epa.gov/climateleadership/scope-3-inventory-guidance. Accessed 10 Oct 2021

Hertwich, E. G., & Wood, R. (2018). The growing importance of scope 3 greenhouse gas emissions from industry. *Environmental Research Letters, 13*(10), 104013. https://doi.org/10.1088/1748-9326/AAE19A

IEA. (2020). *World energy balances 2020.* IEA. https://www.iea.org/data-and-statistics?country=WORLD&fuel=Energytransitionindicators&indicator=TFCShareBySector. Accessed 6 Apr 2021

IEA. (2021a). *IEA world energy statistics and balances.* IEA. https://doi.org/10.1787/enestats-data-en. https://www.oecd-ilibrary.org/energy/data/iea-world-energy-statistics-and-balances_enestats-data-en. Accessed 10 Oct 2021

IEA. (2021b). *World energy balances 2021.* IEA. https://www.iea.org/data-and-statistics/data-product/world-energy-balances. Accessed 25 Oct 2021

ISIC. (2008). *International Standard Industrial Classification (ISIC) of all economic activities revision 4.* United Nations.

Liebreich, M. (2021). *Climate and finance – Lessons from a time machine | BloombergNEF.* BloombergNEF. https://about.bnef.com/blog/liebreich-climate-and-finance-lessons-from-a-time-machine/. Accessed 10 Oct 2021

Lombard Odier. (2021). *Debunking 7 misconceptions on scope 3 emissions.* Lombard Odier. https://www.lombardodier.com/contents/corporate-news/responsible-capital/2021/august/calculating-a-companys-carbon-fo.html

Meinshausen, M., & Dooley, K. (2019). Mitigation scenarios for non-energy GHG. In S. Teske (Ed.), *Achieving the Paris climate agreement goals global and regional 100% renewable energy scenarios with non-energy GHG pathways for +1.5°C and +2°C.* SpringerOpen.

MSCI. (2020). *Global industry classification standard (GICS®) methodology. Guiding principles and methodology for GICS.* https://www.msci.com/documents/1296102/11185224/GICS+Methodology+2020.pdf/9caadd09-790d-3d60-455b-2a1ed5d1e48c?t=1578405935658. Accessed 31 Jan 2021

Teske, S., Pregger, T., Naegler, T., et al. (2019). Energy scenario results. In S. Teske (Ed.), *Achieving the Paris climate agreement goals Sven Teske editor global and regional 100% renewable energy scenarios with non-energy GHG pathways for +1.5°C and +2°C.* SpringerOpen.

WRI & WBCSD. (2013). *Technical guidance for calculating scope 3 emissions, supplement to the corporate value chain (scope 3) accounting & reporting standard.* https://ghgprotocol.org/sites/default/files/standards/Scope3_Calculation_Guidance_0.pdf

WRI & WBCSD. (2021). *Greenhouse gas protocol.* WRI WBCSD. https://ghgprotocol.org/. Accessed 25 Oct 2021

Part IV
Sector-Specific Pathways

Chapter 5
Decarbonisation Pathways for Industries

Sven Teske, Sarah Niklas, and Simran Talwar

Abstract The decarbonisation pathways for the industry sectors are derived. The energy-intensive chemical industry, the steel and aluminium industries, and the cement industry are briefly outlined. The assumptions for future market development used for the scenario calculations are documented, and the assumed development of the energy intensities for product manufacture is presented. An overview of the calculated energy consumption and the resulting CO_2 intensities is given, with the assumed generation mix. The textile and leather industry is also included in this chapter because of its strong ties to the chemical industry and meat production (part of the service sector).

Keywords Net-zero pathways · Industry · Chemicals · Textile and leather · Steel · Aluminium · Cement · Energy intensities · Bottom-up demand projections

The global gross domestic product (GDP) in 2019 was US$87.8 trillion, 3% of which came from agriculture, 26% from industry, 15% from manufacturing, and the remaining 65% from services (World Bank, 2021). The aluminium, steel, and cement industries each had a 1% direct share of the global industry GDP value, and the chemical industry's share was 17%, although the indirect effects of those industries on the GDP were significantly higher. The materials produced by these four industries are essential for the manufacturing and service industries, which generate over 80% of the global GDP. In the next section, the status quo of the aluminium, steel, cement, and chemical industries, and that of the textile and leather industry, is briefly described. Their current production processes and energy intensities (by product unit or GDP value) and their efficiency potentials are documented. The assumptions made for the energy demand of the industry sectors if they are to achieve the OECM 1.5 °C pathways and their energy-related CO_2 emissions are also presented.

S. Teske (✉) · S. Niklas · S. Talwar
University of Technology Sydney – Institute for Sustainable Futures (UTS-ISF),
Sydney, NSW, Australia
e-mail: sven.teske@uts.edu.au

© The Author(s) 2022
S. Teske (ed.), *Achieving the Paris Climate Agreement Goals*,
https://doi.org/10.1007/978-3-030-99177-7_5

The section discusses the development of the energy demand for the industry sector, as defined in the International Energy Agency (IEA) World Energy Balances (IEA, 2020a). The section focuses on the *materials* group (1510) in terms of the *Global Industry Classification Standard* (GICS) classification, plus the textile and leather industry, which is included in the IEA industry statistics, but is classified as *consumer discretionary—textiles* (2520 3030) (see Chap. 4, Sects. 4.1 and 4.2).

5.1 Global Chemical Industry: Overview

The chemical industry is an important intermediate industry, engaged in the conversion of raw materials, such as fossil fuels, minerals, metals, and water, into a variety of chemical products used in other industrial sectors, including pharmaceuticals, fertilisers, pesticides, plastics, dyes, paints, and consumer products. Close overlaps exist between chemical and plastic industries, and many chemical producers are also involved in the manufacture of plastics. Revenue from the global chemical industry increased by 48% to US$3.9 trillion between 2005 and 2019 (Garside, 2020; ACC, 2021). Pharmaceuticals had the largest share in the segment-wise breakdown of the global chemical shipments in 2019 at 26.4%, followed by bulk petrochemicals and intermediates (16.4%), specialty chemicals (16%), plastic resins (12.2%), agricultural chemicals (8.6%), consumer products (8.3%), inorganic chemicals (7.1%), manufactured fibres (4%), and synthetic rubbers (1%). Together, the world's 100 leading chemical companies generated US$1.05 trillion in revenue in 2019 (ACC, 2020).

Basic organic and inorganic chemicals account for the highest shares of production and consumption (by volume) in the global chemical industry (UNEP, 2019):

- *Basic chemicals*, also known as 'commodity chemicals', consist of both organic and inorganic chemicals that are used as feedstock materials for a variety of downstream chemicals. Some of the most frequently used basic chemicals are methanol, olefins (such as ethylene and propylene), and aromatics (such as xylene, benzene, and toluene). Basic chemical production processes are well established, with high capital and energy demands. Among these basic chemicals, petrochemicals and their derivatives, such as organic intermediates, plastic resins, and synthetic fibres, are strongly traded commodities, and the ethylene, propylene, and methanol production capacities account for a vast share of petrochemical production globally.
- *Inorganic chemicals* include acids and bases, salts, industrial gases, and elements such as halogens. Inorganic chemicals are used as intermediate inputs in the manufacture of many specialty chemicals, such as solvents, coatings, surfactants, electronic chemicals, and agricultural chemicals. Nitrogen compounds account for the largest share of inorganic chemical production globally. With the current increases in glass and paper production, the demands for soda ash and caustic soda are increasing rapidly, coupled to the high demand for inorganic chemicals in the food and cosmetics industries.

5.1.1 Major Chemical Industry Companies and Countries

BASF (headquarters [HQ] in Germany), Dow (HQ USA), and Sinopec (HQ China) were some of the world's largest chemical-producing companies (based on sales) in 2018. Each of these three leaders exceeded US\$65,000 million in chemical sales. Eighteen countries were represented in the list of the top 50 chemical companies in 2019, and more than 50% of them were headquartered in the USA (10), Japan (8), and Germany (5) (ACS, 2019). German companies BASF, Bayer, and Linde are the foremost international producers. BASF, for example, owns global operations in the chemical industry and is active across the entire value chain, spanning the manufacture of chemicals, plastics, performance products, functional and agricultural solutions, oil, and natural gas. Bayer is a well-known pharmaceutical and chemical manufacturer, and the Linde group owns large industrial gas and engineering facilities, which produce various gas products, including atmospheric oxygen, nitrogen, and argon.

5.1.2 Chemical Manufacturing and Energy Intensity

The chemical industry uses raw materials from natural gas, ethane, oil-refining by-products (including propylene), and salt to manufacture bulk chemicals, such as sulphuric acid, ammonia, chlorine, industrial gases, and basic polymers, including polyethylene and polypropylene. The manufacturing activity within the chemical industry can be divided into two main categories: basic chemicals and chemical products.

Basic chemicals are those chemicals that feed into the manufacture of other complex chemicals. Petroleum and coal products can be considered basic chemicals because they are used in the manufacture of a variety of polymers, fibres, and other chemicals. The manufacturing processes for basic chemicals, including inorganic chemicals, organic chemicals (such as ethylene and propylene), and agricultural chemicals, are considered energy-intensive industries and require large production facilities.

The second category involves the manufacture of ammonia, polyethylene, and other chemical products. Ammonia production is an energy-intensive process and is considered to be an important contributor to the chemical industry's energy and emission footprints. Ammonium nitrate is used as an agricultural fertiliser and as a blasting explosive in the mining industry. Polyethylene, a by-product of the petrochemical industry, is produced from ethane feedstock and has a variety of uses in the plastic industry. All other chemical products, such as pharmaceuticals, cleaning products and detergents, cosmetics, paints, pesticides and herbicides, fertilisers, and plastic and rubber products, mainly require non-energy-intensive manufacturing processes (USEIA, 2016). Production facilities range from small to large enterprises, with energy supplied by either gas or electricity.

5.1.3   Chemical Industry: Sub-sectors Chosen for the OECM Analysis

To prepare the decarbonisation pathways, we have broken down the chemical industry into the following sub-sectors. These sub-sector classifications are based on the main applications for chemical feedstocks and follow the categorisation based on the American Chemistry Council (ACC, 2020).

1. Pharmaceutical industry
2. Agricultural chemicals
3. Inorganic chemicals and consumer products
4. Manufactured fibres and synthetic rubber
5. Bulk petrochemicals and intermediates, plastic resins

The most important raw materials and chemical products of those five chemical industry subgroups are described below. The division into these subgroups was based on the available economic data required for market projections. An assessment of market development on the basis of the material flow would be more precise but was beyond the scope of this research because of the large variety of products produced by the chemical industry. The analysis focuses on the development of the chemical industry's energy requirements.

5.1.3.1   Sub-sector 1: Pharmaceuticals

Products and materials There are two key stages in pharmaceutical production: (i) the manufacture of the active pharmaceutical ingredient (API) and (ii) the production of the formulation. An API is the part of the drug that generates its effect. The production of APIs is usually chemically intensive, involving reactors specific for the manufacture of specific drug substances. Formulation production is a physical process, in which substances known as 'excipients' are combined with APIs to create consumable products (tablets, liquids, capsules, creams, ointments, and injectables).

Production and processes The world's largest pharmaceutical companies are headquartered in the USA and Europe, although production activities are centred in Asia. Some of the biggest pharmaceutical companies are Pfizer (USA), Roche, Novartis (Switzerland), Merck (USA), and GlaxoSmithKline (UK). Until the mid-1990s, the USA, Europe, and Japan supplied 90% of the world's demand for APIs. However, China's low-cost manufacturing sector and weak environmental regulations have meant that a significant proportion of API production has now shifted, with almost 40% of all APIs currently supplied by China. Together, China and India supply almost 75% of the API demand of pharmaceutical manufacturers in the USA. China's dominance in API production is balanced by India's leadership in global formulation production and its biotechnology sector. India is also the third largest producer of pharmaceuticals, by volume, supplying most of Africa's demand. India hosts the highest number of United States Food and Drug Administration (US FDA)-sanctioned production facilities outside the USA and supplies 40% of the US generic drug market. Despite India's vast pharmaceutical manufacturing industry, the country still imports 70% of its API demand from China.

Uses and applications Pharmaceutical products primarily service the health-care sector, with prescription and over-the-counter drugs, vaccines, and other pharmaceutical applications for human and veterinary use. The biotechnological production of crop seeds, value-added grains, and enzymes is a rapidly growing segment of the industry.

5.1.3.2 Sub-sector 2: Agricultural Chemicals

Products and materials Agricultural chemicals are a type of specialty chemical, and the term refers to a broad variety of pesticide chemicals, including insecticides, herbicides, fungicides, and nematicides (used to kill round worms). Agrichemicals can also include synthetic fertilisers, hormones, and other chemical growth agents, as well as concentrated varieties of raw animal manure (Speight, 2017). The main raw materials for nitrogen fertilisers are natural gas, naphtha, fuel oil, and coal, whereas phosphate fertilisers are based on naturally occurring phosphate rocks or synthetic ammonia.

Production and processes Some of the large agrichemical chemical producers are Syngenta, Bayer Crop Science, BASF, Dow AgroSciences, Monsanto, and DuPont. The fertiliser industry is structured around a few producers who supply the base chemicals to downstream manufacturers. The production facilities usually specialise in single-nutrient or high-nutrient fertiliser products and are located in close proximity to raw material suppliers (petrochemical producers) or agricultural regions (Roy, 2012).

Uses and applications Unsurprisingly, large-scale farming, also referred to as 'industrialised agriculture', is one of the primary users of agrichemicals. In 2010–2011, the global demand for primary plant nutrients was 178 megatonnes (Mt). China (57 Mt), the USA (20 Mt), and India (28 Mt) were the highest consumers.

5.1.3.3 Sub-sector 3: Inorganic Chemicals and Consumer Products

Products and materials Inorganic chemicals are materials derived from metallic and non-metallic minerals, such as ores or elements extracted from the earth (e.g. phosphate, sulphur, potash), air (e.g. nitrogen, oxygen), and water (e.g. chlorine). Other examples include aluminium sulphate, lime, soda ash (sodium carbonate), and sodium bicarbonate. The outputs of the chemical industry are used in the manufacture of consumer products, such as soaps, detergents, bleach, toothpaste and other oral hygiene products, and personal care products, such as hair care, skin care, cosmetics, and perfumes.

Production and processes Basic chemicals are typically produced in large-scale capital-intensive facilities with high-energy demands. Industrial gases, which are also products of the inorganic chemical industry, are heavily used in the production processes associated with steel, other chemicals, electronics, and health-care products. Many global factors influence the production of industrial gases. These factors include high capital intensity, increased consolidation of operations and geographic concentration, service orientation, and innovations in key technologies, such as membrane separation. The chemical conversion processes for consumer products are basic, and the key raw materials include fats, oils, surfactants, emulsifiers, other additives, and basic chemicals. Consumer products are usually formulated in batch-type operations, which involve equipment for mixing, dispersing, and filling (ACC, 2020).

Uses and applications The applications of inorganic chemicals are diverse. For example, chlorine is an important ingredient used to bleach paper pulp and purify drinking water and is used in oil-refining and the steel industry, and caustic soda is used in the production of soaps and detergents. These consumer products are heavily dependent upon vast distribution channels and product segmentation. Therefore, the supply chain and marketing costs are important determinants of the product price, which is also increased by the need for ongoing product development.

5.1.3.4   Sub-sector 4: Manufactured Fibres and Synthetic Rubber

Products and materials Manufactured fibres, also referred to as 'synthetic fibres', consist of *cellulosic fibres*, such as acetate and rayon, and petrochemical-derived *polymeric fibres*, such as acrylics, nylon, polyesters, and polyolefins. There are several types of synthetic rubber, including butyl rubber, ethylene-propylene-diene monomer terpolymers, neoprene, nitrile rubber, styrene-butadiene rubber, and specialty elastomers (ACC, 2020).

Production and processes Synthetic or artificial fibres are derived from polymer industries using processes such as wet spinning (rayon), dry spinning (acetate and triacetate), and melt spinning (nylons and polyesters). Synthetic rubbers have highly flexible material characteristics, and the process of 'vulcanisation' is used to cross-link elastomer molecules.

Uses and applications Plastics, synthetic rubber, and manufactured fibres account for the second highest share (30%) of the total energy consumed by the chemical industry in the USA, preceded by petrochemicals and other basic chemicals, which have a 49% share (ACC, 2020). Synthetic fibres are heavily used in apparel, home furnishings, and automotive and construction industries. Similarly, synthetic rubber is in high demand in automotive manufacturing, construction, and consumer products. Synthetic fibres are increasingly used in textile manufacture because of their durability and abundance and their ability to be processed into long fibres or to be batched and cut for processing. Natural fibres, such as wool, silk, and leather, are most frequently used for high-quality and long-lasting garments, whereas synthetic fibres are popular in the manufacture of fast fashion garments and accessories (ILO, 2021).

5.1.3.5 Sub-sector 5: Petrochemicals

Products and materials Petrochemicals are chemical products derived from petroleum-refining and from other fossil fuels, such as natural gas and coal. The two main classes of petrochemicals are olefins and aromatics. Ethylene, propylene, and butadiene are examples of olefins—ethylene and propylene are used in the manufacture of industrial chemicals and plastic products, whereas butadiene is used to manufacture synthetic rubber. Olefins also form the base compounds in the manufacture of the polymers and oligomers used in plastics, resins, fibres, elastomers, lubricants, and gels.

Benzene, toluene, and xylene isomers are examples of aromatic compounds and are primarily produced from naphtha derived from petroleum-refining. Benzene is used as a raw material in the manufacture of dyes and synthetic detergents, whereas xylene is used to manufacture plastic products and synthetic fibres.

Apart from olefins and aromatics, other chemical products of the petrochemical industry include synthetic gases used to make ammonia and methanol (in steam-reforming plants), methane, ethane, propane, and butanes (in natural gas-processing plants), methanol, and formaldehyde. Ammonia is also used in the manufacture of the fertiliser urea, whereas methanol is used as a solvent and chemical intermediate.

Globally, 190 million tonnes (Mt) of ethylene, 120 Mt of propylene, and approximately 70 Mt of aromatics were produced in 2019.

Production and processes The USA and Western Europe are home to the world's largest petrochemical producers. Some of the most notable petrochemical-manufacturing locations are in the industrial cities of Jubail and Yanbu in Saudi Arabia, Texas and Louisiana in the USA, Teesside in the UK, Rotterdam in the Netherlands, and Jamnagar and Dahej in India. The Middle East and Asia are witnessing increasing investment in new production capacities for petrochemical plants, and a vast majority of the global demand is expected to be met from these regions in the coming decade (Cetinkaya et al., 2018). Some of the fastest-growing petrochemical companies in terms of capacity are PetroChina, Reliance, SABIC, Sinopec, and Wanhua. Both olefins and aromatics can be produced during oil-refining by the fluid catalytic cracking of petroleum fractions or with chemical processes. In chemical plants, the process of steam cracking is used to produce olefins from natural gas liquids, such as ethane and propane. A naphtha catalysis process is used to produce aromatics.

Uses and applications The petrochemical sector supplies materials for the vast majority of chemical industry applications, such as the manufacture of petrochemical derivatives, aromatics from bulk petrochemicals, olefins, and methanol. Seven petrochemicals supply more than 90% of all organic chemicals: benzene, toluene, and xylene (aromatics); ethylene, propylene, and butadiene (olefins); and methanol (ACC, 2020). Bulk petrochemicals are also transformed into intermediate products and downstream derivatives, such as plastic resins, synthetic rubbers, manufactured fibres, surfactants, dyes, pigments, and inks. The end-user industries for petrochemical products are the chemical industry, automotive industry, building and construction, consumer products, electronics, furniture, and packaging.

5.1.4   GDP Projections for the Global Chemical Industry

The economic development of the global chemical industry is significantly more complex than that of the aluminium and steel industries. The product range of the chemical industry is diverse, and the material flow approach used for aluminium and steel is very data-intensive and is therefore beyond the scope of this research. The chemical industry produces materials for almost all parts of the economy—from mining to services—and it is therefore intrinsically connected to overall economic development. Consequently, a GDP-based approach has been used to develop the energy demand projections for the chemical industry over the next three decades.

Table 5.1 provides an overview of the projected economic development of the chemical industry and its five sub-sectors. It is assumed that the chemical industry will follow the trajectory of the global GDP growth and that the chemical industry's share of the global GDP will remain constant until 2050. The sub-sectors are assumed to grow at the same rate as the overall chemical industry, and the market value share of each sub-sector will also remain stable. For example, the pharmaceutical industry had a 26% share of the global chemical industry GDP, just over US$1 billion, in 2019. With this approach, we assume that this share of 26% will remain constant until 2050 and that the growth rate of each sub-sector will develop in line with the global GDP projections. This is a simplification, and the actual development trajectories may vary across all sectors. However, a more nuanced projection of the development of the chemical industry is beyond the scope of this research.

5.1.5   Energy Flows for the Chemical Industry

Natural gas and petroleum products are important energy sources for the chemical industry. Globally, the chemical industry is responsible for 11% of the primary demand for oil and 8% of the primary demand for natural gas (Levi & Pales, 2018). The chemical industry in the USA consumes almost 9% of all petroleum products as feedstock for fuel and power use, natural gas liquids (or liquefied petroleum gases), and heavy liquids (naphtha and gas oil) (ACC, 2020).

Table 5.1 Projected economic development of the chemical industry (ACC, 2020; World Bank, 2021)

Parameter	Units	2019	2025	2030	2035	2040	2050
Global GDP	[bn $GDP]	129,555	142,592	196,715	231,758	266,801	346,236
Total chemical industry	[bn $GDP]	3900	4966	5862	6906	7950	10,317
Global GDP share	[%]	3%	3%	3%	3%	3%	3%
Pharmaceutical industry	[bn $GDP]	1029	1314	1551	1828	2104	2730
GDP—share of the chemical industry market value	[%]	26%	26%	26%	26%	26%	26%
Agricultural chemicals	[bn $GDP]	333.8	424.5	501.1	590.4	679.6	882.0
GDP—share of the chemical industry market value	[%]	9%	9%	9%	9%	9%	9%
Specialties, inorganic chemicals, consumer products	[bn $GDP]	1,225	1558	1839	2167	2495	3237
GDP—share of the chemical industry market value	[%]	31%	31%	31%	31%	31%	31%
Manufactured fibres and synthetic rubber	[bn $GDP]	196.6	249.9	295.0	347.6	400.2	519.3
GDP—share of the chemical industry market value	[%]	5%	5%	5%	5%	5%	5%
Bulk petrochemicals and intermediates, plastic resins	[bn $GDP]	1115.8	1418.9	1674.9	1973.2	2271.6	2947.9
GDP—share of the chemical industry market value	[%]	29%	29%	29%	29%	29%	29%

Petrochemical feedstocks, such as olefins and aromatics, are extracted from hydrocarbons produced with cracking processes. These feedstocks are used in plastics, pharmaceuticals, electronics, and fertiliser industries. *Methanol* is directly converted from methane in natural gas and does not undergo the cracking process. In the USA, natural gas liquids are used in the production of 90% of olefins, whereas naphtha is the main source (70%) of petrochemical production in Europe and Asia.

The IEA (2018a) mapped the flows of fuel feedstocks in the chemical and petrochemical industries in 2015. Most of the oil feedstock was converted to high-value chemicals, and a large proportion of raw materials for the chemical industry were directly supplied by oil refineries. Ammonia and methanol, both chemicals in high demand, require natural gas as the raw material. China also uses coal in the production of ammonia and methanol. Petrochemical production occurs in very large-scale facilities, and a number of related products can be produced at a single petrochemical facility. This differs from the set-ups for commodity chemicals, where specialty chemicals and fine chemicals are manufactured in discrete batch processes. Historically, the accelerating demand for chemical products in these end-use industries has had an inevitable impact on the energy demand and resultant CO_2 emissions of the upstream and overall chemical industry. Together, base chemicals supply the intermediate raw materials for the majority of aforementioned demand industries (IEA, 2018a; Levi & Pales, 2018).

The energy demand in the pharmaceutical industry is largely driven by the critical environmental requirements for temperature, humidity, room pressurisation, cleanliness, and containment. The manufacturing and R&D phases consume a high proportion of the energy demand (>65%), followed by the formulation, packaging, and filling phases (15%). Overall, heating, ventilation, and air conditioning are the highest energy end uses in the industry (>65%), because of the nature of the products manufactured (Centrica, 2021). Another energy-consuming system is the production of compressed air, which has multiple applications and is one of the least energy-efficient functions in a pharmaceutical production facility. There are opportunities for energy and cost savings in this area (Centrica, 2021). In the production of agrochemicals, the energy demand is spread across manufacturing, packaging, and transportation, and the majority of raw materials are derived from the petrochemical industry. The production of nitrogen fertilisers is energy-intensive because the process that converts the fossil-fuel raw materials used to manufacture the usable fertilisers is energy-intensive. In terms of material throughput, 1 tonne of nitrogen fertiliser output consumes 1.5 tonnes of petrol equivalents (Ziesemer, 2007).

5.1.6 Projection of the Chemical Industry Energy Intensity

This brief overview of the energy usage for the sub-sectors analysed has shown that the chemical industry consists of a highly energy-intensive part, which produces the primary feedstock (basic chemicals) and a secondary product manufacturing part, with a relatively low energy intensity, similar to those of other manufacturing industries with energy intensities of < 10 MJ per $GDP.

The energy demands for the five sub-sectors—pharmaceuticals, agricultural chemicals, inorganic chemicals and consumer products, manufactured fibres and synthetic rubber, and petrochemical industry—were calculated with the energy intensities provided in Table 5.2, which are based on the IEA Energy Efficiency extended database (IEA, 2021a) and our own research. The energy intensities for primary feedstock were also considered in estimating the efficiency trajectories of the different sub-sectors. An increase in the efficiency of primary feedstock production of 1% per year over the entire modelling period is required to achieve the assumed efficiency gains for all sub-sectors. However, inadequate data are available for the specific energy intensities of the chemical industry, and no detailed breakdown of the electricity and process heat temperature levels is available in public databases. Therefore, our estimates should be seen as approximate values, and more research, in co-operation with the chemical industry, is required. However, the energy requirements of the entire chemical industry are precisely known and were taken from the IEA statistics *Advanced Energy Balances* (IEA, 2020a)

The energy requirements of the sub-sectors were determined on the basis of market shares and GDP and in discussions with representatives of the chemical industry—specifically members of the Net-Zero Asset Owner Alliance and the Strategic Approach to International Chemicals Management of the United Nations Environment Programme (SAICM UNEP).

Table 5.2 shows the assumed energy intensities per $GDP for the analysed sub-sectors of the chemical industry. The production of primary feedstock is

Table 5.2 Assumed energy intensities for sub-sectors of the chemical industry

Chemical industries—energy intensities		2019	2025	2030	2035	2040	2050
Pharmaceutical industry	[MJ/$GDP]	5.02	4.54	4.36	4.18	4.02	3.70
Assumed annual increase in efficiency	[%/yr]		0.80%	0.80%	0.80%	0.80%	0.80%
Agricultural chemicals	[MJ/$GDP]	8.37	7.56	7.26	6.97	6.69	6.17
Assumed annual increase in efficiency	[%/yr]		0.80%	0.80%	0.80%	0.80%	0.80%
Inorganic chemicals and consumer products	[MJ/$GDP]	4.22	3.81	3.66	3.51	3.37	3.11
Assumed annual increase in efficiency	[%/yr]		0.80%	0.80%	0.80%	0.80%	0.80%
Manufactured fibres and synthetic rubber	[MJ/$GDP]	4.97	4.49	4.31	4.14	3.97	3.66
Assumed annual increase in efficiency	[%/yr]		0.80%	0.80%	0.80%	0.80%	0.80%
Bulk petrochemicals and intermediates, plastic resins	[MJ/$GDP]	4.64	4.15	3.94	3.75	3.56	3.21
Assumed annual increase in efficiency	[%/yr]		1.00%	1.00%	1.00%	1.00%	1.00%
Energy intensities—primary feedstock							
Petroleum-refining	[MJ/$GDP]	54.16	51.45	48.88	46.44	44.11	39.81
Assumed annual increase in efficiency	[%/yr]		1.00%	1.00%	1.00%	1.00%	1.00%
Alkali and chlorine manufacture	[MJ/$GDP]	63.85	60.66	57.62	54.74	52.01	46.94
Assumed annual increase in efficiency	[%/yr]		1.00%	1.00%	1.00%	1.00%	1.00%
All other basic inorganic chemical manufacture	[MJ/$GDP]	54.34	51.62	49.04	46.59	44.26	39.94
Assumed annual increase in efficiency	[%/yr]		1.00%	1.00%	1.00%	1.00%	1.00%
Chemical fertiliser (except potash) manufacture	[MJ/$GDP]	110.41	104.89	99.65	94.66	89.93	81.16
Assumed annual increase in efficiency	[%/yr]		1.00%	1.00%	1.00%	1.00%	1.00%
Chemical industries—average energy intensity	[MJ/$GDP]	4.36	3.64	3.56	3.49	3.42	3.29
Assumed annual increase in efficiency	[%/yr]		1.00%	1.00%	1.00%	1.00%	1.00%

significantly higher than other chemicals owing to the process feedstock used in end products. The share of primary feedstock within a certain production process informs the level of energy efficiency potential. Because no detailed published data are available, the efficiency across all sub-sectors of the industry was assumed to be 1% per year. However, more research and greater access to data are required to allow a more detailed bottom-up energy demand analysis of the chemical industry.

5.1.7 Projection of the Energy Demand and CO₂ Emissions of the Chemical Industry

The projections of the economic development and energy intensities of an industry yield the overall global energy demand projection for that industry. In another step, the share of electricity required to generate thermal process heat has been estimated. Table 5.3 shows the calculated electricity demand and Table 5.4 the process heat demand by temperature level for the chemical industry sub-sectors.

Table 5.3 Projected electricity and process heat demand for the chemical industry to 2050

Sub-sector	Units	2019	2025	2030	2035	2040	2050
Chemical industries							
Chemical industry—electricity demand by sub-sector							
Pharmaceutical industry	[PJ/yr]	1431	1652	1873	2118	2341	2799
	[TWh/yr]	398	459	520	588	650	778
Agricultural chemicals	[PJ/yr]	782	899	1019	1152	1274	1523
	[TWh/yr]	217	250	283	320	354	423
Inorganic chemicals and consumer products	[PJ/yr]	1447	1663	1884	2131	2355	2817
	[TWh/yr]	402	462	523	592	654	782
Manufactured fibres and synthetic rubber	[PJ/yr]	273	314	356	403	445	532
	[TWh/yr]	76	87	99	112	124	148
Bulk petrochemicals and intermediates, plastic resins	[PJ/yr]	1450	1649	1849	2070	2264	2651
	[TWh/yr]	403	458	514	575	629	736
Total chemical industry	[PJ/yr]	5384	6178	6981	7874	8678	10,323
	[TWh/yr]	1496	1716	1939	2187	2411	2867
Heat demand	[PJ/yr]	12,163	15,949	18,024	20,329	22,406	26,653
Heat share	[%]	56%	56%	56%	56%	56%	56%
Heat demand <100 °C	[PJ/yr]	2196	2879	3254	3670	4044	4811
Heat demand 100–500 °C	[PJ/yr]	2722	3570	4034	4550	5015	5965
Heat demand 500–1000 °C	[PJ/yr]	5813	7623	8615	9716	10,709	12,739
Heat demand >1000 °C	[PJ/yr]	1432	1878	2122	2394	2638	3138

Table 5.4 Process and energy-related CO_2 emissions—chemical industry

	Units	2019	2025	2030	2035	2040	2050
CO_2 emissions—power supply	[$MtCO_2$/yr]	761	499	264	115	58	0
CO_2 emissions—heat supply	[$MtCO_2$/yr]	1257	994	707	554	323	0
CO_2 emissions—total energy supply	[$MtCO_2$/yr]	2019	1492	971	669	380	0
Chemical industry total non-energy GHG	[$MtCO_2$eq/ yr]	2520	1852	1220	991	775	682

Finally, energy-related CO_2 emissions have been calculated on the basis of the 1.5 °C energy supply pathway, which is documented in Chap. 12.

5.2 Global Cement Industry

Cement is the second most consumed substance in the world after water and is a central component of the built environment—from civil infrastructure projects and power generation plants to residential houses. Typically made from raw materials such as limestone, sand, clay, shale, and chalk, cement acts as a binder between aggregates in the formation of concrete. Cement manufacture is a resource- and emission-intensive process and is associated with around 7% of the total global CO_2 emissions, according to the Intergovernmental Panel on Climate Change (IPCC; Fischedick et al., 2014, p. 750).

The economic value of the global cement industry was estimated to be US$450 billion in 2015 (McKinsey, 2015). In 2012, the US cement industry's shipment (to support construction projects) was estimated to be US$7.5 billion (Portland Cement Association, 2019), equivalent to 1.6% of the global revenue. In the EU, the cement manufacturing industry's turnover was estimated to be €15.2 billion in 2015, with €4.8 billion in value added (European Commission, 2018).

Beyond the mining of the raw materials, there are five main steps in the cement production process:

1. Raw material preparation—This stage involves the crushing or grinding, classi- fication, mixing, and storage of raw materials and additives. This is an electricity- intensive production step requiring between 25 and 35 kilowatt hours (kWh) per tonne of raw material (shown in Fig. 5.1 as steps 2–3).
2. Fuel preparation—This phase involves optimising the size and moisture content of the fuel for the pyroprocessing system of the kiln (shown in Fig. 5.1 as steps 4–5).
3. Clinker production—The production of clinker involves the transformation of raw materials (predominantly limestone) into clinker (lime), the basic compo- nent of cement, as shown in Fig. 5.1 (step 6). This is achieved by heating the raw materials to temperatures >1450 °C in large rotary kilns. Clinker production is the most energy-intensive stage of the cement manufacturing process, account- ing for >90% of the total energy used in the cement industry.

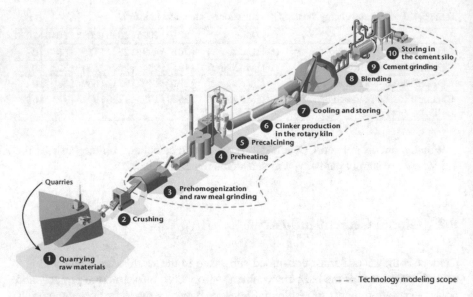

Fig. 5.1 Steps in cement production, from mining to product. (Source: IEA, 2021a)

4. Clinker cooling—After the clinker is discharged from the kiln, it is cooled rapidly (Fig. 5.1, step 7).
5. Finish grinding—After cooling, the clinker is crushed and mixed with other materials (gypsum, fly ash, ground-granulated blast-furnace slag, and fine limestone) to produce the final product, cement (see Fig. 5.1, steps 8–10).

The literature distinguishes between the energy consumed to produce the intermediary product *clinker* (in the form of small rocklike nodules) and the energy consumed for cement production, which is based on clinker.

5.2.1 Major Cement Industry Companies and Countries

Early estimates from the United States Geological Survey (2020) and IEA (2021a) suggest that global cement production reached 4.1 gigatonnes (Gt) in 2019. Over the past decade, global production has averaged close to 4.0 Gt a year, reaching a high of almost 4.2 Gt in 2014 (United States Geological Survey, 2020).

China has become the largest cement producer worldwide, accounting for around 55% of the total global production in 2019 (IEA, 2020b). The second largest producer was India (8%), followed by the USA, Vietnam, Indonesia, and Egypt, with 2% each, and six countries (Iran, Brazil, Russia, Japan, South Korea, and Turkey) each contributed 1% of the global cement production (IEA, 2021b). The remaining 22% of the global production was distributed across all other countries, with production shares of <1% of the global production.

Swiss company LafargeHolcim is the largest single cement producer in the world (responsible for 9% of the global production). Overall, Chinese-owned companies, including the Taiwan Cement Corporation, together account for 13% of the global cement production.

Cement producers in OECD Europe (Switzerland, Germany, and Italy) and OECD America (Mexico) have headquarters in OCED regions but operate cement plants in 50–60 countries worldwide, so cement production-related CO_2 emissions are spread across various countries. It is important to note that the figures on cement production by company are a combination of annual production and production capacity data. Therefore, it is likely that there are discrepancies in the production values (Mt cement per year), because plants often do not meet the plant capacity. The ten largest cement companies produce 32% of the global production (Table 5.5).

Table 5.5 Top ten global cement producers, their headquarters, annual production (Mt), and number of operational cement plants

	Company name	HQ	Production/capacity (Mt/year)		Share of the global production (%)	Number of plants (countries)
1	LafargeHolcim	Zug, Switzerland	386	Production	9	180 (60 countries)
2	Anhui Conch	Wuhu, China	217	Production (2019)	5	32
3	China National Building Materials (CNBM)	Beijing, China	176	Capacity	4	94
4	HeidelbergCement	Heidelberg, Germany	121	Capacity	3	79 (60 countries)
5	Cemex	San Pedro, Mexico	87	Capacity	2	56 (50 countries)
6	Italcementi	Italy	77	Capacity	2	60
7	China Resources Cement	China	78	Capacity	2	24
8	Taiwan Cement Corporation	Taiwan	64	Capacity	2	6
9	Eurocement	Moscow, Russia	45	Production	1	17 (Russia, Ukraine, Uzbekistan)
10	Votorantim Cimentos	Brazil	45	Production	1	34
	Global production (2019)		4100		32	582 production plants

Source: Construction Review Online (2021)

5.2.2   Impact of COVID-19 on Global Cement Production

The global cement demand decreased by 3% in 2020, but this decline varied significantly by region. The largest impacts on the cement industry occurred in Southeast Asia (−10%), Western Europe (−8%), Australia and the Middle East including North Africa (−7% each), and Latin America (−6%) (International Finance Corporation, 2020). The reduction in cement demand due to COVID-19 resulted in a decline in global emissions from the cement industry, estimated at −7–8% globally relative to those in 2019. However, future emission reduction targets for 2025 and beyond are based on 2019 emissions, and it is assumed that the demand for cement will increase to pre-COVID-19 levels by 2022 and 2023. Therefore, the emission targets are based on planned construction projects estimated before the pandemic (International Finance Corporation, 2020).

5.2.3   Energy Efficiency Standards and Energy Intensities for the Cement Sector

5.2.3.1   Thermal Efficiency of Cement Production

In cement manufacturing, a theoretical minimum energy demand of 1850–2800 MJ/t of clinker is determined by the chemical and mineralogical reactions and drying (European Cement Research Academy and Cement Sustainability Initiative, 2017). This demand includes:

- An energy demand of 1650–1800 MJ/t of clinker to heat the raw materials to the required temperature (up to 1450 °C) for the formation of stable clinker phases
- An energy demand of 200–1000 MJ/t of clinker for drying the raw materials

The average global thermal energy intensity of clinker production (grey clinker, excluding the drying of fuels) reduced from 4254 MJ/t clinker in 1990 to 3472 MJ/t clinker in 2017 (GNR, 2021). Table 5.6 (GNR, 2021) shows the average regional

Table 5.6 Selected regional average thermal energy intensities for grey clinker production—excluding drying of fuels (MJ/t clinker)

	North America	Africa	Central America	Europe	Brazil	Middle East	China, Korea, Japan	India	Global average
1990	4944	4612	3933	4056	4214	3973	3476	3907	4254
2000	4591	4056	3700	3726	3413	3453	3444	3145	3753
2010	3888	3740	3588	3700	3675	3366	3397	3130	3581
2016	3894	3743	3627	3675	3560	3382	3206	3086	3519
2017	3821	3660	3641	3584	3489	3378	3194	3058	3472

Source: GNR (2021)

thermal intensity of clinker production (MJ/t clinker). The thermal intensity of clinker production is highest in the Commonwealth of Independent States (CIS; regional intergovernmental organisation in Eastern Europe and Asia), followed by OECD North America and Africa. The average global thermal intensities by kiln type are shown in Table 5.7 (GNR, 2021).

All data in Sect. 5.2.3 are drawn from the 'Getting the Numbers Right (GNR)' database, an independent database of energy performance and CO_2 information for the cement industry. Managed by the Global Cement and Concrete Association (GCCA), GNR compiles uniform data from 877 cement production facilities, which accounted for 19% of the global cement production in 2017.

5.2.3.2 Thermal Efficiency by Kiln Type

There are considerable variations in the thermal efficiency of kiln types, and the best-performing kilns (which dry with preheating and pre-calcining) achieved a weighted average thermal energy intensity of 3350 MJ/t clinker in 2017 and the least-efficient kiln (wet/shaft kiln) a thermal energy intensity of 5900 MJ/t clinker. These data are shown in Table 5.7 (GNR, 2021).

5.2.4 Global Cement Industry: Process- and Energy-Related Emissions

The cement industry is a major source of global CO_2 emissions. However, the data required to estimate the emissions from global cement production are not well documented (Andrew, 2018). Consequently, there is considerable variation between

Table 5.7 Average thermal energy intensity by kiln type—excluding drying of fuels (MJ/t clinker)

	Drying with preheating and pre-calcining	Drying with preheating without pre-calcining	Dry without preheating (long dry kiln)	Semi-wet/ semi-dry kiln	Wet/ shaft kiln
1990	3614	3856	4584	4006	6314
2000	3403	3684	4466	3780	6003
2005	3387	3636	4288	3797	6104
2010	3390	3694	4016	3827	5982
2015	3385	3690	3881	4307	5734
2016	3389	3729	3843	4108	5906
2017	3350	3610	3912	4187	5900
Change 1990–2017	−7.3%	−6.4%	−14.7%	±3%	−6.6%

Source: GNR (2021)

different global estimates. Two main aspects of cement production lead to direct CO_2 emissions:

1. *Energy-related emissions*: Energy is required for the calcination process during clinker production. The combustion of fuels to heat the raw ingredients to >1600 °C in this process accounts for 30–40% of the total emissions associated with cement production. These emissions are commonly referred to as *fuel emissions*.
2. *Process-related emissions*: The calcination of calcium carbonate to calcium oxide is the chemical reaction that takes place when the raw materials (notably limestone) are exposed to high temperatures. The remaining 60–70% of CO_2 emissions from cement production derive from calcination. These emissions are commonly referred to as *process emissions*.

Globally, energy-related (fuel) emissions made up 35% of cement emissions (0.8 $GtCO_2$/yr), and process emissions amounted to 65% (1.5 $GtCO_2$/yr) in 2019 (IEA, 2020c). The energy-related emissions from the cement industry amount to 7% of the global energy emissions in that year (IEA, 2020c). The average emissions associated with the total cement manufacturing process are shown in Fig. 5.2 (McKinsey, 2021).

A comprehensive analysis of the global process emissions from cement production revealed a wide variety of existing datasets (Andrew, 2018). The total global process emissions was 1.5 $GtCO_2$ in 2018 (Andrew, 2018). Table 5.8 outlines the global process emissions ($GtCO_2$) from cement production between 2000 and 2018 (Andrew, 2018).

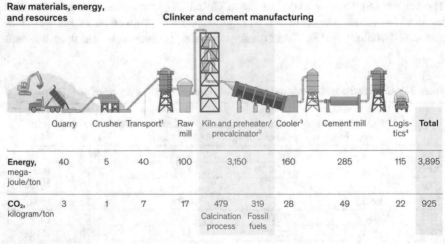

Raw materials, energy, and resources   **Clinker and cement manufacturing**

	Quarry	Crusher	Transport[1]	Raw mill	Kiln and preheater/ precalcinator[2]		Cooler[3]	Cement mill	Logistics[4]	**Total**
Energy, megajoule/ton	40	5	40	100	3,150		160	285	115	3,895
CO_2, kilogram/ton	3	1	7	17	479 Calcination process	319 Fossil fuels	28	49	22	925

[1]Assumed with 1kWh/t/100m.
[2]Assumed global average, data from the Global Cement and Concrete Association, Getting the Numbers Right 2017.
[3]Assumed reciprocating grate cooler with 5kWh/t clinker.
[4]Assumed lorry transportation for average 200km.

Fig. 5.2 Current average energy (MJ/t cement) and emissions (CO_2/t cement) in cement manufacture (Source: McKinsey, 2021)

Table 5.8 Global process emissions from cement production in 2000–2018, in GtCO$_2$

2000	2001	2002	2003	2004	2005	2006
0.72	0.75	0.79	0.85	0.91	0.96	1.05
2007	**2008**	**2009**	**2010**	**2011**	**2012**	**2013**
1.12	1.14	1.17	1.25	1.34	1.37	1.42
2014	**2015**	**2016**	**2017**	**2018**		
1.47	1.43	1.46	1.47	1.5		

5.2.4.1 Reduction of the Clinker/Cement Ratio

The process CO$_2$ emissions released during the production of clinker can be reduced by integrating alternative cement constituents that reduce the clinker/cement ratio. A global clinker/cement ratio of 0.60 is achieved by 2050 under the IEA's 2DS scenario (IEA, 2018b). This represents a fall from 0.65 in 2014, which translates into a reduction in the process CO$_2$ intensity of cement by 30% over that period (the global average carbon intensity for process emissions is projected to reach 0.24 tCO$_2$/t cement by 2050, which will lead to a saving of 364 million tonnes of CO$_2$ (MtCO$_2$) emissions (IEA, 2018b). The OneEarth Climate Model (OECM) also assumes this estimate of the possible decline in process emissions.

Carbonation occurs when CO$_2$ diffuses into the pores of cement-based materials and reacts with hydrated products in the presence of pore water. Carbonation starts at the surface of the concrete or mortar and moves progressively inwards. In contrast to the instantaneous emission of CO$_2$ during the manufacture of cement, carbonation is a slow process that takes place throughout the entire life cycle of cement-based materials (Xi et al., 2016).

Xi et al. (2016) reported that the carbonation of cement materials over their life cycles represents a large and growing net sink of CO$_2$, increasing from 0.10 GtC/yr in 1998 to 0.25 GtC/yr in 2013. In total, they estimated that roughly 43% of the cumulative cement process emissions of CO$_2$ produced between 1930 and 2013 have been reabsorbed by carbonating cement materials. They propose that an average of 44% of the cement process emissions produced each year between 1980 and 2013 has been offset by the annual cement carbonation sink. Moreover, between 1990 and 2013, the annual carbon uptake increased by 5.8% per year on average, slightly faster than the 5.4% per year increase in process cement emissions over the same period (Xi et al., 2016).

5.2.4.2 New Technologies to Reduce Process Emissions in the Cement Industry

The decarbonisation of cement production-related process emissions is being tested and is in various stages of development. These new processes and technologies include clinker displacement by optimising the combination of calcined clay and ground limestone as the cement constituents (European Cement Research Academy and Cement Sustainability Initiative, 2017) and the use of alternative binding materials. Alternative binding materials offer potential opportunities for reducing process CO_2 emissions and involve t mixes of raw materials or alternatives from those used in Portland clinker, although the commercial availability and applicability of the alternatives differ widely.

5.2.4.3   Post-combustion Carbon Capture Technologies

Chemical absorption is the most advanced post-combustion capture technology and allows up to 95% optimum capture yields (European Cement Research Academy and Cement Sustainability Initiative, 2017). A plant began operation in Texas in 2015 to chemically capture and transform 75 $ktCO_2$/yr from a cement plant into sodium bicarbonate, bleach, and hydrochloric acid, which could be sold, so that the sorbents, once saturated, need not be regenerated (IEA, 2018b). The use of membranes as a CO_2 separation technique is another proposed technology, which could theoretically produce a yield of more than 80%. However, membranes have only been proven at small or laboratory scales, at which recovery yields of up to 60–70% were achieved (European Cement Research Academy and Cement Sustainability Initiative, 2017).

    None of the technologies currently under development are assumed for the OECM 1.5 °C pathway because the time of possible commercialisation is yet to be determined.

5.2.5   Global Cement Production and Energy Intensity Projections

Table 5.9 summarises the assumptions of the 1.5 °C OECM cement industry pathway in terms of the projected volume of global cement production, the development of energy intensities for the relevant processes, and the process emissions per tonne of clinker produced. These assumptions are similar, to a large extent, to those made for the IEA Technology Roadmap—Low-Carbon Transition in the Cement Industry projections (IEA, 2018b).

5.2.6   Projections of the Cement Industry Energy Demand and CO2 Emissions

Table 5.10 shows the calculated electricity and process heat demand developments based on the documented assumptions. The breakdown by temperature level is based on the five cement production steps required and their shares of the overall

Table 5.9 Assumed global cement market development and production energy intensities

Parameter	Units	2019	2025	2030	2035	2040	2050
			Projections				
Cement—production volume	[Mt/yr]	4200	4448	4595	4739	4883	5094
Cement—variation compared with 2019	[%]	0	6%	9%	13%	16%	21%
Clinker							
Clinker—production volume	[Mt/yr]	2730	2,869	2,941	3000	3076	3056
Clinker—variation compared with 2019	[%]	0	5%	8%	10%	13%	12%
Clinker/cement ratio	[%]	65.0%	64.5%	64.0%	63.3%	63.0%	60.0%
Energy intensities							
Thermal energy intensity—per tonne of clinker	[GJ/t]	3.5	3.4	3.3	3.25	3.2	3.1
Variation compared with 2019	[%]	0	−3%	−6%	−7%	−9%	−11%
Cement production—electricity intensity	[kWh/t]	116	90	87	85	83	79
Variation compared with 2019	[%]	0	−22%	−25%	−27%	−28%	−32%
Thermal energy intensity—per tonne of cement	[GJ/t]	2.33	2.27	2.20	2.12	2.07	2.01
Variation compared with 2019	[%]	0	−2%	−5%	−9%	−11%	−14%
Process emissions							
Process emissions (calcination process)	[tCO$_2$/t clinker]	0.40	0.40	0.37	0.34	0.30	0.24

Table 5.10 Projected electricity and process heat demand for the cement industry

Parameter	Units		2019	2025	2030	2035	2040	2050
				Projections				
Energy demand—limestone mining	[PJ/yr]		510	526	618	724	829	1034
Energy demand—clinker production	[PJ/yr]		9555	9753	9705	9749	9844	9475
Energy demand—cement production	[PJ/yr]		11,530	11,550	11,552	11,517	11,546	11,670
Electricity demand—cement production	[PJ/yr]		1754	1441	1439	1450	1459	1449
	[TWh/yr]		487	400	400	403	405	402
Heat demand (energy used)	[PJ/yr]		7213	7514	7516	7483	7497	7597
Heat share (final energy)	[%]		81%	88%	88%	87%	87%	88%
Heat demand <100 °C	[PJ/yr]	5%	346	361	361	359	360	365
Heat demand 100–500 °C	[PJ/yr]	2%	146	152	152	152	152	154
Heat demand 500–1000 °C	[PJ/yr]	30%	2189	2280	2281	2271	2275	2305
Heat demand >1000 °C	[PJ/yr]	63%	4532	4721	4722	4701	4710	4773

Table 5.11 Process- and energy-related CO_2 emissions—cement industry

Parameter	Units	2019	2025	2030	2035	2040	2050
			Projections				
CO_2 emissions—power supply	[$MtCO_2$/yr]	248	116	54	21	10	0
CO_2 emissions—eat supply	[$MtCO_2$/yr]	601	425	266	184	97	0
CO_2 emissions—mining	[$MtCO_2$/yr]	38	37	39	25	12	0
CO_2 emissions—total energy supply	[$MtCO_2$/yr]	887	578	360	230	118	0
CO_2 emissions—process-related emissions	[$MtCO_2$/yr]	1,092	1,147	1082	1008	935	734
Total CO_2 emissions	[$MtCO_2$/yr]	848.9	541.1	320.5	204.9	106.5	0.0
Specific energy-related CO_2 emissions per tonne of clinker	[tCO_2/t clinker]	0.220	0.148	0.091	0.061	0.032	0.000
Specific energy-related CO_2 emissions per tonne of cement	[tCO_2/t cement]	0.279	0.174	0.102	0.066	0.033	0.000
Specific CO_2 emissions per tonne of cement (including process emissions)	[tCO_2/t cement]	0.899	0.722	0.561	0.463	0.369	0.240

energy demand. No detailed statistical documentation of the exact breakdown of the process heat demand by temperature level and quantity is available. Table 5.11 shows the energy-related CO_2 emissions—based on the 1.5 °C energy generation pathway—and the expected process emissions.

5.3 Aluminium Industry: Overview

Aluminium is among the most important building and construction materials globally. To understand the opportunities and challenges facing the industry, the global flow of aluminium metal must be considered. Since 1880, an estimated 1.5 billion tonnes of aluminium have been produced worldwide (IAI, 2018a), and about 75% of the aluminium produced is in productive use (IAI, 2018b). In 2019, 36% of aluminium was located in buildings, 25% in electrical cables and machinery, and 30% in transport applications. Aluminium can be recycled, but the availability of scrap is limited by the high proportion of aluminium in use (IAI, 2018a).

5.3.1 Bauxite Production

Primary aluminium production requires bauxite. Bauxite ore occurs in the top soils of tropical and subtropical regions, such as Africa, the Caribbean, South America, and Australia. The largest producers/miners of bauxite include Australia, China, and Guinea. Australia supplies 30% of global bauxite production (M'Calley, 1894). Table 5.12 shows the global distribution of bauxite mine production, aluminium refineries, and production.

Table 5.12 Aluminium resources, bauxite mines, alumina refineries, and aluminium production (in thousand tonnes) by country

	Bauxite mine production		Alumina refineries/ production	Bauxite reserves	Aluminium production	Aluminium production in %
	2018	2019[a] (estimated)	2018	No year	2019	2019
Australia	86,400	100,000	20,400	6,000,000	20,000	15.1%
China	79,000	75,000	72,500	1000,000	73,000	54.9%
Guinea	57,000	82,000	180[c]	7,400,000	No data	No data
Brazil	29,000	29,000	8100	2,000,000	8900	6.7%
India	23,000	26,000	6430	660,000	6700	5.0%
Indonesia	11,000	16,000	1000	1,200,000	No data	No data
Jamaica	10,100	8900	2480	2,000,000	1800	1.4%
Russia	5650	5400	2760	500,000	2700	2.0%
Vietnam	4100	4500	1310	3700,000	No data	No data
Saudi Arabia	3890	4100	1770	200,000	1800	1.4%
The USA	W[a]	W	1570	20,000	1600	1.2%
Canada	No data	No data	No data	No data	1500	1.1%
Other countries	17,000[d]	15,000	11,400[d]	5,000,000[d]	14,600	11.0%
World total	**327,000**[b]	370,000[b]	**131,000**	**30,000,000**	**132,900**	**100%**

Source: United States Geological Survey (2020)
W = Withheld to avoid disclosing company proprietary data
[a]Estimated net exporter
[b]Excludes US production
[c]Only one of the bauxite producers in Guinea refines the raw material in the country; the other aluminium refineries are owned by Russian exporters and Chinese operators
[d]Includes Canada

5.3.2   Aluminium Production

Globally, 63.7 million tonnes of primary aluminium were produced in 2019 (IAI, 2021a). About 32 million tonnes of aluminium is recycled every year (IAI, 2021b). Global primary aluminium production accounts for two-thirds of the total production. However, not all bauxite-rich countries are among the main aluminium-producing nations. China dominates global aluminium production. Overall, nine conglomerates are responsible for global aluminium production (31.5 million tonnes/year), and of those, four have their headquarters in China (Statista, 2021): Chalco, Hongqiao Group, Xinfa, and SPIC Aluminum & Power Investment Co. Ltd. (Statista, 2021). As a result, Chinese aluminium companies produce 17.8 million tonnes per year or 57% of the volume produced by the nine major companies (Statista, 2021). Russian aluminium manufacturer Rusal produces 3.8 million tonnes annually, which is 12% of the amount produced by the nine largest companies. Like China, Russia also owns an aluminium refinery in Guinea (Human Rights Watch, 2018). The Australian/UK mining giant Rio Tinto produces 3.2 million tonnes per year, equivalent to 10.2% of the aluminium produced by the main producers; the UAE aluminium producer EGA produces 2.6 million tonnes per year (8%), the US-owned company Alcoa produces 2.5 million tonnes per year (6.9%), and Norwegian Norsk Hydro produces two million tonnes per year, which is equivalent to 6% of the aluminium produced by the nine top companies (Statista, 2021). Another 1.9 million tonnes per year is produced by other companies.

The proportion of recycled or 'secondary' aluminium production is a key consideration in determining decarbonisation pathways because secondary aluminium production is up to 95% less energy-intensive than its primary production from bauxite (IAI, 2020). The aluminium sector distinguishes between new aluminium scrap (offcuts generated during the manufacture of aluminium) and old scrap (used, discarded, and collected aluminium products). The proportion of aluminium that is recycled can be measured by quantifying the input rate and the efficiency rate:

- The recycling input rate describes the proportion of new and old scrap fed into aluminium production.
- The recycling efficiency rate is the proportion of aluminium available that is recovered from a region.

Once collected, the metal losses from recycling processes are usually <2%, so the net metal yield is >98% (IAI, 2018c; based on a 2005 study). The global recycling input rate has remained constant, at around 32%, since 2000 (IAI, 2020). The most recent data show a global recycling input rate of 32% in 2020, whereas in 2018, the global recycling input rate was 33%, and old scrap accounted for 60% of this.

Globally, up to 30 Mt. of primary aluminium was recycled in 2020, equivalent to a recycling rate of 76% (IAI, 2020):

- Europe has the highest aluminium recycling efficiency rate worldwide, and 81% of scrap available in the region is recovered (IAI, 2020).
- The USA has the highest recycling input rate, at 57%.
- China is the largest producer and consumer of recycled aluminium; it produces ten million tonnes of secondary aluminium from scrap annually or 33% of the global volume (IAI, 2020).

5.3.3 Aluminium Production Processes

An analysis of current and future aluminium production processes is required to understand the decarbonisation opportunities within each process.

Primary aluminium production involves the following processes (excluding mining):

1. *Refining bauxite to produce alumina (Bayer chemical process)*: Bauxite contains ores other than aluminium, including silica, various iron oxides, and titanium

dioxide (The Aluminum Association, 2021). Alumina, an aluminium oxide compound, is chemically extracted with the Bayer process (Scarsella et al., 2015), in which bauxite ore is ground and then digested with highly caustic solutions at elevated temperatures. Approximately 70% of the global bauxite production is refined to alumina with the Bayer process (The Aluminum Association, 2021).

2. *Smelting*: It is the process of refining alumina to pure aluminium metal (Hall–Héroult electrolytic process). Alumina is dissolved at 950 °C (1,750 °F) in a molten electrolyte composed of aluminium, sodium, and fluorine, to lower its melting point, allowing easier electrolysis. An electrical reduction line is formed by connecting several electrolysis cells in series (Haraldsson & Johansson, 2018). Electrolysis separates alumina into aluminium metal at the cathode and oxygen gas at the anode (M'Calley, 1894).

In the *secondary production of aluminium (aluminium recycling process)*, the process of refining the raw material (bauxite) to alumina is not required. Instead, scrap aluminium is re-melted and refined. Therefore, the energy consumption for this process is much lower than for its primary production (Haraldsson & Johansson, 2018; IAI, 2020).

5.3.4 Aluminium Industry: Energy Demand and Energy Intensities

The amount of energy used to generate a unit of GDP is referred to as the 'energy intensity of the economy' (IEA, 2020d). The IEA analyses the energy intensity for different sectors of the economy per GDP, based on US currency. The energy intensities of primary and secondary aluminium production are reported under the sub-sector *basic metals*. In 2018, the production of basic metals was responsible for 27% of the energy consumption in the *manufacturing* sector. The sub-sector *basic metals* includes ferrous metals (22% of energy consumption) and non-ferrous metals, such as aluminium, nickel, lead, tin, brass, silver, and zinc, and accounts for 5% of the *manufacturing* sector's energy consumption (IEA, 2020d). Table 5.13 shows

Table 5.13 Energy intensities (energy consumption per value added) in the manufacturing industry sub-sectors *basic metals* and *non-ferrous metals*, by region (2018 data; MJ per GDP in USD 2015)

	Energy intensity, 2018 data [MJ/GDP in US$, 2015]		
	Basic metals (ferrous and non-ferrous)	Ferrous metals[a]	Non-ferrous metals[a]
Percentage share of the global energy intensity [%]	27	22	5
Global	2724.00	2219.56	504.44
North America [MJ/GDP in US$]	290.00	236.30	53.70
Europe [MJ/GDP in US$]	1568.00	1277.63	290.37

Source: IEA (2020a)
[a]Calculation derived from the total energy consumption of the *basic metals* sector

the energy intensities of the total *basic metal* and *non-ferrous metal* sub-sectors by region.

Compared with aluminium production processes, the energy demand for bauxite mining is relatively small. Bauxite mining requires <1.5 kg of fuel oil (diesel) and < 5 kWh of electricity per tonne of bauxite extracted (IAI, 2018a).

Refining/smelting The global average energy use for the electrolysis cell is 13.4 kWh per kg of aluminium produced. If rectifiers and other cell auxiliaries, such as pollution control equipment, are included, the global average increases to 14.2 kWh per kg of aluminium produced (Haraldsson & Johansson, 2018; IAI data).

Process heat The Bayer process is the most energy-intensive process in primary aluminium production. The energy consumed by the Bayer process varies at 7–21 GJ/tonne (Scarsella et al., 2015). However, the aluminium industry is moving towards more energy-efficient primary production methods. A study of Columbian aluminium-producing companies showed that this energy intensity can be reduced by changing the core elements of the process, including the size, processes, and temperature of the furnaces (Carabalí et al., 2018). That study suggested that energy consumption could be reduced by 32% by installing an oxy-combustion technology, which preheats the combustion air. The costs related to thermal energy could be reduced by 50.5% per tonne of aluminium. However, the investment cost (purchase) of the technology is high, which hinders its widespread application (Carabalí et al., 2018).

5.3.5   Global Aluminium Production and Energy Intensity Projections

The projections for the overall increase in global aluminium production are driven by technology shifts, including in lightweight vehicles and mounting and framing equipment used for solar photovoltaic (PV) panels and large reflectors for concentrated solar power plants (IEA, 2020e). The assumed ratio of primary/secondary aluminium is vital for the calculation of the energy demand, because secondary aluminium production is significantly less energy-intensive than primary production.

The projection of the global energy demand for the aluminium industry until 2050 is based on the projected volume of aluminium production, recycling rates, and energy intensities of the different steps of aluminium production, from bauxite mining to the raw product (aluminium). The IEA Sustainable Development Scenario projects an annual growth rate of around 1.2% until 2030 and 15% overall growth in production between 2018 and 2030 (IEA, 2020e). This is a projected overall increase in global aluminium production from the current 85 million tonnes per year to just under 150 million tonnes per year.

Table 5.14 shows the projected global aluminium production for the OECM 1.5 °C pathway. The global recycling rate is projected to increase from 32% in 2019 to 45% in 2050 (IAI, 2021c). The increased recycling rate will lead to a significant

Table 5.14 Assumed development of the global aluminium market

Parameter	Units	2019	2025	2030	2035	2040	2050
				Projections			
Global aluminium production	[Mt/yr]	84.	96	106	117	127	147
Global primary aluminium production	[Mt/yr]	63.7	68	71	73	76	81
Global aluminium recycling (includes old scrap only, excluding new scarp—re-melted material is listed as part of primary aluminium production)	[Mt/yr]	20.4	28	36	43	51	66
Calculated annual growth rate of the global aluminium market			2.10%	1.90%	1.74%	1.60%	1.38%
Global bauxite mining (estimates based on aluminium growth projections)	[Mt/yr]	325	368	403	438	473	543
Global alumina refineries/production	[Mt/yr]	130	137	139	141	143	147
Bauxite/alumina ratio	[%]	40%	41%	41%	42%	43%	45%
Alumina/primary aluminium ratio	[%]	49%	50%	51%	52%	53%	55%
Global: primary aluminium production Share of the total production	[%]	68%	66%	64%	62%	59%	55%
Global: secondary aluminium production Share of the total production	[%]	32%	34%	36%	39%	41%	45%

decoupling of global bauxite and alumina production from global aluminium production. The efficiency ratio of bauxite to alumina is projected to increase from 40% to 45%, which will lead to a reduction in the energy demand.

Secondary aluminium production occurs through recycling schemes, after which the aluminium is re-melted and refined. The energy consumption involved is much lower than for the primary production of aluminium (Haraldsson & Johansson, 2018). The aluminium sector distinguishes between new or pre-consumer scrap and old or post-consumer scrap (discarded aluminium products). Of the 33 million tonnes of aluminium recycled in 2019, 20 million tonnes was from old scrap, and 14 million tonnes was from new scrap, and the share of new scrap is expected to reach 24 million tonnes in 2050 (IAI, 2021d).

The projected energy intensities for bauxite mining and aluminium production are shown in Table 5.15. The fuel demand per tonne of mined bauxite mainly comprises the fuel consumed by mining vehicles. The projections for the electricity and process heat demand for primary and secondary aluminium reflect the improvements in the industry's efficiency in the past decade and assume incremental improvements based on the efficiency assumptions and opportunities noted above, but with no disruptive new production technologies.

The IEA (2020a) has reported improvements in the energy efficiency (−3% annually) of alumina refining and aluminium smelting between 2010 and 2018. These were due to the highly energy-efficient production in China. Further reductions in global energy intensity (1.2% annually) are required under the IEA Sustainable

Table 5.15 Assumed energy intensities for bauxite mining and aluminium production processes

Parameter	Units	2019 [estimated]	2025	2030	2035	2040	2050
Energy intensities							
Bauxite mining—energy intensity	[PJ/ Mt]	0.721	0.705	0.689	0.674	0.659	0.630
Bauxite mining—fuels for mining machinery (currently 1.5kg of fuel oil per tonne) bauxite/fuel oil = 45.6 kg/ MJ	[PJ/ Mt]	0.068	0.068	0.068	0.068	0.068	0.068
Bauxite mining——alumina electricity demand	[PJ/ Mt]	0.653	0.637	0.621	0.605	0.590	0.561
	[TWh/ Mt]	0.181	0.177	0.172	0.168	0.164	0.156
Bauxite mining and alumina refining—thermal energy	[PJ/ Mt]	10.02	9.77	9.53	9.29	9.06	8.61
Primary aluminium——electricity (anode, electrolysis + ingot)	[PJ/ Mt]	55.8	54.4	53.1	51.7	50.4	47.9
	[TWh/ Mt]	15.5	15.1	14.7	14.4	14.0	13.3
Primary aluminium—thermal (anode, electrolysis + ingot)	[PJ/ Mt]	18.4	17.9	17.5	17.1	16.6	15.8
Secondary aluminium—electricity (anode, electrolysis + ingot)	[PJ/ Mt]	2.8	2.7	2.7	2.6	2.5	2.4
	[TWh/ Mt]	0.8	0.8	0.7	0.7	0.7	0.7
Secondary aluminium—thermal (anode, electrolysis + ingot)	[PJ/ Mt]	0.9	0.9	0.9	0.9	0.8	0.8

Development Scenario, which can be achieved through a shift towards increasing rates of aluminium recycling (Table 5.15). Secondary production must reach 40% by 2030, with a minimum proportion from old scrap of 70% (IEA, 2020e). The IAI projection to 2050, with maximum recycling rates, is 43% secondary production, but material recycled from old scrap will not exceed 70% (IAI, 2021c).

The production and energy intensity data for the aluminium sector were used to calculate the sectorial decarbonisation pathway presented in the following section (5.3.6).

5.3.6 Projection of the Aluminium Industry Energy Demand and CO2 Emissions

Due to the assumed increase in the share of recycled aluminium in global production and the reduced energy intensity per tonne of aluminium produced, a decoupling of the increases in production and energy demand is possible. Between 2019 and 2050, global aluminium production is projected to increase by 75%, whereas

Table 5.16 Projected electricity and process heat demands for the aluminium industry to 2050

Sub-sector	Units	2019	2025	2030	2035	2040	2050
Total electricity demand—aluminium industry	[PJ/yr]	3,694	3860	3924	3982	4035	4125
Total electricity demand—aluminium industry (including re-melting)	[TWh/yr]	1026	1048	1066	1082	1097	1123
Electricity demand—primary aluminium	[TWh/yr]	1005	1027	1040	1051	1062	1079
Electricity demand—secondary aluminium (excluding re-melting)	[TWh/yr]	21	21	26	31	36	44
Total process heat demand—aluminium industry	[PJ/yr]	3110	2581	2590	2597	2601	2601
Process heat demand—primary aluminium	[PJ/yr]	3079	2556	2559	2560	2558	2549
Process heat demand—secondary aluminium	[PJ/yr]	31	25	31	37	42	52
Heat demand <100 °C	[PJ/yr]	261	216	217	218	218	218
Heat demand 100–500 °C	[PJ/yr]	48	40	40	40	40	40
Heat demand 500–1000 °C	[PJ/yr]	569	472	474	475	476	476
Heat demand >1000 °C	[PJ/yr]	2232	1852	1859	1864	1867	1867

Table 5.17 Process- and energy-related CO_2 emissions—aluminium industry

	Units	2019	2025	2030	2035	2040	2050
CO_2 emissions—power supply	[$MtCO_2$/yr]	522	305	145	57	26	0
CO_2 emissions—heat supply	[$MtCO_2$/yr]	191	108	68	47	24	0
CO_2 emissions—total energy supply	[$MtCO_2$/yr]	713	413	213	104	50	0
Process-related emissions	[$MtCO_2$/yr]	210	229	240	250	258	270
Specific energy-related CO_2 emissions per tonne of aluminium	[tCO_2/yr]	4.8	3.5	2.9	2.5	2.2	0
Specific process-related CO_2 emissions per tonne of aluminium	[tCO_2/yr]	2.5	2.4	2.3	2.1	2.0	1.8

the overall energy demand will increase by only 12% (Table 5.16). Due to the already high electrification rates in the aluminium industry—which are projected to increase further—and the decarbonisation of the electricity supply based on renewable power generation, the aluminium industry can halve its specific CO_2 emissions by 2035 (Table 5.17).

5.4   Global Steel Industry: Overview

Steel is an important material for engineering and the construction sector worldwide, and it is also used for everyday appliances at the domestic and industrial levels. About 52% of steel usage is for buildings and infrastructure: 16% is used for

Table 5.18 Global crude steel production data by country (million tonnes per year)

	2017	2018	2019 (IEA prediction +3.4%)	2020
European Union (EU-28)[a]	168.5	167.7	173.4	179.2
Other Europe	42.2	42.4	43.9	
Commonwealth of Independent States (CIS)	101.6	102.1	105.6	
North America (Canada, the USA, Mexico)	114.8	120.3	124.3	100.5
Caribbean	0.6	0.6	0.6	
South America	44.1	44.9	46.5	
Africa	14.8	17.4	18.0	
Middle East	34.5	38.0	39.3	
Asia	1205.5	1278.0	1321.5	
Oceania	6.0	6.3	6.6	
World (Mt)	1732.2	1816.6	1878.4	1878

[a]EU-28 is the abbreviation of European Union (EU) which consists a group of 28 countries (Belgium, Bulgaria, Czech Republic, Denmark, Germany, Estonia, Ireland, Greece, Spain, France, Croatia, Italy, Cyprus, Latvia, Lithuania, Luxembourg, Hungary, Malta, the Netherlands, Austria, Poland, Portugal, Romania, Slovenia, Slovakia, Finland, Sweden, the UK) that operates as an economic and political block

mechanical equipment, such as construction cranes and heavy machinery; 12% is used for automotives (road transport); 10% is used for metal products, including tools; 5% is used for other means of transport, including cargo ships, aeroplanes, and two-wheeler vehicles; 3% is used for electrical equipment; and 2% is used for domestic appliances, such as white goods (World Steel Association, 2020a).

This section provides an overview of global steel production. Table 5.18 shows the data for global crude steel production. The World Steel Association (2020a) production data published in *2020 World Steel in Figures* is not complete for all countries, but is complete for North America (119.2 Mt) and the EU 28 (150.2 Mt) (note: Bulgaria, Croatia, and Slovenia are not included in the report).

5.4.1   Primary and Secondary Steel Production

Steel is produced by various routes. Crude or primary steel is produced from iron ore and secondary steel is produced from recycled steel. These two routes use different technologies and different energy sources. The share of secondary steel production increased by 25% globally in 2013 and by 28% in 2018 (IEA, 2020f).

Secondary steel production is limited by the availability of scrap. Currently, the total global scrap steel collection rate is 85% (IEA, 2020f), i.e. on average, 85% of steel consumed or utilised will be collected and recycled (Gauffin & Pistorius, 2018). However, the scrap collection rate varies for different steel applications: for structural reinforcement, it is as low as 50%, whereas for industrial equipment, it is as high as 97% (IEA, 2020f). Secondary steel production is up to 74% less

Table 5.19 Share of scrap (%) in crude steel production, by region, 2018

	2018 (%)	Change from previous year (%)
EU-28	55.9	+0.3
The USA	69.4	+2.2
Japan	35	+2.1
Russia	42.5	+5.5
Turkey	80.7	−0.4
South Korea	41.4	−2.3

Source: Bureau of International Recycling (2019)

energy-intensive than making steel from iron ore (primary production) (ISRI, 2019). Altogether, scrap input accounts for about 35% of the total primary steel production.

By 2030, this share should increase to 40% under the IEA Sustainable Development Scenario (IEA, 2020f). The share of scrap in primary steel production varies among countries and from year to year (Table 5.19):

- In EU-28, the proportion of recycled steel in crude steel production was 55.9% in 2018.
- In the USA, the proportion of steel scrap in crude steel production was 69.4% in 2018.

Global steel production is highly concentrated, and 12 companies are responsible for >50% of the global steel production. Steel companies with headquarters in China dominate the sector (Fig. 5.3). Seven corporations based in China are responsible for 30% of the global steel production. European steel manufacturers produce 9% of the global steel, Japanese companies 7%, South Korean companies 4%, and Indian steel manufacturers 3%.

Regional age profiles show that production capacity (manufacturing plants) in the steel sector differs among world regions. The average age profile of steel plants in the Asia Pacific region, including China, is among the youngest (IEA, 2020g); as a result, energy efficiency improvement is significant. Considering this region is responsible for one-third of the global production, energy efficiency improvement had an effect at the global scale.

Impact of COVID-19 on global steel production Global crude steel production decreased by 1.4% in the first 3 months of 2020 compared with that in the same period of the previous year, and in March, a reduction of 6% was reported (World Steel Association, 2020b). The largest declines in steel production in the first quarter of the year (Q1) occurred in the EU (−10%), Japan (−9.7%), South Korea (−7.9%), and North America (−4%) (World Steel Association, 2020a). The long-term consequences of COVID-19 for the steel sector are unclear. During the Global Financial Crisis (GFC) in 2009, steel production in Europe alone dropped by 30% compared with that in previous years.

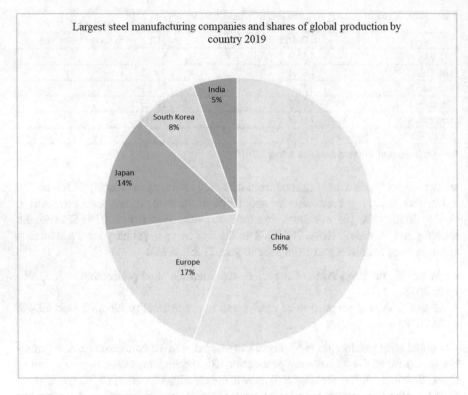

Fig. 5.3 Largest steel manufacturing companies and shares of global production, 2019

5.4.2 Technological Overview of Steel Production

On average, 20 GJ of energy is consumed to produce 1 tonne of crude steel globally (World Steel Association, 2021). The IEA's Tracking Industry Report (, 2020c) showed a gradual decline in energy intensity between 2009 and 2018. The largest year-to-year fall was in 2017–2018, when energy intensity declined by 3.6%. As mentioned earlier, there are two routes by which steel is produced (Table 5.20). Primary or crude steel is produced by the coal- or natural gas-based blast-furnace-basic oxygen furnace (BF-BOF) route, in which iron ore is reduced at very high temperatures in a blast furnace. The iron ore is melted to a liquefied form (pig iron or direct reduced iron [DRI]) and then oxidised and rolled (Table 5.21). Coal or natural gas is required to generate high temperatures of up to 1650 °C. In the secondary production route, scrap steel is melted in electric arc furnaces (EAFs). The EAF route has the lowest emission intensities. In the EAF (gas-fuelled) process, scrap is usually blended at a rate of about 10% with DRI. A more energy-efficient pathway for primary production is to use scrap steel with ore-based inputs in BF-BOF production, usually at a rate of 15–20% scrap (IEA, 2020f).

Table 5.20   Steel production—main processes

	Blast furnace-basic oxygen furnace (BF-BOF), 75% of steel is made with this process		Electric arc furnace (EAF), 25% of steel is made with this process	
Energy indirectly consumed (mining, preparation, and transport of raw materials)	9%		6%	
Energy input from	Used as energy in	Energy use (%) in the total process	Used as energy in	Energy use (%) in the total process
Coal	Blast furnaces (BFs), sinter, and coking plants	89%	Coke production, BF pulverised coal injection	11%
Electricity	EAFs, rolling mills, and motors	7%	Melting steel scrap	50%
Natural gas	Furnaces, power generators	3%	BF injection, DRI production	38%
Other gases and sources (%)	Steam production	1%	BF injection	1%

BF-BOF: production of primary steel from iron ore (oxygen is blown through liquid pig iron, increasing its temperature and releasing carbon)
EAF: production of secondary steel from scrap metal

Table 5.21   Steel production—main processes and energy requirements

	Energy use (GJ/tonne)			
Process	Absolute minimum	Practical minimum	Actual average requirement	% Over practical minimum
Liquid metal 'pig iron' (iron ore is reduced to iron)	9.8	10.4	13.5	23%
Liquid hot metal: basic oxygen furnace (iron ore is converted to steel)	7.9	8.2	11	25%
Liquid hot metal: electric arc furnace	1.3	1.6	2.25	29%
Hot rolling flat (after rolling, steel is delivered as strips, plates, bars, etc.)	0.03	0.9	2.2	59%
Cold rolling flat	0.02	0.02	1.2	98%

Emission benchmarks for the steel industry   Table 5.22 shows the emission values allowed for the manufacture of steel under the emission trading scheme of the EU (EU-ETS). The manufacture of secondary steel with EAFs is significantly less carbon-intensive—in tonnes of CO_2 per tonne of steel (tCO_2/tonne)—than the production of primary steel by the iron ore-based route, in which hot metal is produced in blast furnaces (BF-BOF route).

Table 5.22 EU-ETS benchmark values for iron and steel manufacture, as of February 2020

Material	Benchmark (tCO_2e/tonne product)[a]
Hot metal	1.328
Sintered ore	0.171
Iron casting	0.325
Electric arc furnace (EAF) high-alloy steel	0.352
Electric arc furnace (EAF) carbon steel	0.283
Coke (excluding lignite coke)	0.286

Source: EU (2020)

[a]tCO_2e = tonne carbon dioxide emission equivalent—a term that describes a unique global warming impact, includes all GHG emissions (CO_2 and non-CO_2 emissions)

5.4.3 Projections for the Global Steel Industry: Production and Energy Intensity

To calculate the future energy demand for the global steel industry requires a range of assumptions—from the actual market volume to the recycling rates and energy intensities, to the actual production process itself. Unlike the aluminium industry, steel manufacturing involves GHG emissions that are not related to energy generation but to the process itself. The emission intensity of the steel sector, specifically steel plants, depends upon the production route (BF-BOF or EAF) and the energy source (Table 5.23). Both routes can, for example, be fuelled by natural gas (IEA, 2020f). The actual process emissions per tonne product for each of the production process options are assumed to remain at current levels.

Table 5.23 shows the assumed development of global iron ore and steel production in million tonnes per year and the shares of primary and secondary steel production for the 1.5 °C OECM steel pathway. All assumed energy intensities, which are dependent on the production technologies used and process emissions that are used for the energy demand projections, are provided.

The global steel market is estimated to grow by 1–1.5% throughout the entire modelling period. The recycling rates are assumed to increase so that the share of secondary steel will grow from 35% in 2019 to 48% in 2050. The shares of electricity for primary and secondary steel in the overall production process are projected to remain at the current levels. Secondary steel production is, to a large extent, based on electricity, whereas primary steel production is 98% dependent upon process heat for the melting processes. The energy and electricity intensities per tonne of manufactured volume for both secondary and primary steel production are based on IEA projections (IEA, 2020f). Table 5.23 shows all the assumed market and energy intensity developments for the global steel industry according to the production process.

Table 5.23 Assumed market and energy intensity developments for the global steel industry according to the production process

Parameter	Units	2019	2025	2030	2035	2040	2050
Global iron ore production—estimates based on steel growth projections	[Mt/yr]	2339	2377	2511	2676	2851	3289
Global: annual production volume—iron and steel industry	[Mt/yr]	1,869.6	1904.2	2018.4	2159.7	2310.9	2695.4
Calculated annual growth rate for global steel market	[%/yr]		0.95%	1.13%	1.31%	1.31%	1.48%
Development of production structures (primary and secondary)							
Primary steel production	[%]	65%	63%	61%	59%	56%	52%
Secondary steel production (share of scrap)	[%]	35%	37.2%	39.3%	41.5%	43.7%	48%
Share of electricity in *primary* steel production	[%]	2%	2%	2%	2%	2%	2%
Share of electricity in *secondary* steel production	[%]	91%	91%	91%	91%	91%	91%
Energy intensities							
Energy intensity for iron ore mining	[PJ/Mt]	0.069	0.067	0.066	0.064	0.062	0.059
Global: average energy intensity for steel production	[GJ/t]	18.6	12.81	12.4	12.2	12.0	11.4
Global range: average energy intensity for primary steel production	[GJ/t]	21	16	16	16	16	16
Global range: average energy intensity for secondary steel production	[GJ/t]	9.1	8.26	7.65	7.55	7.45	7
Primary steel production—electricity demand	[GJ/t]	0.42	0.31	0.31	0.31	0.31	0.31
Primary steel production—process heat demand	[GJ/t]	15.57	11.51	11.51	11.51	11.51	11.51
Secondary steel production—electricity demand	[GJ/t]	8.28	7.52	6.96	6.87	6.78	6.37
Secondary steel production—process heat demand	[GJ/t]	6.75	6.13	5.68	5.61	5.53	5.20
Electricity intensities							
Electricity intensity—*primary* steel production	[TWh/Mt]	0.12	0.09	0.09	0.09	0.09	0.09
Electricity intensity—*secondary* steel production	[TWh/Mt]	2.30	2.09	1.93	1.91	1.88	1.77
Development of process-related emissions							
Specific process emissions—assumption in the OECM for the global average	[tCO_2/t crude steel]	1.06	0.92	0.60	0.37	0.23	0.08

(continued)

Table 5.23 (continued)

Parameter	Units	2019	2025	2030	2035	2040	2050
Basic oxygen furnace (BOF)—production share	[%]	65%	58%	35%	20%	10%	0%
Basic oxygen furnace (BOF)—emission factor	[tCO$_2$/t steel]	1.46	1.46	1.46	1.46	1.46	1.46
Open hearth furnace (OHF)—production share	[%]	5%	3.0%	2.5%	1.0%	1.0%	0%
Open hearth furnace (OHF)—emission factor	[tCO$_2$/t steel]	1.72	1.72	1.72	1.72	1.72	1.72
Electric arc furnace (EAF)—production share	[%]	30%	40%	63%	79%	89%	100%
Electric arc furnace (EAF)—emission factor	[tCO$_2$/t steel]	0.08	0.08	0.08	0.08	0.08	0.08

Table 5.24 Projected electricity and process heat demands for the steel industry to 2050

Sub-sector	Units	2019	2025	2030	2035	2040	2050
Steel industry							
Total electricity demand—iron and steel industry	[PJ/yr]	4559	5691	5906	6550	7245	8676
Total electricity demand—iron and steel industry	[TWh/yr]	1266	1581	1641	1819	2012	2410
Electricity demand—primary steel	[TWh/yr]	83	103	105	109	112	121
Electricity demand—secondary steel	[TWh/yr]	1184	1478	1535	1711	1900	2289
Total process heat demand—iron and steel industry (final energy)	[PJ/yr]	17,451	18,146	18,639	19,603	20,604	22,900
Process heat demand—primary steel	[PJ/yr]	13,269	13,797	14,120	14,569	15,011	16,163
Process heat demand—secondary steel	[PJ/yr]	4183	4349	4518	5034	5593	6738
Heat demand	[PJ/yr]	13,060	18,146	18,639	19,603	20,604	22,900
Heat share	[%]	74%	76%	76%	75%	74%	73%
Heat demand <100 °C	[PJ/yr]	595	2341	2405	2529	2658	2955
Heat demand 100–500 °C	[PJ/yr]	211	336	345	363	382	424
Heat demand 500–1000 °C	[PJ/yr]	2489	5038	5175	5442	5720	6358
Heat demand >1000 °C	[PJ/yr]	9765	10,431	10,714	11,268	11,844	13,164

5.4.4 Projection of the Steel Industry Energy Demand and CO2 Emissions

The assumed division between primary and secondary production rates and the assumed production process technologies are key to the energy demand projections. Whereas secondary steel production requires significantly more electricity per tonne, its demand for high-temperature process heat is significantly lower (Table 5.24).

Table 5.25 Process- and energy-related CO_2 for the steel industry

	Units	2019	2025	2030	2035	2040	2050
CO_2 emissions—power supply	[$MtCO_2$/yr]	645	459	223	96	48	0
CO_2 emissions—heat supply	[$MtCO_2$/yr]	1073	762	489	353	187	0
CO_2 emissions—total energy supply	[$MtCO_2$/yr]	1717	1221	712	449	235	0
Process-related emissions	[$MtCO_2$/yr]	1980	1757	1219	804	542	216
Specific energy-related CO_2 emissions per tonne of steel	[$tCO2$/t steel]	0.6	0.4	0.2	0.2	0.1	0.0

Furthermore, as the share of primary steel will be reduced, demand for iron ore mining (volumes) that is required will decrease with higher recycling rates.

The energy-related CO_2 emissions and estimated process emissions are shown in Table 5.25. Whereas the energy-related emissions are projected to be phased out by 2050, the process-related emissions are not, although they will be significantly reduced due to the predominant use of EAF ovens and the phase-out of high-emitting BOF ovens.

5.5   Textile and Leather Industry: Overview

The international fashion industry is estimated to be worth US$2.4 trillion, and the textile and leather industry constitutes a large proportion of it (valued at US$818.19 billion in 2020) (SC, 2019; GNW 2021). 'Textiles' refer to natural and synthetic materials used in the manufacture of clothing (including finished garments and ready-to-wear clothing), furniture and furnishings, automotive accessories, and decorative items. Therefore, the textile industry spans activities related to the design, manufacture, distribution, and sale of yarn, cloth, and clothing. We refer to the textile and leather industry and the fashion industry interchangeably, because some data are available for the fashion industry as a whole, to which textiles and leather contribute almost 35% (SC, 2019; GNW, 2021).

The textile and leather industry has close links with the agricultural and chemical industries. Agricultural output provides the raw materials for the textile industry in the form of natural fibres; similarly, the chemical industry outputs are used as synthetic raw materials in the textile industry. Chemical industry products are also used in the processing of fibres into textiles, especially during dyeing processes. Some of the commonest chemical products used in textile production include spinning oils, lubricants, solvents, adhesives, binders, detergents, bleaches, acids, dyes, pigments, and resins (ChemSec, 2021).

Over 60% of textiles are used in the manufacture of apparel. Natural fibre crops, such as cotton, jute, kenaf, industrial hemp, sun hemp, and flax, are used in the manufacture of yarn for textiles, paper, and rope. Natural fibres can also be extracted from animals (sheep, goats, rabbits, and silkworms) and minerals (asbestos). Synthetic fibres are increasingly used in textile manufacture because of their durability and abundance and as by-products of the chemical and petro-chemical industries.

Cotton is the most commonly grown natural fibre. The main processes involved include cultivation and harvesting, spinning (yarn), weaving (fabric), and finishing (textiles). Most natural fibres are short (only few centimetres) and generally have a rough surface. In contrast, synthetic fibres have the ability to be processed as long fibres or batched and cut to be processed like natural fibres. Synthetic or artificial fibres are derived from polymer industries using processes such as wet spinning (rayon), dry spinning (acetate and triacetate), and melt spinning (nylons and polyes-ters). Natural fibres such as wool, silk, and leather most often result in high-quality and long-lasting garments, whereas synthetic fibres are popular in the manufacture of fast fashion garments and accessories (ILO, 2021).

The fashion industry's vast scale has raised international alarm about the envi-ronmental effects and social equity of many offshore production facilities. In addi-tion to glaring issues like child labour, unsafe working conditions, and inequitable wages, the industry's increasing dependence on energy, non-renewable synthetic fibres, and water is an issue of global concern. Estimates suggest that textile dyeing and treatment processes are responsible for almost 20% of all water pollution from industrial effluent (Ellen MacArthur Foundation, 2017). The fashion industry's environmental impact is spread across the value chain, although the manufacturing process is the most energy-, water-, and chemical-intensive, with high volumes of toxic chemical effluent and wastewater ending up in marine systems. Some of the estimated environmental impacts of the industry are:

- The consumption of 79 trillion litres of water.
- An 8–10% share of global emissions.
- 20% of water pollution from industry is from textile treatment and dyeing.
- The generation of 92 million tonnes (Mt) of waste.
- 35% (190,000 tonnes) of all oceanic primary microplastic pollution.

Note: All estimates are calculated annually.

(Kant, 2012; GFA, 2017; Quantis, 2018; UNFCCC, 2018; Niinimäki et al., 2020)

The International Labour Organization (ILO, 2021) noted that the stages of yarn and fabric production in textile manufacture consume significant quantities of water, chemicals, and energy. These stages are also responsible for a large share of GHG emissions from the textile industry. In the leather value chain, 63–68% of emissions are generated during the manufacture of products such as footwear, whereas the production of raw materials accounts for only 20–29% of emissions (Cheah et al., 2013; Quantis, 2018). The United Nations Framework Convention on Climate Change (UNFCCC) Fashion Industry Charter for Climate Action (2018) aims to achieve a 30% reduction in GHG emissions by 2030.

Although the use of recycled fibres in new textiles is gaining momentum, Dahlbo et al. (2017) have cautioned that more research and empirical evidence is required to determine the impact of recycled fibres on the replacement of virgin fibres in the textile value chain and the rebound effects of the reuse and resale of textiles on the demand for new production. However, the present analysis focuses on the energy demand and supply of the industry and the resulting GHG emissions.

5.5.1 Global Textile and Leather Production: Major Companies and Countries

The textile, clothing, leather, and footwear (TCLF) industry is characterised by geographically dispersed production and high volatility to factors external to the market, driven by rising fuel and material prices, low agricultural yields of natural fibres, escalating geopolitical tensions around offshore manufacturing, and higher costs of labour and capital in erstwhile havens for textile manufacturing, such as China, Sri Lanka, and Bangladesh. Niinimäki et al. (2020) mapped the environmental impacts of the fashion industry (energy demand, chemical use, water demand, waste output) across various value chain activities and the countries that lead in each stage of the value chain (Fig. 5.4). It is evident that the different stages of yarn and textile manufacture have environmental impacts across all categories (other than GHG emissions). Despite the fashion industry's global footprint, a vast proportion of fibre production and garment manufacture occurs in developing countries (Niinimäki et al., 2020).

In terms of consumer spending, the Asia Pacific region accounted for 37% of the global sales of apparel and footwear in 2018. China had the largest share of demand at US\$380 billion, followed by the USA at US\$370 billion (Lissaman, 2019). Despite the fashion industry's highly fragmented production and sales operations, it is reported that just 20 multinational companies own 138% of the sector's profits. In 2018, fashion brands such as Nike, Adidas, H&M, Uniqlo, Zara, Levi's, Old Navy,

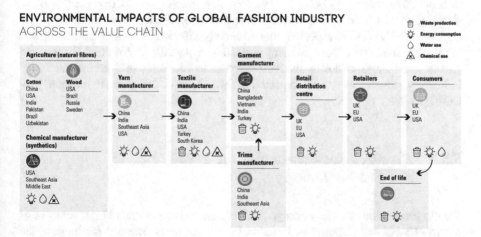

Fig. 5.4 Environmental impacts of global fashion industry across the value chain. (Source: Niinimäki et al., 2020)

and Ralph Lauren owned 8% of the global sales. Given the highly competitive industry dynamics and the low-profit margins in most of the upstream value chain activities, the industry is faced with mounting international pressure to incorporate sustainable resource management practices. Compounded by the impact of COVID-19, triggering closures and retail degrowth, the industry is struggling, because of its global labour- and resource-intensive operations.

5.5.1.1   Volume of Global Textile Production

In terms of the raw material demand of the textile industry, cotton had the highest value in 2019 (US$378.6 billion). However, in terms of volume, polyester recorded a 28% share of the textile demand, as a result of the diversity of its applications in textiles and apparel. Unsurprisingly, China leads global textile production and exports of both raw textiles and finished garments. Within Asia, the Indian textile industry constitutes 6.9% of the global textile production, valued at US$150 billion. India is the second largest textile producer, after China, in terms of production volume, and the textile industry contributed 15% to India's export earnings in 2018–2019. The USA leads global production and exports for raw cotton and is also a strong importer of raw textiles and finished garments (BV, 2020). It is also the third largest textile producer, with its industry valued at US$76.8 billion in 2018.

5.5.2   Impact of COVID-19 on Global Textile Production

Global textile and apparel exports were valued at US$750 billion in 2017 and were projected to grow at a compound annual growth rate of 18.7% to US$971—38 billion by 2021, before COVID-19. Most of this growth is still expected in Asia, although it will be dependent on the recovery of individual economies from the impacts of COVID-19, especially the adversely affected local manufacturing and retail sales industries. Because many countries, especially in Asia and Europe, are still experiencing lockdowns and a slow return to economic resurgence, the TCLF industry's growth trajectory is expected to take at least a few years to return to pre-COVID levels.

5.5.3   Resource Requirements of the Textile and Leather Industry

Textile production is water-, energy-, and chemical-intensive, and high volumes of liquid effluent are disposed of in natural water systems. Beyond production, the impact of textile and leather products at the end of the value chain is problematic because they generate high volumes of waste and the lifespans of many synthetic materials are short.

Reputable fashion events are increasingly promoting the theme of sustainability, and regenerated materials and accessories are being adopted by leading fashion designers. Whereas such initiatives are mainly targeted at material waste streams, there has also been a conscious effort to stimulate the use of natural and regeneratable materials in fashion. For the fashion industry to reduce its energy and emission intensities, systemic shifts must work in tandem. These shifts range from innovation in product design; the use of regenerative materials; more efficient technologies for processing and manufacture; decentralised production; reduced chemical use and dyeing; water cycling; common effluent treatment, especially in developing countries, which are major producers of fast fashion; and business models that accelerate longer use, reuse, sharing, and recovery.

The water footprint of the textile industry is one of the primary resource challenges for the environmental sustainability of the processing and production phases of this sector. The industry has one of the greatest demands for fresh water in the world, arising from the high water consumption across different stages: farming (especially cotton farming), washing and cleaning, textile processing, printing, dyeing, and finishing (ILO, 2021). The demand for water in the fashion industry is estimated to be 1.5–2.5 trillion gallons annually. In terms of the most polluting processes, textile dyeing accounts for the greatest shares of water use and pollution (SC, 2019).

5.5.4 Textile and Leather Industry: Energy Intensities and Emissions

Energy consumption in the textile industry is significantly high in the wet processing stages of dyeing and finishing, where it is used to generate steam, heat water, and dry fabrics. Alkaya and Demirer (2014) found that almost 46% of the energy demand was for the conversion of natural gas to steam, most of which is used to heat water for wet processing. The energy demand for the drying process was 30% in the same cotton mill (Alkaya & Demirer, 2014). The ILO (ILO, 2021) reported that energy use was the major contributor to the textile industry's GHG emissions, other than the emissions associated with agriculture and farming, manufacturing in the chemical industry, and livestock breeding for leather production.

The textile industry's carbon intensity relies on the type of energy source and production processes used. For example, hard coal and natural gas are the primary sources of industrial heat in India, thus raising the carbon footprint of apparel manufactured in India. China's textile industry accounts for almost 17% of the industrial sector's overall energy demand; in Bangladesh, the textile industry's energy use accounts for 9% of this demand. The type of input material also affects the energy demand over a product's lifespan. For example, a cotton t-shirt may have a higher-energy demand during the consumption phase than during its production, whereas the energy demand is highest during the production of a viscose garment (Allwood et al., 2006).

The various stages of textile production have different energy intensities, and these also vary significantly across regions. Therefore, the assumptions made about energy intensity must be simplified for any global analysis. Dyeing and finishing processes are most energy-intensive and, because they are currently supplied with predominantly fossil fuels, have the highest energy-related emissions (36%), followed by yarn production (28%), fibre production (15%), and fabric manufacture (12%) (Quantis, 2018). Despite the longevity and reuse characteristics of natural fibres, GFA (GFA, 2017) found that leather, silk, and wool processing generates the highest emissions per kilogram of material. In contrast, synthetic materials, such as polypropylene and acrylic fibres, record the lowest emissions, although post-use issues, such as microplastic pollution and the difficulty in recycling composite fibres, make natural fibres more sustainable.

5.5.5 Projections for the Global Textile and Leather Industry: Production and Energy Intensities

Table 5.26 shows the assumed economic development and energy intensities for the textile and leather industry used to calculate the 1.5 °C OECM pathway. The energy intensities per product volume (e.g. in tonnes per year) are not available, so the

Table 5.26 Projected economic development and energy intensities of the textile and leather industry

Parameter	Units	2019	2025	2030	2035	2040	2050
			Projection				
Textile industries—economic value	[bn $GDP]	1275	1632	1927	2270	2614	3392
Variation compared with 2019	[%]	0%	28%	51%	78%	105%	166%
Leather industry—economic value	[bn $GDP]	252	323	381	449	516	670
Variation compared with 2019	[%]	0%	28%	51%	78%	105%	166%
Total textile and leather value	[bn $GDP]	1527	1955	2308	2719	3130	4062
Variation compared with 2019	[%]	0%	28%	51%	78%	105%	166%
Textile and leather—sector share (global/total GDP)	[%]	1.2%	1.2%	1.2%	1.2%	1.2%	1.2%
Textile industry—energy intensities							
Textile mills	[MJ/$GDP]	4.5	4.4	4.3	4.2	4.1	3.9
Textile product mills	[MJ/$GDP]	4.6	4.5	4.4	4.2	4.1	3.9
Clothing industries	[MJ/$GDP]	0.9	0.8	0.8	0.8	0.8	0.7
Textile industry—average energy intensity	[MJ/$GDP]	2.0	1.9	1.9	1.8	1.8	1.7
Variation compared with 2019	[%]	0%	−2%	−5%	−7%	−10%	−14%
Leather industry—energy intensities							
Leather and allied product industries	[MJ/$GDP]	1.49	1.45	1.42	1.38	1.35	1.28
Variation compared with 2019	[%/yr]	0%	−2%	−5%	−7%	−10%	−14%

energy demand is calculated as a product of the assumed economic development in $GDP and the average energy units required per dollar. This simplification was necessary because the level of detail in the available energy demand data for the textile and leather industry on the global level did not allow a more exact approach. Textile mills have a significantly higher energy intensity than the clothing industry, which manufactures the clothing in downstream processes. The assumed average energy intensity for both textile and leather sections of the industry is estimated on the basis of the overall energy demand for both industries according to the IEA World Energy Statistics and the GDP shares.

5.5.6 Projection of the Textile and Leather Industry Energy Demand and CO2 Emissions

Analogous to the previous industry energy and emission projections, Tables 5.27 and 5.28 show the results for the textile and leather industry. All values are calculated on the basis of the documented assumptions. Based on the production processes typical of the industry, it is assumed that the process heat demand does not exceed the temperature level of 100 °C. The 1.5 °C OECM pathway requires that the global textile and leather industry decarbonises the required energy demand entirely by 2050, whereas a reduction by almost 50% seems achievable by 2030.

Table 5.27 Projected electricity and process heat demands for the textile and leather industry to 2050

Parameter	Units	2019	2025	2030	2035	2040	2050
			Projections				
Energy demand—textile industry	[PJ/yr]	2474	3134	3607	4143	4650	5737
Variation compared with 2019	[%]	0%	27%	46%	67%	88%	132%
Energy demand—leather industry	[PJ/yr]	425	469	539	620	696	858
Variation compared with 2019	[%]	0%	10%	27%	46%	64%	102%
Energy demand—textile and leather industry	[PJ/yr]	2899	3603	4146	4763	5346	6595
Variation compared with 2019	[%]	0%	24%	43%	64%	84%	128%
Electricity demand—textile and leather industries	[PJ/yr]	1277	1569	1805	2074	2328	2872
	[TWh/yr]	355	436	501	576	647	798
Heat demand	[PJ/yr]	2899	3603	4146	4763	5346	6595
Heat share	[%]	56%	56%	56%	56%	56%	56%
Heat demand <100 °C	[PJ/yr]	1622	2034	2341	2689	3018	3723
Heat demand 100–500 °C	[PJ/yr]	0	0	0	0	0	0
Heat demand 500–1000 °C	[PJ/yr]	0	0	0	0	0	0
Heat demand >1000 °C	[PJ/yr]	0	0	0	0	0	0

Table 5.28 Process- and energy-related CO_2 emissions for the textile and leather industry

Parameter	Units	2019	2025	2030	2035	2040	2050
			Projections				
CO_2 emissions—power supply	[million tCO_2/ yr]	181	127	68	30	15	0
CO_2 emissions—heat supply	[million tCO_2/ yr]	178	151	109	87	51	0
CO_2 emissions—total energy supply	[million tCO_2/ yr]	359	278	177	117	67	0
CO_2 emissions—process-related emissions	[million tCO_2/ yr]	0	0	0	0	0	0
Total CO_2 emissions	[million tCO_2/ yr]	359	278	177	117	67	0
Textile and leather: Specific CO_2 emissions per $GDP	[kgCO_2/$GDP]	0.12	0.08	0.05	0.03	0.02	0.00

5.6 Energy Demand Projections for the Five Industry Sectors Analysed

The industry sectors analysed, aluminium, steel, cement, chemical industries, and the textile and leather sector, consume more than half the electricity and process heat demand of the combined industry sectors (Table 5.29). The remaining large energy consumers are in *machinery*, including the manufacturing industry, food processing, mining, and construction. The aim of this sectorial pathway analysis is to inform the finance industry, which uses industry and service classification systems such as GICS. GICS differs from the IEA in the IEA sectors *industry* and *services*, as described in Chap. 4. The energy demand of *food processing*—a subgroup of the IEA industry sector—in the OECM is part of the demand analysis and projections for the *services* sector, whereas the IEA industry sector *construction* is part of the *buildings* analysis. Furthermore, the *transport equipment* sector has been analysed as part of the OECM 1.5 °C pathway for global transport.

Table 5.30 shows that the high-temperature process heat (>500 °C) accounts for two-thirds of the total process heat demand. Consequently, the generation of process heat for specific industries, such as in arc furnace ovens for steel, aluminium smelters, and process heat plants for chemical processes, is key to the decarbonisation of the global industry sector.

Therefore, the challenge is less the generation of carbon-free renewable power than the implementation of applications and manufacturing equipment especially designed for the cement, steel, and chemical industries. Timely investments in new manufacturing equipment may lead to the early retirement of existing industrial plants. The 1.5 °C global carbon budget of 400 GtCO_2 between 2020 and 2050, identified by the Intergovernmental Panel on Climate Change (IPCC; see Chap. 2),

Table 5.29 Total electricity demand of the industries analysed

Sector	Units	2019	2025	2030	2035	2040	2050
Electricity demand			Projections				
Chemical industry	[PJ/yr]	5384	6178	6981	7874	8678	10,323
Cement industry	[PJ/yr]	1754	1441	1439	1450	1459	1449
Aluminium industry	[PJ/yr]	3694	3860	3924	3982	4035	4125
Steel industry	[PJ/yr]	4559	5691	5906	6550	7245	8676
Textile and leather industry	[PJ/yr]	1277	1569	1805	2074	2328	2872
Total: five industry sectors	[PJ/yr]	16,668	18,739	20,055	21,930	23,745	27,445
	[TWh/yr]	4630	5205	5571	6092	6596	7624

Table 5.30 Total process heat demand of the industries analysed

Sector	Units	2019	2025	2030	2035	2040	2050
Process heat demand			Projections				
Chemical industry	[PJ/yr]	12,163	15,949	18,024	20,329	22,406	26,653
Cement industry	[PJ/yr]	7213	7514	7516	7483	7497	7597
Aluminium industry	[PJ/yr]	3110	2581	2590	2597	2601	2601
Steel industry	[PJ/yr]	17,451	18,146	18,639	19,603	20,604	22,900
Textile and leather industry	[PJ/yr]	2899	3603	4146	4763	5346	6595
Total: five industry sectors	[PJ/yr]	42,836	47,793	50,915	54,775	58,454	66,346
Heat demand <100 °C	[PJ/yr]	5020	7831	8578	9465	10,298	12,072
Heat demand 100–500 °C	[PJ/yr]	3127	4098	4571	5105	5589	6583
Heat demand 500–1000 °C	[PJ/yr]	11,060	15,413	16,545	17,904	19,180	21,878
Heat demand >1000 °C	[PJ/yr]	17,961	18,882	19,417	20,227	21,059	22,942

Table 5.31 Total energy-related CO_2 emissions of the industries analysed

Sector	Units	2019	2025	2030	2035	2040	2050
Chemical industry	[MtCO₂/yr]	2019	1492	971	669	380	0
Cement industry	[MtCO₂/yr]	887	578	360	230	118	0
Aluminium industry	[MtCO₂/yr]	713	413	213	104	50	0
Steel industry	[MtCO₂/yr]	1717	1221	712	449	235	0
Textile and leather industry	[MtCO₂/yr]	359	278	177	117	67	0
Total: five industries	[MtCO₂/yr]	5695	3982	2433	1569	850	0

has set a clear and hard limit for future emissions, and industries must be supported by government policies to implement the required transition to decarbonisation.

The five main industry sectors are responsible for about 85% of the energy-related CO_2 emissions of the entire *industry* sector and for almost 20% of all global energy-related CO_2 emissions (Table 5.31).

5.7 OECM 1.5 °C Pathways for Major Industries: Limitations and Further Research

The development of energy and emission pathways for industry sectors requires an energy model with high technical resolution. Compared with regional and global energy scenarios, sectorial pathways for industries are based on significantly more statistical data and must be developed in close co-operation with industry partners. Furthermore, the estimation of carbon budgets for specific industry sectors (based on GICS) requires a holistic approach, and all sectors must be considered in order to capture the interactions between the different industries and with the energy sector. To estimate a carbon budget based on current emissions for a single sector, such as the aluminium industry, will inevitably lead to inaccurate results because this approach does not consider the possible technical developments in other individual industries. The current discussions of net-zero targets for specific industries are often developed for a single industrial sector in isolation. This means that the total of all sub-concepts for certain industries may exceed the actual CO_2 emitted, and/or the responsibility for the reduction of CO_2 may be shifted to other sectors. In this research, bottom-up projections of the energy demand for the chemical, aluminium, and steel industries formed the basis for the supply scenarios for electricity, process heat, and fuels. The supply of carbon-free electricity is the key to the decarbonisation of all industry sectors. Furthermore, the electricity demand will increase with the electrification of process heat to replace fuels. Therefore, power utilities will play a crucial role in those industries reaching their decarbonisation targets. The decarbonisation of process heat will require changes in specific production processes and is therefore the core responsibility of the industry itself. We found that it is technically possible to decarbonise the energy supply of the analysed industries with available technologies. However, the OECM 1.5 °C pathway is not a prognosis, but a backcasting scenario that shows what must be done to achieve the carbon target. More detailed analyses for specific industry locations, e.g. China or India, are required because our global analysis simplifies processes and calculates energy demand projections on the basis of average global energy intensities. Moreover, energy demand was calculated with energy intensities (e.g. for steel production) derived with a literature search. Energy statistics, especially for the chemical industry, are sparse, and all the energy demand for sub-sectors are based on GDP projections. More research is required for industries in specific GICS classes, in terms of both statistical data and the current and future energy intensities of industry-specific processes. A central database of energy intensities and energy demand for each GICS class would significantly enhance the level of detail available for the calculation of net-zero pathways in the future.

References

ACC. (2020). *Guide to the business of chemistry.*

ACC. (2021). *American Chemistry Council (ACC).* Off. website. https://www.americanchemistry. com/default.aspx

ACS. (2019). *C&EN's global top 50 chemical companies of 2018.*

Alkaya, E., & Demirer, G. N. (2014). Sustainable textile production: A case study from a woven fabric manufacturing mill in Turkey. *Journal of Cleaner Production, 65*, 595–603. https://doi. org/10.1016/j.jclepro.2013.07.008

Allwood, J., Laursen, S., Rodriguez. C. M. de, & Bocken, N. (2006). *Well dressed? The present and future sustainability of clothing and textiles in the United Kingdom.*

Andrew, R. M. (2018). Global CO2 emissions from cement production, 1928–2017. *Earth System Science Data, 10*, 2213–2239.

Bureau of International Recycling. (2019). *World steel recycling in figures 2014-2018. Steel scrap – A raw material for steelmaking* (10th ed.). Bureau of International Recycling (BIR) Ferrous Division.

BV. (2020). *Global textile industry factsheet 2020: top 10 largest textile producing countries and top 10 textile exporters in the world.* BizVibe (BV). https://blog.bizvibe.com/blog/top-10-largest-textile-producing-countries. Accessed 14 Oct 2021.

Carabalí, D. M., Forero, C. R., & Cadavid, Y. (2018). Energy diagnosis and structuring an energy saving proposal for the metal casting industry: An experience in Colombia. *Applied Thermal Engineering, 137*, 767–773. https://doi.org/10.1016/j.applthermaleng.2018.04.012

Centrica. (2021). *Pharma companies cutting energy consumption to gain a competitive advantage.* Blog. https://www.centricabusinesssolutions.com/us/blogpost/pharma-companies-cutting-energy-consumption-gain-competitive-advantage. Accessed 5 Oct 2021.

Cetinkaya, E., Liu, N., Simons, T. J., & Wallach, J. (2018). *Petrochemicals 2030: Reinventing the way to win in a changing industry.*

Cheah, L., Ciceri, N. D., Olivetti, E., et al. (2013). Manufacturing-focused emissions reductions in footwear production. *Journal of Cleaner Production, 44*, 18–29. https://doi.org/10.1016/j. jclepro.2012.11.037

ChemSec. (2021). *The textile process.* Chem. Secr. https://textileguide.chemsec.org/find/get-familiar-with-your-textile-production-processes/. Accessed 12 Nov 2021.

Construction Review Online. (2021). Top 10 cement producers in the world. https://construc-tionreviewonline.com/top-companies/top-10-cement-producers-in-the-world/. Accessed 15 Dec 2021.

Dahlbo, H., Aalto, K., Eskelinen, H., & Salmenperä, H. (2017). Increasing textile circulation – consequences and requirements. *Sustain Prod Consum, 9*, 44–57. https://doi.org/10.1016/j. spc.2016.06.005

Ellen MacArthur Foundation. (2017). *A new textiles economy: Redesigning fashion's future.*

EU. (2020). *Financing a sustainable European economy.*

European Cement Research Academy, Cement Sustainability Initiative. (2017). *Development of state of the art-techniques in cement manufacturing: Trying to look ahead.* Duesseldorf.

European Commission. (2018). *Competitiveness of the European cement and lime sectors final report.*

Fischedick, M., Roy, J., Abdel-Aziz, A., et al. (2014). Industry. In O. Edenhofer, R. Pichs-Madruga, Y. Sokona, et al. (Eds.), *Climate change 2014: Mitigation of climate change. Contribution of Working Group III to the Fifth Assessment Report of the Intergovernmental Panel on Climate Change* (pp. 739–810). Cambridge University Press.

Garside, M. (2020). *Chemical industry worldwide.*

Gauffin, A., & Pistorius, P. C. (2018). The scrap collection per industry sector and the circulation times of steel in the U.S. between 1900 and 2016, calculated based on the volume correlation model. *Metals (Basel), 8*. https://doi.org/10.3390/met8050338

GFA. (2017). *Pulse of the fashion industry.*

GNR. (2021). *Reporting CO2*. GNR Proj. https://gccassociation.org/gnr/. Accessed 2 Oct 2021.

GNW. (2021). *Worldwide apparel and leather products industry to 2030*. Globe News Wire (GNW), Res. Mark.

Haraldsson, J., & Johansson, M. T. (2018). Review of measures for improved energy efficiency in production-related processes in the aluminium industry – From electrolysis to recycling. *Renewable and Sustainable Energy Reviews, 93*, 525–548. https://doi.org/10.1016/j.rser.2018.05.043

Human Rights Watch. (2018). *"What do we get out of it?" The human rights impact of bauxite mining in Guinea*.

IAI. (2018a). *Global metal flow*. Alum. Futur. Gener. https://recycling.world-aluminium.org/review/global-metal-flow. Accessed 23 Aug 2021.

IAI. (2018b). *Sustainability*. Alum. Futur. Gener. https://recycling.world-aluminium.org/review/sustainability/. Accessed 3 Aug 2021.

IAI. (2018c). *Recycling indicators*. Int. Alum. Inst. https://recycling.world-aluminium.org/review/recycling-indicators/. Accessed 2 Aug 2021.

IAI. (2020). *Aluminium recycling*. Int. Alum. Inst. https://aluminium.org.au/wp-content/uploads/2020/10/IAI-Recycling-Factsheet.pdf. Accessed 8 Aug 2021.

IAI. (2021a). *Primary aluminium production*. Stat. Int. Alum. Inst. https://international-aluminium.org/statistics/primary-aluminium-production/. Accessed 1 Mar 2021.

IAI. (2021b). *Aluminium sector greenhouse gas pathways to 2050*. https://international-aluminium.org/resource/aluminium-sector-greenhouse-gas-pathways-to-2050-2021/. Accessed 10 Oct 2021.

IAI. (2021c). *Aluminium sector greenhouse gas pathways to 2050*.

IAI. (2021d). *Material flow model – 2021 Update*. https://international-aluminium.org/resource/iai-material-flow-model-2021-update/

IEA. (2018a). *The future of petrochemicals: towards a more sustainable chemical industry (technology report)*.

IEA. (2018b). *Technology roadmap – low-carbon transition in the cement industry*.

IEA. (2020a). *World energy balances 2020*. IEA. https://www.iea.org/data-and-statistics?country=WORLD&fuel=Energytransitionindicators&indicator=TFCShareBySector. Accessed 6 Apr 2021.

IEA. (2020b). *Tracking cement 2020*.

IEA. (2020c). *Low-carbon transition in the cement industry: Technology roadmap*.

IEA. (2020d). *Energy efficiency indicators*.

IEA. (2020e). *Aluminium tracking report*.

IEA. (2020f). *Iron and steel*.

IEA. (2020g). *Age profile of global production capacity for the steel sector (blast furnaces and DRI furnaces)*.

IEA. (2021a). *Energy efficiency indicators extended database*. Updat. 21th June 2021.

IEA. (2021b). Driving energy efficiency in heavy industries.

ILO. (2021). *Reducing the footprint? How to assess carbon emissions in the garment sector in Asia*.

International Finance Corporation. (2020). *The impact of COVID-19 on the cement industry*.

ISRI. (2019). *2019 recycling industry year book*.

Kant, R. (2012). Textile dyeing industry: An environmental hazard. *Natural Science, 4*, 22–26. https://doi.org/10.4236/ns.2012.41004

Levi, P., Pales, A. F. (2018). *From energy to chemicals*. Int. Energy Agency.

Lissaman, C. (2019). *The size of the global fashion retail market*. Common Object.

M'Calley, H. (1894). Bauxite mining. *Science*, (80-) ns-23:29–30. https://doi.org/10.1126/science.ns-23.572.29.

McKinsey. (2015). *The cement industry at a turning point: A path toward value creation*. McKinsey Insights. https://www.mckinsey.com/industries/chemicals/our-insights/the-cement-industry-at-a-turning-point-a-path-toward-value-creation

McKinsey. (2021). *Laying the foundation for zero-carbon cement.*

Niinimäki, K., Peters, G., Dahlbo, H., et al. (2020). The environmental price of fast fashion. *Nature Reviews Earth & Environment, 1*, 189–200. https://doi.org/10.1038/s43017-020-0039-9

Portland Cement Association. (2019). *Cement industry overview.* https://archive.epa.gov/epa-waste/nonhaz/industrial/special/web/pdf/chap-2.pdf

Quantis. (2018). *Measuring fashion: insights from the environmental impact of the global apparel and footwear industries.*

Roy, A. H. (2012). Fertilizers and food production. In J. A. Kent, T. Bommaraju, & S. D. Barnicki (Eds.), *Handbook of industrial chemistry and biotechnology* (12th ed., pp. 757–804). Springer International Publishing AG.

SC. (2019). *Global garment and textile industries: workers, rights and working conditions.*

Scarsella, A., Noack, S., Gasafi, E., et al. (2015). Energy in alumina refining: Setting new limits. In M. Hyland (Ed.), *Light metals 2015.* Springer.

Speight, J. G. (2017). Sources and types of organic pollutants. In J. G. Speight (Ed.), *Environmental organic chemistry for engineers* (pp. 153–201). Butterworth-Heinemann.

Statista. (2021). *World's leading primary aluminum producing companies 2020.* https://www.statista.com/statistics/280920/largest-aluminum-companies-worldwide/

The Aluminum Association. (2021). *Bauxite.* Production. https://www.aluminum.org/industries/production/bauxite. Accessed 1 Apr 2021.

UNEP. (2019). *Global chemicals outlook II.*

UNFCCC. (2018). *UN helps fashion industry shift to low carbon.* United Nations Framew. Conv. Clim.

United States Geological Survey. (2020). *Mineral Commodity Summaries.*

USEIA. (2016). *International energy outlook.*

World Bank. (2021). *World development indicators.* http://wdi.worldbank.org/table/4.2#. Accessed 3 Jul 2021.

World Steel Association. (2020a). *World steel in figures.*

World Steel Association. (2020b). *March 2020 crude steel production.* Woldsteel Stat.

World Steel Association. (2021). *Raw materials.*

Xi, F., Davis, S. J., Ciais, P., et al. (2016). Substantial global carbon uptake by cement carbonation. *Nature Geoscience, 9*, 880–883. https://doi.org/10.1038/ngeo2840

Ziesemer, J. (2007). *Energy use in organic food systems.*

Chapter 6
Decarbonisation Pathways for Services

Sven Teske, Kriti Nagrath, and Sarah Niklas

Abstract The decarbonisation pathways for the service sector are derived. Brief outlines of the agriculture—food and forestry—wood product sectors, fishing industry, and water utilities are presented. The projected development of product quantities or GDP and the assumed development of energy intensities are given. The industry-specific energy consumptions and CO_2 emission intensities are provided in tables. The non-energy-related CO_2 emissions for all sectors analysed in this chapter are discussed and quantified.

Keywords Decarbonisation pathways · Service industry · Agriculture · Food · Forestry · Wood products · Water utilities · Fisheries · Energy intensities · Bottom-up demand projections

The service sector contributes 65% of the global gross domestic product (GDP in 2019, US\$ 56.9 trillion (World Bank, 2021). In this analysis, we use the IEA World Energy Balances as the basis for the energy statistics which defines three main sub-sectors: 'industry', 'transport', and 'other sectors'.

While 'industry' and 'transport' overlap with their respective GICS classification used for the 1.5 °C OECM sectoral pathways to a large extent, the service sector is scattered across several GICS sectors and the IEA 'other sectors' and 'industry' group (see Chap. 4). In this section, we describe four service sectors that supply essential goods:

1. Agriculture and food processing
2. Forestry and wood products
3. Fisheries
4. Water utilities

The combined share of global energy demand of these sectors at about 7.5% is relatively minor. Even though the energy demand is low and current energy-related CO_2 emissions contribute only 6% to global CO_2 emissions, the non-energy GHG

S. Teske (✉) · K. Nagrath · S. Niklas
Institute for Sustainable Futures, University of Technology Sydney, Sydney, NSW, Australia
e-mail: sven.teske@uts.edu.au; kriti.nagrath@uts.edu.au; Sarah.Niklas@uts.edu.au

© The Author(s) 2022
S. Teske (ed.), *Achieving the Paris Climate Agreement Goals*,
https://doi.org/10.1007/978-3-030-99177-7_6

emissions are significant. Agriculture and forestry are among the main emitter of non-energy CO_2, methane (CH_4), and nitrous oxide (N_2O)—emissions referred to as *AFOLU* (agriculture, forestry, and other land uses) in climate science.

6.1 Overview of the Global Agriculture and Food Sector

The agriculture and food sector is an essential economic sector contributing to food security, livelihoods, and well-being. Valued at 3.5 trillion USD, agriculture, forestry, and fisheries (AFF)[1] accounted for 4% of the global GDP in 2019, with the largest contributions from China and India. The value added[2] in agriculture[3] alone was 0.2 trillion USD (FAO, 2021b; The World Bank, 2019). Value is also added in some of the manufacturing sectors supported by AFF. In 2018, the manufacture of food and beverages contributed 1.5 trillion USD, and the manufacture of tobacco products contributed 167 billion USD (UNIDO, 2020). The corresponding GICS sectors addressed in this section are listed in Table 6.1 (ISIC, 2008).

The most widely produced commodities in the world are cereals, sugar crops, vegetables, and oil crops. The area under agricultural use has been increasing since the 1960s, until it started to plateau at the beginning of this century, with almost 5 billion hectares under cultivation by 2018. China, the United States, and Australia

Table 6.1 Relevant Global Industry Classification Standard (GICS) sectors

Sector	Industry group	Industry	Sub-industry	Description
Consumer staples 30	Food, beverages, and tobacco 3020	Food products 302020	Agricultural products 30202010	Producers of agricultural products. Includes crop growers, owners of plantations, and companies that produce and process foods but do not package or market them. Excludes companies classified in the forest products sub-industry and those that package and market the food products classified in the packaged food sub-industry
			Packaged foods and meats 30202030	Producers of packaged foods, including dairy products, fruit juices, meats, poultry, fish, and pet foods
		Tobacco 302030	Tobacco 30203010	Manufacturers of cigarettes and other tobacco products

[1] Corresponds to ISIC divisions 1–3 and includes forestry, hunting, and fishing, as well as cultivation of crops and livestock production.

[2] Net output of a sector after all the outputs are summed and the intermediate inputs subtracted.

[3] Includes crop and animal production, hunting, and related service activities (ISIC division A_01).

have the largest areas of agricultural land (FAO, 2021b). Besides land and energy (discussed in the next section), other major inputs to agriculture are fertilisers and pesticides, which have been increasing progressively over time.

The impacts of agriculture, forestry, and other land uses (AFOLU) can be both positive and negative. The IPCC describes AFOLU emissions as follows: 'Plants take up carbon dioxide (CO_2) from the atmosphere and nitrogen (N) from the soil when they grow, re-distributing it among different pools, including above and below-ground living biomass, dead residues, and soil organic matter. The CO_2 and other non-CO_2 greenhouse gases (GHG), largely methane (CH_4) and nitrous oxide (N_2O), are in turn released to the atmosphere by plant respiration, by decomposition of dead plant biomass and soil organic matter, and by combustion' (Smith et al., 2014).

6.1.1   Energy Demand Projection for the Global Agriculture and Food Sector

Although energy is an important input to agriculture, the sector accounts for only 2.2% of the total final energy consumption globally, with oil and oil products meeting most of this demand (IEA, 2020). Generally, as agriculture is industrialised, this energy consumption increases. In regions where most agricultural systems are industrialised, efficiency gains may have plateaued (in the United States, after a peak in 2006 [FAO, 2021a)]), and the sectoral final energy consumption may even have decreased (in EU, 10.8% decrease since 1998 [Eurostat, 2020]).

However, the global food system is estimated to account for almost one third of the world's total final energy demand. In high-GDP countries, approximately 25% of the total sectoral energy is consumed behind the farm gate (agriculture including in fisheries): 45% in food processing and distribution and 30% in retail, preparation, and cooking (Sims et al., 2015). In low-GDP countries, a smaller share is spent on the farm and a greater share on cooking (FAO, 2011).

In this study, projections of the future energy demand for the agriculture and food processing sector are based on GDP development projections. The assumed global GDP projections until 2050 are based on the World Bank and IEA projections (IEA, 2019). It is anticipated that both agriculture and food and processing industries will grow in proportion to the global economy and that their share of the global GDP will remain between 3.5% and 4%. The production volumes for cereals, pulses, and other agricultural products for 2019, shown in Table 6.2, are taken from the Food and Agriculture Organization (FAO) database (FAO, 2021b).

The estimated global population growth is based on UN population projections (UN DESA, 2019) and will decrease evenly from about 1% per year in 2020 to 0.5% per year in 2050. The food production volumes for each product shown will develop accordingly. No dietary or lifestyle changes are assumed in estimating the future energy demand of the agriculture and food processing sector. In addition to

Table 6.2 Economic development—agriculture and food processing: 2019 and projections towards 2050

Parameter	Unit	2019	2025	2030	2035	2040	2050
Global GDP	[bn $]	129,555	142,592	196,715	231,758	266,801	346,236
Agriculture—economic value	[bn $]	3887	4687	5533	6518	7504	9738
Food and processing industry	[bn $]	1010	1326	1565	1844	2123	2755
Global GDP share	[%]	3.8%	4.2%	3.6%	3.6%	3.6%	3.6%
Total volume—main food products	[Million tonnes]	9609	10,068	10,392	10,689	10,953	11,415
Cereals, total	[Million tonnes]	2979	3159	3285	3400	3502	3680
Pulses, total	[Million tonnes]	88	94	97	101	104	109
Vegetables, primary	[Million tonnes]	1130	1199	1246	1290	1329	1396
Roots and tubers, total	[Million tonnes]	861	913	950	983	1012	1064
Sugar crops, primary	[Million tonnes]	2229	2242	2253	2265	2276	2299
Oil crops	[Million tonnes]	1101	1168	1215	1257	1295	1360
Milk, total	[Million tonnes]	883	937	974	1008	1039	1091
Meat, total	[Million tonnes]	337	357	371	384	396	416

Source: Food and Agriculture Organisation of the United Nations (FAOSTAT: Production)

food for human consumption, agricultural products are also needed for animal feed. However, the impacts of diets on agricultural product demand and emissions are discussed in the next section.

According to the IEA's *Advanced Energy Balances* database structure, the food processing industry is part of the *industry* sector, whereas agriculture is part of *other sectors* group. Furthermore, the statistical data for the relevant energy demand are provided as 'food and tobacco', and separate data for the food processing industry are not available. Similarly, the IEA database provides the energy demand for agriculture and forestry, but no further separation of the two industries is available.

To calculate the energy demand for each sub-sector, the economic values in $GDP energy for agriculture, forestry, food processing, and tobacco industry are divided by the average energy intensities (in MJ per $GDP) for each of those sectors. Table 6.3 shows a selection of energy intensities taken from the IEA database for different agricultural products. To calibrate the model and to understand the development in the past, statistical data for the years 2005–2019 are used. To project the future energy demand for each of the sub-sectors, the calculation method then

Table 6.3 Energy intensities for selected food processing industries

Energy intensities		2019	2025	2030	2035	2040	2050
Bakery product industry	[MJ/$GDP]	3.32	3.28	3.24	3.20	3.16	3.08
Assumed efficiency increase per year	[%/yr]		0.25%	0.25%	0.25%	0.25%	0.25%
Fruit and vegetable industries	[MJ/$GDP]	7.26	7.17	7.08	6.99	6.90	6.73
Assumed efficiency increase per year	[%/yr]		0.25%	0.25%	0.25%	0.25%	0.25%
Dairy product industry	[MJ/$GDP]	4.02	3.97	3.92	3.87	3.82	3.73
Assumed efficiency increase per year	[%/yr]		0.25%	0.25%	0.25%	0.25%	0.25%
Meat product industries	[MJ/$GDP]	3.49	3.45	3.40	3.36	3.32	3.24
Assumed efficiency increase per year	[%/yr]		0.25%	0.25%	0.25%	0.25%	0.25%
Average food processing industry	[MJ/$GDP]	3.49	2.82	2.81	2.8	2.78	2.67
Assumed efficiency increase per year	[%/yr]		1%	0.1%	0.1%	0.1%	0.1%
Average agriculture and farming	[MJ/$GDP]	1.74	1.53	1.39	1.27	1.15	0.96
Assumed efficiency increase per year	[%/yr]		1%	0.8%	0.8%	0.8%	0.8%

changes, and the projected GDP development (Table 6.2) is multiplied by the average sector-specific energy intensities, incorporating an assumed efficiency factor, giving the projected energy demand. For more details of the OECM methodology, see chap. 3.

The average energy intensity of the food processing industry for 2019 has been calculated to be around 3.5 MJ/GDP, and it is assumed that the annual efficiency gain is 0.25% on average (Table 6.3). The main energy demand for food processing is for heating processes in the range of 100–500 °C. Based on the study of Ladha-Sabur et al. (2019), the share of thermal energy is estimated to be 75% of the final energy demand on average for food processing and the remaining 25% for electricity. Transport energy is not included in this approach because the transport sector is analysed separately (see the Methodologies for *Scopes 1, 2,* and *3* section).

Based on the methodology described above, the energy demand for the agriculture and farming sector is calculated with an energy intensity of 1.74 MJ per $GDP for the base year 2019. The majority of the energy demand is estimated to be for fuel for agricultural machinery, such as tractors and harvesters, whereas 30% of the energy is electricity. Efficiency gains for the agriculture sector are assumed to be higher—0.8–1% per year—than for the food processing industry.

Table 6.4 shows the calculated energy demand broken down according to the electricity, heat, and fuel requirements for the agriculture and food processing sector. The energy-related CO_2 emissions for the calculated demand are based on the 1.5 °C OECM supply scenario (see Chap. 12) (Table 6.5).

Table 6.5 Energy-related CO_2 emissions for agriculture and food processing

Parameter	Unit	2019	2025	2030	2035	2040	2050
			Projection				
Agriculture: total energy-related CO_2 emissions	[Mt CO_2/ yr]	542	384	215	117	70	0
Agriculture: emissions—heat and fuels	[Mt CO_2/ yr]	195	152	99	69	47	0
Agriculture: emissions—electricity	[Mt CO_2/ yr]	346	232	116	48	23	0
Food processing: total energy-related CO_2 emissions	[Mt CO_2/ yr]	772	506	284	148	88	0
Food processing: emissions—heat and fuels	[Mt CO_2/ yr]	148	109	77	59	44	0
Food processing: emissions—electricity	[Mt CO_2/ yr]	624	397	206	89	44	0

Table 6.4 Energy demand projection for agriculture and food processing

Parameter	Unit	2019	2025	2030	2035	2040	2050
			Projection				
Agriculture							
Energy demand—agriculture	[PJ/yr]	7803	8655	9297	9967	10,442	11,221
Agriculture: electricity demand	[PJ/yr]	2450	2873	3087	3309	3467	3725
	[TWh/ yr]		681	798	857	919	963
Agriculture: heat and fuel demand	[PJ/yr]	5352	5781	6210	6658	6975	7496
Food processing							
Energy demand—food processing	[PJ/yr]	6071	6381	7498	8795	10,079	12,549
Food processing: electricity demand	[PJ/yr]	1931	2000	2349	2755	3156	3932
	[TWh/ yr]	536	556	653	765	877	1092
Food processing: heat and fuel demand	[PJ/yr]	4140	4381	5149	6040	6923	8617
Agriculture and food processing							
Energy demand—agriculture and food processing	[PJ/yr]	13,873	15,036	16,795	18,762	20,520	23,770
Agriculture and food processing: electricity demand	[PJ/yr]	4382	4873	5436	6064	6622	7657
	[TWh/ yr]	1217	1354	1510	1684	1840	2127
Agriculture and food processing: heat and fuel demand	[PJ/yr]	9492	10,162	11,359	12,698	13,898	16,113

6.1.2   Food Demand and Implications

Food Equity

The FAO estimates that sufficient global aggregate food is produced for nearly everyone to be well fed. However, income inequalities and resource constraints in different parts of the world mean that everyone is not well fed. Progress towards eliminating hunger and malnutrition is still lagging, with 821 million people under-nourished in 2017 (FAO, 2018). However, while we recognise the need for the redistribution of available food calories and a discussion of nutrition, in this research, we take a global aggregate view of food production, rather than a nuanced view of food security and nutritional equity in the local context.

Demand for Agricultural Products

The key drivers of food (and consequently feed) demand are population growth and changes in consumption patterns, which are driving a shift to a more meat-based diet. The demand for commodities, such as food grains, is primarily driven by increases in population because the per capita food demand is stagnant or even decreasing in several high-income countries (although the demand for coarse grains for use as feed will increase as meat and dairy consumption increases). Income, individual preferences, and changes in lifestyle and consumption patterns will play a greater role in the demand for vegetable oils, sugar, meat, and dairy products (OECD-FAO, 2020). The use of cereals for feed is projected to grow at 1.2% per year over the coming decade as livestock production expands and intensifies in low- and middle-income countries, compared with the projected growth of 1% per year for food use (OECD-FAO, 2021).

The average dietary energy supply per person per day in low- and middle-income countries is around 2750 kilocalories, whereas in high-income countries, it is around 3350 kilocalories. Both these figures exceed the minimum requirement of around 1950 kilocalories per person per day (FAO, IFAD, & WFP, 2015). It is expected that overall per capita consumption will increase globally, including in developed countries, even as concerns around obesity increase (Alexandratos & Bruinsma, 2012).

The global demand for food for human consumption is the main component of the overall demand for agricultural products. However, non-food uses of several commodities, mainly animal feed and fuel, are important and have experienced faster growth than food for human consumption over the last decade(s). It is anticipated that in the coming decade, the relative importance of food, feed, and biofuel use will remain constant, because no major structural shifts in the demand for agricultural commodities are expected (OECD-FAO, 2020). The global demand for agricultural commodities (including for non-food uses) is projected to grow at 1.2% per year over the coming decade, which is well below the 2.2% per year growth experienced over the last decade. This projected slowdown is due to a lower global demand for biofuels, especially as many high-income and emerging countries achieve saturation levels (OECD-FAO, 2021).

6.1.3 Meeting Global Food Demand While Reducing the Environmental Impact of Food Production

As noted above, a major source of emissions from the agricultural sector is associated with land use. Key complementary strategies for increasing food production while reducing the impact on land use are discussed below, followed by a discussion of the environmental impacts and emissions specifically related to animal protein production, including enteric emissions. These impacts are fundamentally driven by the overall demand for agricultural products.

Crop Yield

The substantial additional amounts of food required in the coming decades will mainly be produced through yield increases, rather than any major expansion of cultivated areas (FAO, 2017). The FAO expects 77% of this increased production to come from increased yields, compared with 9% from the expansion of cultivated land and 14% from increased cropping intensities (Alexandratos & Bruinsma, 2012). A review of the scientific literature showed that most of the focus on how to feed the world is on increasing food production through technological advances, whereas attention on reducing the food demand through dietary changes to less-intensive patterns has remained constant and low (Tamburino et al., 2020).

In either case, crop yields must increase to meet the needs of the growing population without increasing croplands. Agricultural yields have increased without a significant increase in agricultural land use in the past. For example, between 1961 and 2000, the global population more than doubled, and the per capita cereal consumption increased by 20%. However, the area of harvested cereals increased by only 7%, largely because cropping intensities increased (Piesse, 2020). Mueller et al. (2012) found that by maximising crop yields (i.e. closing yield gaps), the global crop production could increase by 45–70% with the same land use.

Food Waste

Another important consideration to improve the efficiency of food systems is the reduction of food waste. The energy embedded in global food losses is 38% of the total final energy consumed by the whole food supply chain. This means that more than 10% of the world's total energy consumption is food that is lost and wasted. By one estimate, the food losses and waste that occur every year generate more than 3.3 gigatonnes of CO_2 equivalents (FAO, 2013), equal to the combined annual CO_2 emissions of Japan and the Russian Federation (FAO, 2017).

Kummu et al. (2012) determined that an additional one billion people could be fed if food waste was halved, from 24% to 12%. The World Resources Institute reported that a 25% reduction in food waste would push food production 12% closer to the level necessary to feed the world in 2050 and would reduce the amount of increased agricultural land needed by 27%, inching closer to fully closing the land gap (Ranganathan et al., 2018).

Dietary Changes

Most developed countries have largely completed the transition to livestock-based diets, although it is unlikely that all developing countries—including India—will shift to levels of meat consumption typical of western diets in the foreseeable future (Alexandratos & Bruinsma, 2012).

The FAO 2030 Agriculture Outlook suggests that near-saturation levels of meat consumption, as well as health and sustainability concerns, might limit the growth of animal protein consumption in high-income countries, particularly reducing the demand for beef. However, the demand for poultry is expected to increase in high-income countries in the move to a more sustainable and healthy diet and in middle- and lower-income countries because it is the most economic animal protein (this will also circumvent religious reasons for the non-consumption of meat, such as the consumption of beef and pork in India and Muslim countries, respectively). However, it is estimated that over the next decade, any gains (emission-wise) made from the reduced demand for animal products in developed countries due to increases in vegetarianism or veganism will be offset by the increased consumption of meat in middle-income countries due to lifestyle changes and increasing per capita caloric consumption.

The projected improvements in production efficiency will be insufficient to meet the future food demand without increasing the total environmental burden posed by food production. By contrast, transitioning to less impactful diets would, in many cases, allow production efficiency to keep pace with the growth in human demand while minimising the environmental burden of the food system (Davis et al., 2016). Changing diets to a globally adequate diet of 3000 kcal per capita per day, with 20% animal kcal would allow an additional 2.1–3.1 billion people to be fed in 2050 if yield gaps are closed (Davis et al., 2014). Another study showed that a transition towards more sustainable production and consumption patterns could support 10.2 billion people within the planetary boundaries given if cropland is spatially redistributed, water and nutrient management improved, food waste reduced, and dietary changes imposed (Gerten et al., 2020).

Environmental Impacts

Increased meat production impacts land use in terms of increased pastureland and increased cropland. To accommodate the increasing ruminant production (especially sheep and goats) in sub-Saharan Africa, pastureland is expected to expand by 1.2 Mha. The projected expansion in livestock production in North America will require additional pastureland (+3.22 Mha), with the conversion of marginal croplands (OECD-FAO, 2021).

The other main contributor to agricultural emissions is methane emissions from the enteric fermentation in livestock. Diets rich in meat, particularly that from ruminants such as cattle, are associated with higher environmental costs and higher emissions of GHGs: methane, from enteric fermentation; CO_2, which is released from the clearing of forests for pasture; and nitrous oxide (N_2O), which is generated in feed production (FAO, 2017) . Diets with a smaller meat component have significantly lower emission intensities. The FAO 2030 Agriculture Outlook projects predict that agricultural GHG emissions will grow by 4% between 2018–2020 and 2030, with livestock accounting for more than 80% of this global increase (OECD-FAO, 2021).

Non-energy-related carbon emissions are calculated with the Generalized Equal Quantile Walk (GQW) method, the land-based sequestration design method, and the carbon cycle and climate model (*Model* for the Assessment of Greenhouse Gas Induced Climate Change, MAGICC) (Meinshausen & Dooley, 2019). The model also accounts for other GHG gas emissions arising from the enteric fermentation of livestock (CH_4), crop residues and fertilisers, and manure management (N_2O).

An industry sub-sector share has been assigned for each GHG, as explained in the attached supplementary material. Only a small part (20%) of the CO_2 emissions attributable to changes in land use are assigned to the agriculture sub-sector, with 80% assigned to forestry. Table 6.6 shows the breakdown of the different emission sources in agriculture. These emissions are multiplied by the global warming potential of other GHG gases to obtain the total CO_2 equivalents (CO_2e) for the sector.

6.2   Overview of Global Forestry and Wood Sector

Forestry contributes to food security, livelihoods, and well-being; supports terrestrial ecosystems and biodiversity; provides (human) life-sustaining ecosystem services; and acts as a carbon sink. Value is also added by some of the manufacturing sectors supported by forestry. In 2018, wood and wood products contributed 183 billion USD, and paper and paper products contributed 324 billion USD to the global economy. Together with agricultural manufacturing, this is about 18% of the value added in total manufacturing globally (UNIDO, 2020). The corresponding GICS sectors addressed are listed in Table 6.7.

Globally, 30% of all forests are used for production. Of this 30%, about 1.15 billion ha of forest are primarily used for the production of wood and non-wood forest products, and another 749 million ha are designated for multiple uses. In contrast, only 10% is allocated for biodiversity conversation, although more than half of total forests have management plans (FAO, 2020a).

Table 6.6 Non-energy emissions from the agriculture sector

Parameter	Unit	2019	2025	2030	2035	2040	2050
Agriculture—AFOLU	[Mt CO_2/yr]	662	409	232	224	216	200
Agriculture: synthetic and organic fertiliser	[kt N_2O/yr]	7827	6849	6300	6091	6126	6047
Agriculture	[Mt CH_4/yr]	154	119	96	88	87	80
Agriculture: ammonia	[Mt NH_3/yr]	22	21	20	21	21	20
Agriculture—total non-energy GHGs	[Mt CO_2e/ yr]	6837	5413	4515	4243	4205	3994

Table 6.7 Relevant Global Industry Classification Standard (GICS) sectors

Sector	Industry group	Industry	Sub-industry	Description
Materials 15	Materials 1510	Paper and forest products 151050	Forest products 15105010	Manufactures timber and related wood products. Includes lumber for the building industry
			Paper products 15105020	Manufactures all grades of paper. Excludes companies specialising in paper packaging, which is classified in the paper packaging sub-industry

Table 6.8 Global economic development of the forestry, wood, and wood products industry

Parameter	Unit	2019	2025	2030	2035	2040	2050
Forestry industry—economic value	[bn $]	155	187	221	261	300	390
Wood industry—economic value	[bn $]	143	183	216	255	293	381
Pulp and paper industry—economic value	[bn $]	117	150	177	209	240	312
Round wood	[Million m^3]	3969	3993	4013	4033	4053	4094
Variation compared with 2019	[%]	0.0%	0.6%	1.1%	1.6%	2.1%	3.1%
Sawn wood	[Million m^3]	489	492	494	497	499	504
Variation compared with 2019	[%]	0.0%	0.6%	1.1%	1.6%	2.1%	3.1%
Pulp for paper	[Million tonnes]	194	195	196	197	198	200
Variation compared with 2019	[%]	0.0%	0.6%	1.1%	1.6%	2.1%	3.1%
Paper and paperboard	[Million tonnes]	404	429	446	461	475	499
Variation compared with 2019	[%]	0%	6%	10%	14%	18%	24%

6.2.1 Energy Demand Projection for the Global Forestry, Wood, and Wood Product Sector

The sectoral final energy consumption of forestry has remained stable over the last three decades, and half of this demand is met by oil products.

The energy demand of the forestry and wood sector was calculated with the same methodology as for the agricultural and food processing sector (Table 6.8). The IEA Advanced Energy Balances show the wood and wood products separately but combine the energy demand for forestry with that for agriculture. The energy demand for forestry was calculated both as the energy intensity (Table 6.10) multiplied by the global GDP for this sector, as shown in Table 6.9, and by subtracting the calculated energy for agriculture (see previous section) from the combined energy demand for agriculture and forestry provided by the IEA. With this repeated calculation, the energy intensity for forestry, taken from the literature, was evaluated again. The economic values for forestry were taken from FAO 2015 (Lebedys, 2015).

Selected energy intensities of the wood products and paper industry, as well as the average energy intensities, were used to calculate the energy demand for the

Table 6.9 Assumed energy intensities for the forestry, wood, and wood product industry

Energy intensities	Unit	2019	2025	2030	2035	2040	2050
			Projection				
Forestry	[MJ/$GDP]	3	3.34	3.30	3.25	3.21	3.13
Assumed efficiency increase per year	[%/yr]		0.25%	0.25%	0.25%	0.25%	0.25%
Average energy intensity forestry, wood, and paper industry	[MJ/$GDP]	26	23.38	22.21	21.10	20.04	19.25
Assumed efficiency increase per year	[%/yr]		1.00%	1.00%	1.00%	1.00%	1.00%

Table 6.10 Energy demand for the forestry and wood product industry

Energy demand	Unit	2019	2025	2030	2035	2040	2050
Forestry							
Energy demand—forestry	[PJ/yr]	832	923	992	1063	1114	1197
Forestry: electricity demand	[PJ/yr]	74	5	11	22	44	176
	[TWh/yr]	20	2	3	6	12	49
Forestry: heat and fuel demand	[PJ/yr]	759	918	981	1041	1070	1021
Wood and wood products							
Energy demand—wood and paper	[PJ/yr]	7039	7791	8737	9779	10,695	13,330
Wood and paper: electricity demand	[PJ/yr]	2165	2259	2534	2836	3102	3866
	[TWh/yr]	602	628	704	788	862	1074
Wood and paper: heat and fuel demand	[PJ/yr]	4873	5532	6204	6943	7593	9464
Forestry and wood products							
Total energy demand	[PJ/yr]	7871	8715	9729	10,842	11,809	14,526
Electricity	[PJ/yr]	2239	2265	2545	2858	3146	4042
	[TWh/yr]	622	629	707	794	874	1123
Heat and fuels	[PJ/yr]	5632	6450	7184	7984	8663	10,484

forestry industry and the wood and wood product industry. For forestry, it is assumed that the improvement in energy efficiency per year will be relatively small, at only 0.25% per year, because this industry is already highly automated (Ringdahl, 2011).

The wood and wood product industry, as defined in the IEA statistic, includes the manufacture of wood and of products made of wood and cork, except furniture, and the manufacture of articles of straw and plaiting materials, as classified under the United Nations International Standard Industrial Classification of All Economic Activities (ISIC, 2008).

The calculated total final energy demand, further broken down to the electricity and heat/fuel demand for the forestry and wood product industry, is shown in Table 6.10. The processing of wood to wood products requires considerably more energy than forestry activities. For this reason, in developing the 1.5 °C energy pathway, the energy efficiency in this area is given greater importance than that for timber harvesting.

Table 6.11 Energy-related CO_2 emissions of the analysed sectors under the 1.5 °C energy pathway

Energy-related CO_2 emissions	Unit	2019	2025	2030	2035	2040	2050
			Projection				
Forestry: total energy-related CO_2 emissions	[Mt CO_2/yr]	58	18	12	9	6	0
Forestry: emissions—heat and fuels	[Mt CO_2/yr]	21	16	11	7	5	0
Forestry: emissions—electricity	[Mt CO_2/yr]	37	2	1	1	1	0
Wood products: total energy-related CO_2 emissions	[Mt CO_2/yr]	630	396	220	120	73	0
Wood products: emissions—heat and fuels	[Mt CO_2/yr]	175	139	94	69	49	0
Wood products: emissions—electricity	[Mt CO_2/yr]	455	257	126	51	24	0
Total energy-related CO_2 emissions	[Mt CO_2/yr]	688	414	232	128	79	0

Based on the 1.5 °C OECM supply scenario documented in Chap. 12, the energy-related CO_2 emissions for the analysed forestry and wood product sector are provide in Table 6.11. To decarbonise the energy supply of the forestry requires to switch machinery such as chainsaws and other heavy-duty tools from combustion engines to electric motors, and all-terrain vehicles need to be electrified.

6.2.2 Land-Use Demand for Forestry

There is potential for 'nature-based solutions' to remove CO_2 from the atmosphere at the gigatonne scale, with potentially significant co-benefits (Meinshausen & Dooley, 2019) (see also Chap. 14). Simulations of nature-based approaches, such as forest restoration, reforestation, reduced harvest, agroforestry, and silvopasture, were combined and found to sequester an additional 93 Gt carbon by 2100. This would require an additional 344 million ha of land for reforestation (Littleton et al., 2021). The key pathway for managing land-use change is reforestation, which is limited to biomes that will naturally support forests, by identifying previously forested land in close proximity to intact or degraded natural forests. This comprises of 274 Mha of land in proximity to intact forests in subtropical and tropical forest biomes and another 70 Mha identified in temperate biomes.

Decarbonisation pathways are being developed at the global level. At this level, there is little conflict between the competing uses of cropland, pastureland, and forests for carbon removal. Adopting nature-based approaches, such as agroforestry or silvopasture, where trees are integrated into cropland or grazing lands, will help to increase the carbon stock while meeting the increasing demand for forestry and agricultural products. It should be noted that a lot of deforestation and the capacity

Table 6.12 Non-energy GHG emissions in the forestry industry

Emissions		2019	2025	2030	2035	2040	2050
Non-energy GHG emissions							
Forestry—AFOLU	Mt CO/yr	2648	1164	−619	−1241	−835	−1359
Change to 2019	[%]		−56%	−123%	−147%	−132%	−151%

and demand for increased agricultural and livestock products will occur in tropical and subtropical regions, often in developing countries. At the local level, there must be a more nuanced approach to addressing the balance between environmental, economic, and well-being outcomes.

The OECM model also calculates the non-energy GHG emissions from the forestry sector, as shown in Table 6.12. The OECM 1.5 °C net-zero pathway is based on efficient energy use and renewable energy supply only—leading to full-energy decarbonisation by 2050. No negative emission technologies are used and the OECM leads to zero energy-related carbon emissions. The model assumes no net deforestation from 2030 onwards and the adoption of nature-based approaches to land-use management. Therefore, from 2030 onwards, there will be carbon removal or negative emissions.

6.3   Overview of the Global Fisheries Sector

About 7% of the total protein intake globally comes from seafood (FAO, 2020b). Over 200 million tonnes of fish and seafood are produced annually (Ritchie & Roser, 2021). According to the OECD, the fisheries industry employs over 10% of the world's population (OECD, 2020b). While the overall food fish consumption expanded by 122% between 1990 and 2018, the global capture fisheries—fish that has been caught from natural environments by various fishing methods—only grew by 14%. The main rise of fish 'production' came from aquaculture, which increased output by factor five. However, the percentage of fish stocks caught in the open ocean within biologically sustainable levels decreased from 90% in 1909 to only 65.8% in 2018 (FAO, 2020b).

The economic (first sale) value of the global fishing industry in 2018 was estimated at USD 401 billion, of which USD 250 billion came from aquaculture production (FAO, 2020b).

The Fishing Industry and Their Relevance Within the Energy Sector
While the fishing industry plays a significant role in food supply and economic income for a large part of the global coastal population, its share on global energy demand is minor with less than 0.1% of the global energy demand (IEA, 2020). The IEA World Energy Statistics lumps the energy demand of agriculture, forestry, and fisheries in one category. And even within this category, the energy demand of fisheries only makes up 3% within that group. The energy demand of the agricultural/

forestry sector is with 8900 PJ per year—compared to around 300 PJ annually for fisheries—about 25 times higher (IEA, 2020).

However, the OECM decided to develop a specific scenario for fisheries due to its importance for small island states. Subsistence fishing is a key economic pillar for island nations in the Pacific, the Indian Ocean, and the Caribbean. Over the past decades, large fishing vessels have been in dispute with the traditional fish grounds of local indigenous people.

Marine and aquatic ecosystems are under stress—from climate change, overfishing and unsustainable fishing, and aquaculture practices in some areas, as well as pollution from various other human activities, which lead to ocean acidification and declining biodiversity. Furthermore, illegal, unreported, and unregulated (IUU) fishing continues in many parts of the world, adding excessive pressure on fish stocks, harming law-abiding fishers through unfair competition, and thereby reducing their profitability, in addition to limiting employment opportunities throughout the value chain (OECD, 2020b).

Among the most unsustainable fishing methods is bottom trawling with large vessels which accounts for about one quarter of fish catch globally. Traditional artisanal fishing boats which are either entirely unpowered or with small outboard engines cannot compete with industrial fishing vessels. Increasing fuel costs make it increasingly uneconomic for the fisherman as fuel costs can often outweigh income from fish. Besides, most island states still rely on expensive diesel generators to provide electricity for households and cooling equipment for food preservation.

6.3.1 Fisheries: Projection of Economic Development and Energy Intensities

The economic value of the fishery industry is assumed to maintain its current global GDP share of 0.2% and to increase from US$ 272 billion in 2019 according to growth projection for global GDP to over US$700 billion in 2050. However, the shares between marine fishing, aquaculture, and inland fishing change significantly in favour of aquaculture. Table 6.13 shows all key assumptions used of the 1.5 °C pathway for fisheries.

The projected development of produced fish in million tonnes per year is certainly arguable, and forecasts of fish production volumes over the next 30 years are not available—thus the assumption that the volume of wild fish catch and fish from aquaculture plateaus on 2020 level, while the market value steadily increases. The rationale behind this is that marine fishing will not be able to increase fishing volumes, while costs and economic values per tonne of fish continue to increase. The catch per unit effort (CPUE)—the amount of energy per tonne—is assumed to remain stable. In this case, longer distances and sailing time to catch 1 tonne of fish can be compensated by increased energy efficiency of fishing vessels.

Table 6.13 Key assumption for the energy demand projection of the global fisheries industry

Parameter	Unit	2019	2025	2030	2035	2040	2050
			Projection				
Fishing (marine)—economic value	[bn $]	194	317	315	313	320	346
Fishing (aquaculture)—economic value	[bn $]	65	150	157	185	267	346
Fishing (inland)—economic value	[bn $]	13	17	20	23	27	35
Fishing—total economic value	[bn $]	272	483	492	521	614	727
Total volume—fish consumption	[Million tonnes]	159	159	159	159	160	160
	[%]	0%	0%	0%	1%	1%	1%
Marine landings	[Million tonnes]	47	47	47	47	46	46
Variation compared to 2019	[%]	0%	0%	0%	−1%	−1%	−1%
Aquaculture	[Million tonnes]	106	107	107	107	107	108
Variation compared to 2019	[%]	0%	1%	1%	1%	1%	2%
Inland fisheries	[Million tonnes]	6	6	6	6	6	6
Variation compared to 2019	[%]	0%	1%	1%	1%	1%	2%
Fishing fleet—number of vessel (powered)	[Million]	2.07	2.26	2.33	2.40	2.47	2.62
Unpowered	[Million]	1.16	1.16	1.16	1.16	1.16	1.16
Powered artisanal	[Million]	1.63	1.81	1.91	2.02	2.13	2.36
Powered, industrial (including aquaculture)	[Million]	0.43	0.45	0.42	0.38	0.35	0.26
Fishing fleet—total motor power	[GW]	144	154	151	147	144	135
Artisanal motor power	[GW]	57	63	67	71	74	83
Industrial motor power	[GW]	87	90	84	77	69	52
Catch per unit effort (CPUE)—energy units: petajoule per million tonnes of fish	[PJ/MtF]	6	6	6	6	6	6

The 1.5 °C OECM pathway for the fishing industry suggests moving away from large-scale fish trawlers towards a more decentralised fleet of fishing boats.

In regard to the fishing vessel fleet, 2.07 million vessels were registered in 2019, 1.16 unpowered, 1.63 million powered artisanal vessels, and 0.43 million industrial vessels (Rousseau et al., 2019). The overall motor power of the global fishing fleet is estimated with a capacity of 144 GW, 87 GW of which are from industrial vessels. The 1.5 °C pathways assume that the power artisanal fishing vessels steadily increase in numbers on the expense of industrial vessels which lose market shares in a stable fish market by volume.

The average motor power of artisanal vessel is estimated with 35 kW that operate with around 500 full load hours per year. The electricity share for fishing vessels increases from 0% in 2020 to 2% in 2025, to 4% in 2030, to 16% in 2040, to 64% in 2050.

Table 6.14 shows the resulting energy demand under those documented assumptions and Table 6.15 the expected energy-related CO_2 emissions. However, the available data about energy demand of fishing vessels is scarce and the results are indicative. More research is required in order to develop more detailed scenarios for and around the fishing industry, their vessels, and electrification concepts for artisanal fishing boats.

Decarbonising the energy and electricity supply of island nations away from diesel generators for electricity generation and gasoline-fuelled outboard engines to renewable-powered—mainly battery solar systems—mini- and micro-electricity grids will afford the island energy independence from expensive fuel supply via boat and planes. While the electrification of road vehicles for passenger and freight transport is already progressing worldwide, the electrification of ships and fishing vessels is still in its very first developments. Electric outboard engines, supplied with batteries charged with renewable electricity, can support subsistence fishing and help moving away from destructive fishing practices. However, electric outboard engines are still significantly more expensive than two-stroke or four-stroke outboarder, and the market is small. Economies of scale are required to make electric outboard engines—preferably in the range of 30–50 kW—cost-competitive.

Table 6.14 Projected energy demand for global fisheries industry

Parameter	Unit	2019	2025	2030	2035	2040	2050
			Projection				
Energy demand—fisheries	[PJ/yr]	300	309	315	327	349	483
Variation compared to 2019	[%]	0%	3%	5%	9%	16%	61%
Fuel demand—fishing fleet	[PJ/yr]	272	276	276	276	276	277
Variation compared to 2019	[%]	0%	1%	1%	2%	2%	2%
Electricity and synthetic fuel demand fishing fleet	[PJ/yr]	27	32	37	50	72	205
	[TWh/yr]	8	9	10	14	20	57

Table 6.15 Energy-related CO_2 emissions of the fisheries industry under the 1.5 °C energy pathway

Parameter	Unit	2019	2025	2030	2035	2040	2050
			Projection				
Fishing industry: total energy-related CO_2	[Mt CO_2/yr]	33	31	26	22	16	0
Fishing industry: emissions—heat and fuels	[Mt CO_2/yr]	1	1	1	0	0	0
Fishing industry: fuels for vessels	[Mt CO_2/yr]	28	27	24	20	15	0
Fishing industry: emissions—electricity	[Mt CO_2/yr]	4	3	1	1	0	0
Variation compared to 2019	[%]	0%	−8%	−21%	−36%	−52%	−100%

6.4 Overview of the Global Water Utilities Sector

Water is important for basically every process that supports human life on Earth. Keeping potable drinking water of high quality is therefore a basic requirement for the health of humans, for the environment, and for an intact economy. Thus, the economic value of water utilities is far beyond the monetary values of this industry. While the projection of future energy demand for various sectors in the analysis is based on economic values, the energy demand projection for water utilities must be based on production volumes.

The 1.5 °C OECM pathways are developed according to sectors as defined in the Global Industry Classification Standard (GICS). Water utilities (5510 40) are a subsector of the GICS sector 55 *utilities* together with electric utilities (5510 10), gas utilities (5510 20), multi-utilities (5510 30), and independent power and renewable electricity producers (5510 50). According to the GICS definition, water utilities are companies that purchase and redistribute water to the end consumer, including large-scale water treatment systems.

Only a fraction of water utilities globally has been privatised. The global market value of privatised water utility companies in 2020 was USD 158.79 billion (Statistica, 2021). Globally, the largest privatised water utilities are located in China and in the United States and are worth between USD22 and USD33 billion (Fig. 6.1).

However, the majority of member countries of the European Community decided against a privatisation of the water sector. The European Economic and Social Committee called for a stop of water utility privatisation (EESC, 2018), and the controversial debate kept the sector predominately in public ownership. Therefore, US American, Chinese, and companies from the United Kingdom dominate the overview due to their high share of privatisation.

To ensure that drinking water is of high quality, stricter water regulations have been implemented, and treatment practises have been intensified. As a consequence, energy consumption of wastewater treatment plants increased (Rothausen & Conway, 2011). The energy intensity for wastewater treatment depends on the

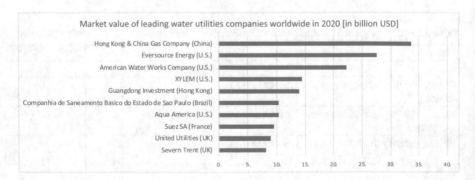

Fig. 6.1 Market value of leading water utilities companies worldwide in 2020, by country, in billion USD. (Source: Statistica (2021))

process and/or technology and the scale of the treatment plant (Paul et al., 2019). As electricity consumption in the water sector grows, the carbon footprint of the sector becomes larger and more significant if fossil fuel-based electricity is used. If this electricity is purchased from power utilities, energy costs might be significant. In developed countries, water utilities, on average, spend 15–30% of their budget on energy—this is for large wastewater plants—costs for small wastewater treatment plants are higher and make up 30–40% of their budget (Paul et al., 2019). For drinking water plants, the largest energy use (80%) is used to operate motors for pumping (Copeland & Carter, 2017).

6.4.1   Water Utilities: Commodity Demand Projections and Usage

There are a several international organisations that oversee water governance and also offer comprehensive databases; the most relevant organisations are:

- Food and Agriculture Organization (FAO)—a specialised agency of the United Nations that leads international efforts to defeat hunger
- Organisation for Economic Co-operation and Development (OECD)
- World Bank
- International Energy Agency (IEA)—data only for energy-related water usage

For this analysis, we use the FAO AQUASTAT database for all water-related data, which contains detailed data on water withdrawal, usage, and treatment (FAO, 2021a). The OECD database is most useful for OECD regions (North America and Europe) but not as comprehensive for global data. A comparison of the different data is shown in Table 6.13. IEA data on water extraction aligns best with the FAO data for global and for OECD regional. The OECD and World Bank data is particularly patchy; historical data is therefore displayed as averaged values for different timeframes. Considering the diversity of databases and approaches to compile data, the OECM project decided to use the FAO database for global analysis.

The FAO defines *total water withdrawal* as the 'annual quantity of water withdrawn for agricultural, industrial and municipal purposes. It can include water from renewable freshwater resources, as well as water from over-abstraction of renewable groundwater or withdrawal from fossil groundwater, direct use of agricultural drainage water, direct use of (treated) wastewater, and desalinated water. It does not include in-stream uses, which are characterised by a very low net consumption rate, such as recreation, navigation, hydropower, inland capture fisheries, etc.'.

The FAO water extraction data is based on the following calculation:

Total water withdrawal   Municipal water withdrawal

Industrial water withdrawal   Agricultural water withdrawal

Table 6.16 Total water withdrawal in billion cubic meters per year including extraction from desalination

	2017 (FAO AQUASTAT)	OECD average values (2010–2019)	IEA (2007–2015)	Desalination (FAO data)
OECD North America	567.7	505,896	543.16	0.6
OECD Europe	253.7	270.4	281.3	1.1
World [total]	4012.4	1047.6	3771.9	10.7

Data source: FAO AQUASTAT (2021a) and OECD (2020c) stats (most recent values)

Table 6.17 Assumed global water withdrawal quantities for the energy demand projection for water utilities

Parameter	Unit	2019	2025	2030	2035	2040	2050
			Projection				
Water withdrawal—total	[Billion m³]	4134	4388	4608	4838	5080	5601
Variation compared to 2019	[%]	0%	6%	11%	17%	23%	35%
Of which is saltwater	[Billion m³]	11	11	11	12	12	13
Saltwater share (of total water withdrawal)	[%]	0.3%	0.3%	0.2%	0.2%	0.2%	0.2%
Agricultural water	[Billion m³]	2956	3138	3295	3459	3632	4005
Variation compared to 2019	[%]	0%	6%	11%	17%	23%	35%
Municipal water	[Billion m³]	475	505	530	556	584	644
Variation compared to 2019	[%]	0%	6%	11%	17%	23%	35%
Industrial water	[Billion m³]	703	746	783	822	864	952
Variation compared to 2019	[%]	0%	6%	11%	17%	23%	35%

In addition to total extraction, the database allows to break down the data into water withdrawal by sector and by industry, which links the water sector with energy consumption in agriculture in form of irrigation.

According to the OECD, 70% of all water abstractions is used for agriculture (OECD, 2020a, p. 35). While freshwater extractions dominate total water extractions, desalinisation plants are an important parameter considering their high-energy consumption. However, water extraction through desalination plants only makes up 0.2% of the global water extraction (Table 6.16). Globally, about one third of all countries representing 80% of global population (OECD, 2020a) are connected to sewerage treatment plants. Table 6.17 shows the assumed global water withdrawal quantities—broken down by usage sector—which form the basis for the projection of the energy demand projection for water utilities.

6.4.2   Energy Efficiency Standards and Energy Intensities of Water Utilities

The following processes require the use of energy for water utilities:

- Water sourcing
  - Surface water pumping or
  - Groundwater pumping

- Wastewater treatment—energy demand dependent on the level of pollution
- Water distribution—energy for pumping, dependent on required distances

In addition, the topography of a region and the climatic conditions—especially seasonal temperature differences and rainfall pattern—affect energy use in the water sector (Copeland & Carter, 2017). In dry regions such as California, 19% of the state's electricity consumption is used for pumping, treating, collecting, and discharging water and wastewater (ibid). The following provides a brief overview of the technical processes and their energy intensities.

Water Extraction To lift 1000 litre (1 m^3) on metre requires 0.0027 kW/h—at 100% efficiency (Rothausen & Conway, 2011). But, in practice, the value is higher and dependents on the quality and efficiency of water pumps.

Wastewater Collection Wastewater is collected from domestic, commercial, or industrial use and processes. In general, the composition of wastewater by weight consists of 99.9% wastewater and 0.1% contaminants, including organic or inorganic matter, or microorganisms that need to be removed (ERC, 2019). Wastewater must be collected and transported; this process requires water pumps.

Wastewater Treatment Plants (WWTP) There are four types of wastewater treatment plants: (1) sewage treatment plants (STPs), (2) effluent treatment plants (ETPs), (3) activated sludge plants (ASPs), and (4) common and combined effluent treatment plants (CEPTs).

For water utilities, only sewage treatment plants (STPs) and activated sludge plants (ASPs), which are part of STPs, are important. Effluent treatment plants (ETPs) are typically used to clean industrial wastewater (ERC, 2019)—most of these are integrated into industrial parks for manufacturing and/or the chemical sector.

Sewage Treatment Plant (STP) The STP receives wastewater from domestic and commercial use and industrial processes. It also collects rainwater, storm water, and associated debris. The main processes include a basic filtering procedure to remove debris, dirt, grit, and sand:

- *Primary treatment*—settling: In the primary treatment, heavier and lighter organic solids are separated in a clarification tank which promotes sinking of heavier and floating of lighter solids, so they can be removed. This primary sludge is then moved into aeration basins, where the secondary treatment takes place.

- *Secondary treatment*—Secondary treatment involves aerobic aeration, which consists of ceramic or rubber membranes which have holes for aeration. The inflow of oxygen (compressed air) initiates a biological process in which the bacteria in the sewage digest organic matter. Aerobic aeration can remove chemicals, with the exception of nitrates (there are additional processes which can remove NO_3). After this process, the sludge moves into the dewatering tank to remove any water from the activated sludge.
- *Tertiary treatment—disinfection*: The tertiary process combines mechanical and photochemical processes. This step of the wastewater treatment process is required for sanitary sewage with microorganisms which require disinfection.

The secondary treatment is the most energy-intensive process for wastewater treatment plants; aeration—the introduction of air into the biological tank—consumes about 60% of the plant's total energy. There are ways to improve energy efficiency, e.g. by removing the aeration process through enhanced primary solid removal, based on advanced micro sieving and filtration processes (Oulebsir et al., 2020).

For the calculation and projection of energy demand for global water utilities, the energy intensities need to be simplified, and average values are used. Table 6.18 shows the values used for the 1.5 °C OECM pathway for water utilities. It is assumed the water withdrawal quantities (Table 6.17) will have to be pumped and distributed and—after usage—go back into wastewater treatment. The average energy intensity is provided. A small share of the water withdrawal—around 0.2%—will come from desalination plants which have a relatively high energy intensity per cubic metre.

Sewage plants often have onside electricity generation from biological material collected during wastewater treatment. The 1.5 °C pathway assumes that 5% of all sewage plants will utilise this potential in 2020 and that the share increase by 1% annually to 35% in 2050.

Furthermore, water utilities have significant non-energy GHG emissions from sewers, biological wastewater treatment, and sludge—mainly CH_4 and N_2O. Table 6.18 shows the assumed values in CO_2 equivalent per cubic metre with the conservative assumption that those specific values will remain on 2020 level until 2050.

6.4.3 Projection of the Energy Demand and CO_2 Emission for Water Utilities

The projected global energy demand for water utilities was calculated with the documented assumed global quantities of required water and energy intensities (Table 6.19). Based on the required energy and the 1.5 °C energy supply scenario,

Table 6.18 Assumed global energy intensities for process relevant for water utilities

Parameter	Unit	2019	2025	2030	2035	2040	2050
			Projection				
Energy intensity—water utilities							
Water pumping and distribution	[kWh/m^3]	0.18	0.18	0.18	0.18	0.18	0.17
Desalination	[kWh/m^3]	4.50	4.44	4.39	4.33	4.28	4.17
Wastewater treatment	[kWh/m^3]	0.06	0.06	0.06	0.05	0.05	0.05
Average electricity intensity	[kWh/m^3]	0.22	0.20	0.20	0.19	0.19	0.18
Average heat intensity	[MJ/m^3]	0.52	0.48	0.47	0.46	0.45	0.44
Energy intensity across all processes	**[kWh/m^3]**	**0.36**	**0.33**	**0.33**	**0.32**	**0.31**	**0.30**
Variation compared to 2019	[%]	0%	−8%	−9%	−11%	−13%	−16%
Wastewater treatment—electricity generation with syngas (from waste water)	[kWh/m^3]	−0.21	−0.21	−0.21	−0.21	−0.21	−0.21
Non-energy-related GHG—specific emissions							
Water utilities: specific non-energy GHG emissions	[kg CO$_2$eq/m^3]	0.20	0.20	0.20	0.20	0.20	0.20
CH$_4$ from sewers or biological wastewater treatment and sludge	[kg CO$_2$eq/m^3]	0.17	0.17	0.17	0.17	0.17	0.17
N$_2$O from sewers or biological wastewater treatment and sludge	[kg CO$_2$eq/m^3]	0.03	0.03	0.03	0.03	0.03	0.03
Water utilities—specific GHG emission	[kg CO$_2$eq/m^3]	0.33	0.27	0.24	0.22	0.21	0.20

Table 6.19 Projected global energy demand for water utilities

Energy demand	Unit	2019	2025	2030	2035	2040	2050
			Projection				
Water utilities: total energy demand	[PJ/yr]	5358	5284	5510	5745	5992	6518
Variation compared to 2019	[%]	0%	−1%	3%	7%	12%	22%
Water utilities: process heat energy demand	[PJ/yr]	2143	2098	2164	2232	2303	2451
Variation compared to 2019	[%]	0%	−2%	1%	4%	7%	14%
Water utilities: electricity demand	[PJ/yr]	3215	3186	3346	3513	3688	4066
	[TWh/yr]	893	885	929	976	1025	1130
Variation compared to 2019	[%]	0%	−1%	4%	9%	15%	26%

Table 6.20 Global energy-related CO_2 emissions and non-energy GHG for water utilities under the 1.5 °C energy pathway

Emissions	Unit	2019	2025	2030	2035	2040	2050
			Projection				
Water utilities: total energy-related CO_2 emissions	[$MtCO_2$/yr]	104	67	40	25	16	0
Water utilities: emissions—heat	[$MtCO_2$/yr]	77	53	33	22	15	0
Water utilities: emissions—electricity	[$MtCO_2$/yr]	27	14	7	3	1	0
Water utilities: non-energy GHG emissions	[$MtCO_{2\,equi.}$/yr]	830	881	925	971	1020	1125

the global energy-related CO_2 emissions have been estimated (Table 6.20). However, the main GHG emissions from water utilities do not originate from energy-related CO_2, but from methane and N_2O (nitrous oxide or 'laughing gas') which have a significant greenhouse potential (see Chap. 11).

6.5   Energy Demand Projection for the Four Analysed Service Sectors

The combined energy demand for the analysed sectors represented 7.5% of the global demand in 2019. The results of the energy demand projection suggest that demand will continue to grow even with energy efficiency measures as the volume of their produced commodities—especially food and water—will have to increase to meet the demands of a growing population by 165% by 2050. The two main drivers for the increased energy demand are agriculture and food processing and forestry and wood products. Due to electrification of machinery and (process) heat, the overall electricity demand increases significantly by 162% in 2050 in comparison to 2019. Especially the electricity demand for fisheries with the projected electrification of marine fishing vessel increases by factor 7 between 2019 and 2050 (Table 6.21).

6.6   The OECM 1.5 °C Pathways for Major Industries: Limitations and Further Research

We have shown that the four analysed sectors can phase out their energy-related CO_2 emissions (Table 6.22) with a combination of energy efficiency and a shift to a renewable energy supply. Key technologies for the decarbonisations are the following:

Agriculture and Forestry Heavy-duty machinery for harvesting food products, such as crops, or timber is currently almost entirely based on fossil fuel-driven com-

Table 6.21 Projected global energy demand for water utilities

Parameter	Unit	2019	2025	2030	2035	2040	2050
			Projection				
Agriculture and food processing							
Energy demand—agriculture and food processing	[PJ/yr]	13,873	15,036	16,795	18,762	20,520	23,770
Agriculture and food processing: electricity demand	[PJ/yr]	4382	4873	5436	6064	6622	7657
	[TWh/yr]	1217	1354	1510	1684	1840	2127
Agriculture and food processing: heat and fuel demand	[PJ/yr]	9492	10,162	11,359	12,698	13,898	16,113
Forestry and wood products							
Total energy demand	[PJ/yr]	7871	8715	9729	10,842	11,809	14,526
Electricity	[PJ/yr]	2239	2265	2545	2858	3146	4042
	[TWh/yr]	622	629	707	794	874	1123
Heat and fuels	[PJ/yr]	5632	6450	7184	7984	8663	10,484
Fisheries							
Energy demand—fisheries	[PJ/yr]	300	309	315	327	349	483
Electricity and synthetic fuel demand fishing fleet	[PJ/yr]	27	32	37	50	72	205
	[TWh/yr]	8	9	10	14	20	57
Fuel demand—fishing fleet	[PJ/yr]	272	276	276	276	276	277
Water utilities							
Water utilities: total energy demand	[PJ/yr]	5358	5284	5510	5745	5992	6518
Water utilities: electricity demand	[PJ/yr]	3215	3186	3346	3513	3688	4066
	[TWh/yr]	893	885	929	976	1025	1130
Water utilities: process heat energy demand	[PJ/yr]	2143	2098	2164	2232	2303	2451
Total—service sector							
Energy demand	[PJ/yr]	27,403	29,344	32,349	35,676	38,670	45,297
Electricity demand	[PJ/yr]	9862	10,356	11,364	12,485	13,529	15,971
	[TWh/yr]	3031	3134	3382	3656	3921	4492
Heat and fuel demand	[PJ/yr]	17,539	18,985	20,983	23,191	25,141	29,325

Table 6.22 Global energy-related CO_2 emissions and non-energy GHG for water utilities under the 1.5 °C energy pathway

Parameter	Unit	2019	2025	2030	2035	2040	2050
			Projection				
Total—service sector							
Agriculture and food processing	[Mt CO_2/yr]	1314	890	499	265	157	0
Forestry and wood products	[Mt CO_2/yr]	540	339	202	118	75	0
Fisheries	[Mt CO_2/yr]	4	3	1	1	0	0
Water utilities	[Mt CO_2/yr]	104	67	40	25	16	0
Total energy-related CO_2 emissions	[Mt CO_2/yr]	1963	1298	742	409	248	0

bustion engines. However, biofuels and—after 2030—electric vehicles are assumed to be available to reduce energy-related CO_2 emissions to zero by 2050.

The management of forests, croplands, and pastures can lead to both emission and sequestration of CO_2 and other GHGs. The need to feed a population of nine billion in 2050 will exert significant demands on the global agriculture and food systems. Advances in technology, particularly the increasing role of renewable energy in the agri-food sector, will help to reduce the energy emissions of the sector. However, given the crop intensification and agricultural expansion required to meet these food demands, it is expected that the agriculture sector will be unable to achieve zero emissions of non-energy GHGs by 2050. Improving soil management, reducing the yield gap, and initiating substantial shifts in dietary and nutritional patterns will help to reduce emissions. However, an increase of agricultural land at the expense of forests and/or their expansion in order to achieve negative emissions is likely if crop yield efficiencies cannot be improved. Further research is required on the individual contributions of each of these pathways to the complete decarbonisation of the sector.

Nature-based approaches, particularly reforestation, also offer offset options. With an increasing focus on saving and regenerating forests, the forestry sector can become not only carbon-neutral but also carbon-negative, as early as 2030. The abolition of carbon emissions or the achievement of negative emissions between 2030 and 2050 will compensate for the unavoidable process emissions in other sectors, such as the cement and steel industries.

The authors found a lack of policy mechanisms to unlock the large potential for nature-based solutions to create carbon sinks, although the scientific literature confirms the significant role of land-use emissions in climate mitigation pathways (IPCC 2021). More research is required into the compensation mechanisms for process emissions and their potential roles in the implementation of nature-based solutions (see also Chap. 11).

Food Processing Food processing, in particular, requires process heat, most of which was supplied by fossil fuel-based technologies in 2019. A significant increase in the electrification of process heat generation is assumed to occur. To achieve the overall CO_2 emission targets, the electricity generation under the OECM pathway will increase the average global renewable electricity share from 25% in 2019 to 74% in 2030. Although the transition to renewables under the OECM 1.5 °C pathways that phase out energy-related Scopes 1 and 2 emissions is ambitious, the implementation of the assumed Scope 3 emission pathways is significantly more challenging.

Wood and Paper Products   The wood processing and pulp and paper industry can use organic residuals and biomass as fuel for onside power and heat generation which is already a common practice especially in Scandinavia and Canada. An increase of those applications is assumed in the OECM.

Water Utilities   Similar to the wood and paper industry, water utilities can use organic residuals and especially methane from sludge to fuel onside power and heat generation to supply their own demand. Those technologies are assumed to become mainstream in the OECM to reduce 'behind the meter' demand and to capture methane emissions which have a high global warming potential (GWP)—see Chap. 11.

Fisheries   The transition to sustainable fisheries includes to move away from industrial fishing trawlers towards a more decentralised fishing fleet. The electrification of marine artisanal vessels via electric outboard engines seems a promising way to reduce emissions from inefficient diesel ship engines. However, the energy intensity for aquaculture farms is diverse, and a global average value in energy units per tonne of fish is not available. The literate suggests that it is entirely dependent on the region and the fish species. Thus, the calculated energy demand for the global fishery industry is fraught with very great uncertainties, and more research is needed.

We found that industry-specific data for energy intensities, although available (especially for the food sector), are often incomparable because they are based on different assumptions and/or methodologies. Therefore, we recommend the standardisation of the calculation and reporting methodologies for industry-specific energy intensities for the various technical processes. Furthermore, industry-specific energy statistics, including those for the sub-sectors of industries classified under the GICS system, would significantly enhance the level of detail available for setting net-zero targets in the future.

References

Alexandratos, N., & Bruinsma, J. (2012). *World agriculture towards 2030/2050: The 2012 revision* (No. 12; 03). World Agriculture. http://www.fao.org/3/ap106e/ap106e.pdf

Copeland, C., & Carter, N. T. (2017). *Energy-water nexus: The water sector's energy use.* https://fas.org/sgp/crs/misc/R43200.pdf

Davis, K. F., D'Odorico, P., & Rulli, M. C. (2014). Moderating diets to feed the future. *Earth's Future, 2*(10), 559–565. https://doi.org/10.1002/2014EF000254

Davis, K. F., Gephart, J. A., Emery, K. A., Leach, A. M., Galloway, J. N., & D'Odorico, P. (2016). Meeting future food demand with current agricultural resources. *Global Environmental Change, 39*, 125–132. https://doi.org/10.1016/J.GLOENVCHA.2016.05.004

EESC. (2018). *No more water privatisation, says EESC*. European Economic and Social Committee.

ERC. (2019). *Types of wastewater*. Blog. https://www.ercofusa.com/blog/types-of-wastewater-treatment/

Eurostat. (2020, September). *Agri-environmental indicator—Energy use*. Eurostat. https://ec.europa.eu/eurostat/statistics-explained/index.php/Agri-environmental_indicator_-_energy_use#Data_sources

FAO. (2011). *Energy-smart food for people and climate. Issue paper*. https://www.fao.org/3/i2454e/i2454e00.pdf

FAO. (2013). *Food wastage footprint: Impacts on natural resources—Summary report*. https://www.fao.org/3/i3347e/i3347e.pdf

FAO. (2017). *The future of food and agriculture: Trends and challenges*. http://www.fao.org/3/i6583e/i6583e.pdf

FAO. (2018). *The future of food and agriculture alternative pathways to 2050*. http://www.fao.org/3/I8429EN/i8429en.pdf

FAO. (2020a). *Global forest resources assessment 2020 Main report*. https://www.fao.org/3/ca9825en/ca9825en.pdf

FAO. (2020b). *The state of world fisheries and aquaculture 2020*. https://doi.org/10.4060/ca9229en

FAO. (2021a). *AQUASTAT*.

FAO. (2021b). *FAOSTAT*. FAO. http://www.fao.org/faostat/en/#data

FAO, IFAD, & WFP. (2015). *Achieving Zero Hunger: The critical role of investments in social protection and agriculture*. www.fao.org/contact-us/licence-request

Gerten, D., Heck, V., Jägermeyr, J., Bodirsky, B. L., Fetzer, I., Jalava, M., Kummu, M., Lucht, W., Rockström, J., Schaphoff, S., & Schellnhuber, H. J. (2020). Feeding ten billion people is possible within four terrestrial planetary boundaries. *Nature Sustainability, 3*(3), 200–208. https://doi.org/10.1038/s41893-019-0465-1

IEA. (2019). *World energy outlook 2019*. https://iea.blob.core.windows.net/assets/98909c1b-aabc-4797-9926-35307b418cdb/WEO2019-free.pdf

IEA. (2020). *World energy balances 2020*. IEA. https://www.iea.org/data-and-statistics

IPCC. (2021). Climate Change 2021: The Physical Science Basis. Contribution of Working Group I to the Sixth Assessment Report of the Intergovernmental Panel on Climate Change. Cambridge University Press.

ISIC. (2008). *International Standard Industrial Classification of All Economic Activities Revision 4*. https://unstats.un.org/unsd/demographic-social/census/documents/isic_rev4.pdf

Kummu, M., de Moel, H., Porkka, M., Siebert, S., Varis, O., & Ward, P. J. (2012). Lost food, wasted resources: Global food supply chain losses and their impacts on freshwater, cropland, and fertiliser use. *Science of the Total Environment, 438*, 477–489. https://doi.org/10.1016/J.SCITOTENV.2012.08.092

Ladha-Sabur, A., Bakalis, S., Fryer, P. J., & Lopez-Quiroga, E. (2019). Mapping energy consumption in food manufacturing. *Trends in Food Science & Technology, 86*, 270–280. https://doi.org/10.1016/J.TIFS.2019.02.034

Lebedys, A. (2015). *Forest products contribution to GDP*. https://unece.org/fileadmin/DAM/timber/meetings/20150318/2015jwpfsem-item6-roundtable-2-lebedys.pdf

Littleton, E. W., Dooley, K., Webb, G., Harper, A. B., Powell, T., Nicholls, Z., Meinshausen, M., & Lenton, T. M. (2021). *Dynamic modelling shows substantial contribution of ecosystem restoration to climate change mitigation* (No. 02). https://www.exeter.ac.uk/media/universityofexeter/globalsystemsinstitute/documents/Littleton_et_al-Dynamic_modelling_shows.pdf

Meinshausen, M., & Dooley, K. (2019). Mitigation scenarios for non-energy GHG. In S. Teske (Ed.), *Achieving the Paris climate agreement goals global and regional 100% renewable energy scenarios with non-energy GHG pathways for +1.5°C and +2°C.* SpringerOpen. https://link. springer.com/chapter/10.1007/978-3-030-05843-2_4

Mueller, N. D., Gerber, J. S., Johnston, M., Ray, D. K., Ramankutty, N., & Foley, J. A. (2012). Closing yield gaps through nutrient and water management. *Nature, 490*(7419), 254–257. https://doi.org/10.1038/nature11420

OECD-FAO. (2020). *OECD-FAO agricultural outlook 2020–2029.* OECD. https://doi. org/10.1787/1112c23b-en.

OECD-FAO. (2021). *OECD-FAO agricultural outlook 2021–2030.* https://doi. org/10.1787/19428846-en.

OECD. (2020a). *Environment at a glance 2020.* OECD Publishing.

OECD. (2020b). *Sustainable fisheries and aquaculture policies for the future.* OECD. https:// www.oecd.org/agriculture/topics/fisheries-and-aquaculture/

OECD. (2020c). *Water abstractions.* OECD Stat.

Oulebsir, R., Lefkir, A., Safri, A., & Bermad, A. (2020). Optimization of the energy consumption in activated sludge process using deep learning selective modeling. *Biomass and Bioenergy, 132*, 105420. https://doi.org/10.1016/j.biombioe.2019.105420

Paul, R., Kenway, S., & Mukheibir, P. (2019). How scale and technology influence the energy intensity of water recycling systems-An analytical review. *Journal of Cleaner Production, 215*, 1457–1480. https://doi.org/10.1016/j.jclepro.2018.12.148

Piesse, M. (2020). *Global food and water security in 2050: Demographic change and increased demand.* Future Directions International. https://www.futuredirections.org.au/publication/ global-food-and-water-security-in-2050-demographic-change-and-increased-demand/

Ranganathan, J., Waite, R., Searchinger, T., & Hanson, C. (2018). *How to sustainably feed 10 billion people by 2050, in 21 charts.* https://www.wri.org/insights/ how-sustainably-feed-10-billion-people-2050-21-charts

Ringdahl, O. (2011). *Automation in forestry development of unmanned forwarders* [Department of Computing Science, Umeå University]. https://www.researchgate.net/ publication/235676406_Automation_in_Forestry_Development_of_Unmanned_Forwarders

Ritchie, H., & Roser, M. (2021). *Fish and overfishing.* OurWorldInData.Org. https://ourworldin-data.org/fish-and-overfishing#

Rothausen, S. G. S. A., & Conway, D. (2011). Greenhouse-gas emissions from energy use in the water sector. *Nature Climate Change, 1*(4), 210–219. https://doi.org/10.1038/nclimate1147

Rousseau, Y., Watson, R. A., Blanchard, J. L., & Fulton, E. A. (2019). Evolution of global marine fishing fleets and the response of fished resources. *Proceedings of the National Academy of Sciences of the United States of America, 116*(25), 12238–12243. https://doi.org/10.1073/ pnas.1820344116

Sims, R., Flammini, A., Puri, M., & Bracco, S. (2015). *Opportunities for agri-food chains to become energy-smart.* http://www.fao.org/3/i5125e/i5125e.pdf

Smith, P., Bustamante, M., Ahammad, H., Clark, H., Dong, H., Elsiddig, E. A., Haberl, H., Harper, R., House, J., Jafari, M., Masera, O., Mbow, C., Ravindaranath, N. H., Rice, C. W., Robledo Abad, C., Romanovskaya, A., Sperling, F., & Tubiello, F. (2014). 2014: Agriculture, forestry and other land use (AFOLU). In O. Edenhofer, R. Pichs-Madruga, Y. Sokona, E. Farahani, S. Kadner, K. Seyboth, A. Adler, I. Baum, S. Brunner, P. Eickemeier, B. Kriemann, J. Savolainen, S. Schlömer, C. von Stechow, T. Zwickel, & J. C. Minx (Eds.), *Climate change 2014: Mitigation of climate change contribution of working group III to the fifth assessment report of the Intergovernmental Panel on Climate Change* (pp. 811–922). Cambridge University Press. https://www.ipcc.ch/site/assets/uploads/2018/02/ipcc_wg3_ar5_chapter11.pdf

Statistica. (2021). *Market value of leading water utilities companies world-wide in 2020.* Water & Wastewater. https://www.statista.com/statistics/1182423/ leading-water-utilities-companies-by-market-value-worldwide/

Tamburino, L., Bravo, G., Clough, Y., & Nicholas, K. A. (2020). From population to production: 50 years of scientific literature on how to feed the world. *Global Food Security, 24*, 100346. https://doi.org/10.1016/J.GFS.2019.100346

The World Bank. (2019). *Agriculture, forestry, and fishing, value added (% of GDP) | Data*. The World Bank. https://data.worldbank.org/indicator/NV.AGR.TOTL. ZS?end=2019&most_recent_value_desc=true&start=2008&view=chart

UN DESA. (2019). *World population prospects 2019*. United Nations Department of Economic and Social Affairs. https://population.un.org/wpp/Download/Standard/Population/

UNIDO. (2020). *INDSTAT 2 2020*. UNIDO Statistics Data Portal.https://stat.unido.org/

World Bank. (2021). *World development indicators*. http://wdi.worldbank.org/table/4.2#.

Chapter 7
Decarbonisation Pathways for Buildings

Souran Chatterjee, Benedek Kiss, Diana Ürge-Vorsatz, and Sven Teske

Abstract This section documents the development of four different energy demand pathways on the basis of the high-efficiency buildings (HEB) model of the Central European University. The assumptions and the scenario narratives are derived and the results provided in numerous graphs and tables. Of the four derived scenarios, two are selected for the OECM and the selection criteria are justified. The results in terms of the global energy demand and energy-related CO_2 emissions are provided in tables.

Keywords Decarbonisation pathways · Buildings · Residential · Commercial · High-efficiency buildings (HEB) model · Energy intensities · Floor area · Bottom-up demand projections

The developments of the regional and global energy demand for the building sector are described in this chapter. Sections 7.1 and 7.2 document the development of the bottom-up energy demand projections for buildings with the methodology described in Sect. 3.2 and are authored by Prof. Dr. Diana Ürge-Vorsatz, Dr. Souran Chatterjee, and Benigna Boza-Kiss of the Central European University Budapest, Hungary. The last section describes the implementation of this research in the wider OneEarth Climate Model (OECM) to generate a single 1.5 °C energy pathway for buildings and construction.

S. Chatterjee · B. Kiss · D. Ürge-Vorsatz
Department of Environmental Sciences and Policy, Central European University,
Budapest, Hungary
e-mail: ChatterjeeS@ceu.edu; KissB@ceu.edu; Vorsatzd@ceu.edu

S. Teske (✉)
Institute for Sustainable Futures, University of Technology Sydney, Sydney, NSW, Australia
e-mail: sven.teske@uts.edu.au

© The Author(s) 2022
S. Teske (ed.), *Achieving the Paris Climate Agreement Goals*,
https://doi.org/10.1007/978-3-030-99177-7_7

7.1   Buildings

The building sector is responsible for 39% of process-related greenhouse gas (GHG) emissions globally and accounts for almost 32% of the global final energy demand, making the building sector pivotal in reducing the global energy demand and climate change (Ürge-Vorsatz et al., 2015a, 2020). The building sector is often suggested to have the largest low-cost climate change mitigation potential, achievable by reducing the energy demand (Ürge-Vorsatz & Tirado Herrero, 2012a; Güneralp et al., 2017a). However, with the increasing rates of population growth and urbanisation, the building stock is projected to double in developing regions by 2050, so reducing the global energy demand will become challenging (EIA, 2015a). Along with these challenges, new building stocks in developing regions will simultaneously provide opportunities for energy-efficient construction, which could substantially reduce the global energy demand. In developed regions, opportunities to reduce the energy demand will predominantly involve renovating the existing building stock (Prieto et al., 2019a; Chatterjee & Ürge-Vorsatz, 2020).

The IPCC's fifth assessment report makes clear that the energy demand must be reduced substantially by 2050 to limit the global temperature rise to 1.5 °C (Rogelj et al., 2018a). However, today, most mitigation pathways still rely on supply-side solutions, and little effort has been made to understand the demand-side potential (Creutzig et al., 2018a). More precisely, understanding the global energy demand for the building sector by assessing the future growth in floor area and the corresponding energy demand is crucial in the context of the 1.5 °C target. Therefore, different models of the building energy demand are used to understand the future energy consumption and emission potential of the building sector under different policy scenarios.

7.2   The High-Efficiency Buildings (HEB) Model: Energy Demand Projections for the Building Sector

To develop detailed energy demand projections for the regional and global building sectors, the *high-efficiency buildings* (HEB) *model* was used. The HEB methodology is documented in Sect. 3.2 and is among the most detailed models for this sector. The key output of the HEB model consists of floor area projections for different types of residential and tertiary buildings in different regions and countries, the total energy consumption of residential and tertiary buildings, the energy consumption for heating and cooling, the energy consumption for hot water energy, the total CO_2 emissions, the CO_2 emissions for heating and cooling, and the CO_2 emissions for hot water energy. The HEB is based on a bottom-up approach, and it includes rather detailed technological information for one sector of the economy. However, it also uses certain macroeconomic and socio-demographic data, including population growth rates, urbanisation rates, and floor areas per capita. The HEB model uses four different scenarios to understand the dynamics of energy use and to explore the potential of the building sector to mitigate climate change by exploiting various opportunities. The four scenarios are:

1. *Deep efficiency scenario*: The *deep efficiency* scenario demonstrates the potential utility of state-of-the-art construction and retrofitting technologies, which can substantially reduce the energy consumption of the building sector and therefore CO_2 emissions while also providing full thermal comfort in buildings. In this scenario, exemplary building practices are implemented worldwide for both new and renovated buildings.
2. *Moderate efficiency scenario*: The *moderate efficiency* scenario incorporates present policy initiatives, particularly the implementation of the Energy Performance of Building Directive (EPBD) in the EU and building codes for new buildings in other regions.
3. *Frozen efficiency scenario*: This scenario assumes that the energy performance of new and retrofitted buildings does not improve relative to the baseline. Retrofitted buildings will consume around 10% less energy for space heating and cooling than standard existing buildings, whereas most new buildings have a lower level of energy performance than that in the *moderate efficiency* scenario due to their lower compliance with building codes.
4. *Nearly net-zero scenario*: The last scenario models the potential of deploying 'nearly net-zero energy buildings' (buildings that can produce as much energy locally through the utilisation of renewables as they consume, on annual balance) around the world. It differs from other three scenarios in that it not only calculates the energy consumption but already incorporates the local energy supply to arrive at the final energy demand. In other aspects, it uses the same parameters as the *deep efficiency* scenario.

The aim of the scenario analysis is to determine the importance of different policies for building energy-efficiency measures and to show how much the final energy consumption of the building sector can be reduced across the world. Table 7.1 summarises the actual parameters of the four scenarios.

7.2.1   Regional Breakdown of the High-Efficiency Buildings (HEB) Model

The end-use demand and its corresponding emissions are produced until 2060 at yearly resolution for 11 key regions, which include 28 member states of the European Union and 3 key countries (India, China, and the USA), and cover the world. Those 11 regions shown in Fig. 7.1 differ from the 10 IEA regions used for the regional transport demand analysis. The main differences are as follows: OECD Europe (IEA) is broken down into Western and Eastern Europe (HEB); Africa (IEA) and the Middle East (IEA) are grouped into Middle East and Northern Africa (HEB)

Table 7.1 Parameters of the four scenarios

Parameter	Deep efficiency scenario	Moderate efficiency scenario	Frozen efficiency scenario	Nearly net-zero scenario
Initial retrofit rate	1.4%	1.4%	1.4%	1.4%
Accelerated retrofitting rate	3% in developed countries and 1.5–1.6% in developing countries after 2027	3% in developed countries and 1.5–1.6% in developing countries after 2027	No accelerated retrofitting rate is assumed	3% in developed countries and 1.5–1.6% in developing countries after 2027
Energy-efficiency measures for new buildings	New buildings are built to regional standards	New buildings are built to regional standards	New buildings do not improve relative to the existing stock	New buildings are built to regional standards
Energy-efficiency measures for renovated buildings	Renovations reduce the energy demand by approximately 30%	Renovations reduce the energy demand by approximately 30%	Renovations reduce the energy demand by approximately 10%	Renovations reduce the energy demand by approximately 30%
Share of advanced buildings within the new and retrofitted stock	All new and retrofitted buildings have a very low energy demand after 2030 in EU, NAM, and PAO and after 2037 in other parts of the world	Advanced buildings (new buildings) are only introduced in Western Europe after 2035, and after 2045, all retrofitted buildings have a very low-energy design	Advanced buildings are only introduced in Western Europe after (1% of the new and retrofitted building stock)	All new and retrofitted buildings have a net-zero energy demand after 2030 in EU, NAM, and PAO and after 2037 in other parts of the world

Fig. 7.1 Global coverage of HEB model

and sub-Saharan Africa (HEB); India (IEA) is part of the South Asia (HEB) region, which includes the neighbouring countries Bangladesh, Bhutan, Sri Lanka, Nepal, and Pakistan—countries that are part of the IEA region non-OECD Asia; China (IEA) is part of the group Centrally Planned Asia, which includes Cambodia, Lao, Mongolia, North Korea, and Vietnam, all of which are part of the IEA region non-OECD Asia; Pacific Asia (HEB) is the remaining part of the non-OECD Asia (IEA) region and all Pacific Island states.

7.2.2 HEB: Data and Assumptions

Similar to any bottom-up energy demand model, the HEB model is very data-intensive. Therefore, it relies on a broad variety of input sources, including statistical databases and the scientific peer-reviewed and grey literature, to incorporate the most up-to-date data. The HEB model largely depends on four sources for its basic input data:

World Bank Databases Both present and historical data on population and real gross domestic product (GDP) figures are obtained from the World Bank databases. The GDP forecast data play a particularly crucial role because they determine the growth in floor area of non-residential buildings. The HEB model calculates future GDP values based on historic and present GDP growth rate data obtained from the World Bank database. The future real GDP is predominantly calculated for non-OECD countries for which future forecasts of real GDP are not available. However, for the OECD member states, this model uses the OECD database of real GDP projections. In addition to the forecast GDP and real GDP databases, the HEB model uses the population forecast database of the World Bank to calculate the future population growth for different countries and regions.

United Nations Development Programme (UNDP), UN-Habitat, and United Nations Conference on Trade and Development (UNCTAD) Population Databases To calculate the growth in floor area and therefore the final energy consumption for heating and cooling, population projection data are required. Together with the World Bank database, the HEB model uses the UNDP population projection database to calculate future populations. Furthermore, because the HEB uses rural and urban classifications, urbanisation rate data are obtained from the UNCTAD database. However, none of these databases contains data on slums or the informal settlement of different regions. Therefore, urban populations living in slums are calculated based on UN-Habitat projections.

In addition to population and GDP data, other important data points used in HEB, such as building stock data and energy intensity data, have been collected from several project reports and datasets of the European Commission, as well as the Eurostat database in the case of the EU, the US Energy Information Administration (EIA) database, and various literature sources. Further information on the data collection can be found in the previous report of HEB in Urge-Vorsatz et al. (2012a). In some cases, data for some of the parameters are unavailable, and in those cases, the HEB model relies on expert judgement. For instance, the energy intensity (specific energy consumption) of advanced buildings mainly utilises the 'passive house' principle, meaning that the useful energy demand may not exceed 15 kWh/m²/year for heating. This concept has been shown to be applicable throughout the world, and various other measures are used to reduce the cooling and dehumidification demands. The total useful demand can be supplied by increasingly efficient heat pumps, which results in very low final energy demands in such advanced buildings. In the *nearly net-zero* scenario, the energy consumption of advanced buildings is even more reduced by potential local energy production, which is calculated with the Better Integration for Sustainable Energy (BISE) model at the building level. The basic input data used in the HEB model are presented, together with their sources, in Table 7.2.

The key assumptions of the model are presented in Table 7.1, and the sources of the key input data are documented in Table 7.2. Assumptions, such as the retrofitting rate, the share of advanced buildings within the new and retrofitted stock, and the energy performance of buildings of different vintages, are based on expert judgements and the authors' experience in the field of modelling building energy. Because data for these parameters are not available, the authors have made several assumptions related to their magnitudes (Table 7.1). Moreover, because the HEB model provides a realistic evaluation of the building energy demands under different policy scenarios, different scenario-specific assumptions are also used to define the scenarios.

The findings are presented in Sects. 7.2.3 and 7.2.4. First, the findings of the study show the future floor area projections under different scenarios, and then it presents future space-heating and space-cooling demand of the different regions. Space-heating and space-cooling demand largely depends on the floor area growth,

Table 7.2 Key input data used in the HEB model and their sources

Description	Sources
GDP forecast	World Bank (2020) and OECD (2021a)
Population forecast	World Bank (2021a, b) and UN DESA (2019)
Urbanisation rate forecast	Our World in Data (2021a), United Nations Conference on Trade and Development (2021), and UN DESA (2018)
Urban populations living in slums (forecast)	Our World in Data (2021a) and UN-Habitat (2021a)
Shares of building types (residential and commercial) within a region	Eurostat (2021a), Hong et al. (2014a), EIA (2015a), and European Commission (2021a)
Demolition rate, retrofitting rate, floor area per capita/GDP	European Commission (2021b), EIA (2012, 2015a), and ENTRANZE (2014a), and literature (Chatterjee & Ürge-Vorsatz, 2020)
Specific energy use for heating	Schnieders et al. (2015a), Mantzos et al. (2015), Hotmaps (2021a), and Heat Roadmap and experts' judgement

and hence, the results of floor area are presented first. To calculate floor area and final energy demand, HEB model first calculates region-specific population and GDP with the help of Eqs. 3.1–3.9 (Sect. 3.2). Based on the region-specific populations and GDP growth rates, then region-specific floor area and final energy consumption for space heating and cooling are calculated.

7.2.3 Floor Area

The floor area for each of the regions is calculated with Eqs. 3.10 and 3.11 (Sect. 3.2). In accordance with the HEB modelling assumptions, the growth in floor area in the residential sector depends predominantly on the population growth, whereas the growth in non-residential or commercial floor area depends on the GDP growth of the region. Based on these equations and assumptions, the findings of the HEB model show that the global floor area will increase by 77% from 2022 to 2060 and the global floor area growth will be dominated by the growth in the Asian, Middle Eastern, and African regions. Precisely, substantial growth in floor area will be observed in the Middle East and Africa (180%), followed by Pacific Asia (174%), Africa (131%), and Latin America (130%) (refer to Fig. 7.2).

Significant population and GDP growth is projected for regions such as the Middle East and Africa, Africa, and Pacific Asia in the future, so the floor area growth in these regions will be substantial. If the global growth in floor area is further analysed according to different building categories and classifications, it can be seen that the substantial increase in floor area will be dominated by urban floor area (99% growth is projected by 2060 relative to 2022), which will mainly be caused by an increasing rate of urbanisation. As a result of the increasing urbanisation rate, urban slums are projected to increase significantly to 176% by 2060. However, the floor area of slums constitutes only a small proportion of the global floor area (2.4%

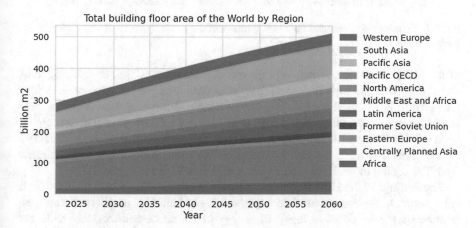

Fig. 7.2 Growth of the total floor area and its distribution among the regions of the world

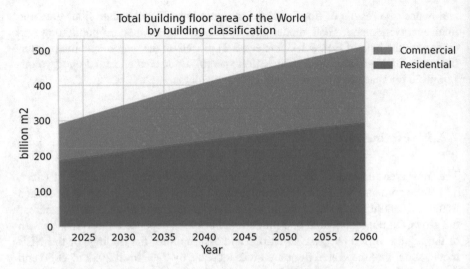

Fig. 7.3 Total building floor area in the world by building classification

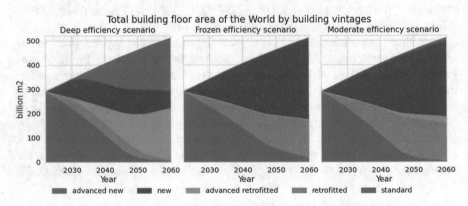

Fig. 7.4 Distribution of the total floor area throughout the world by building vintage across the modelling period

of the global floor area, which is projected to increase to 3.7%), so the growth of slum areas will have little impact on the global floor area growth. Moreover, if floor area growth is analysed per building classification, substantial growth can be projected for both residential and commercial buildings. More precisely, the global residential building sector is projected to grow from 186 billion m^2 in 2022 to 292 billion m^2 by 2060, and the global commercial building sector is projected to grow from 102 billion m^2 in 2022 to 217 billion m^2 by 2060 (refer to Fig. 7.3).

The findings of the HEB model are summarised in Figs. 7.4 and 7.5. However, it is important to understand the future proportions of buildings of different vintages, because they have different levels of energy performance and therefore different energy consumption patterns. The floor area growth for buildings of different

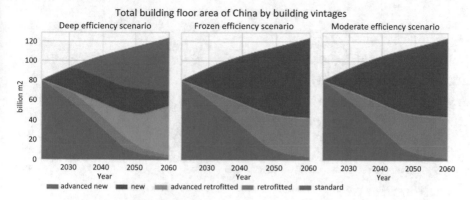

Fig. 7.5 Distribution of the total floor area in China by building vintages across the modelling period

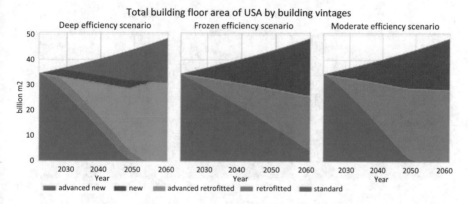

Fig. 7.6 Distribution of the total floor area in the USA by building vintages across the modelling period

vintages is presented in Figs. 7.4, 7.5, 7.6, 7.7, 7.8 and 7.9, which shows the share of each vintage and its change over the modelling period across the different scenarios in each of the regions. It is important to note that the total floor area remains the same in all scenarios.

The findings show that the growth in total floor area is mainly dominated by growth in China and India. More precisely, China's share of the global total floor area in 2022 will be around 28%, and by 2060, it will increase by 54%, whereas India's share in 2022 will be 14% and will increase by 96% by 2060. Furthermore, significant growth in floor area can be observed by 2060 in key regions, such as the USA (41%), Pacific OECD (25%), and EU-28 (22%).

The results of the HEB model also show that a very small amount of today's building stock will remain as it is until 2060. Therefore, to reduce the energy demand and the impact of the energy demand of the building sector on climate change, it

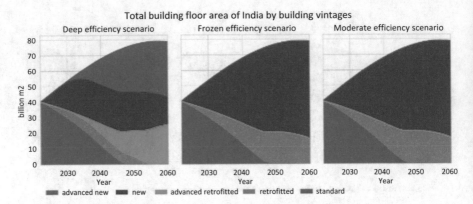

Fig. 7.7 Distribution of the total floor area in India by building vintages across the modelling period

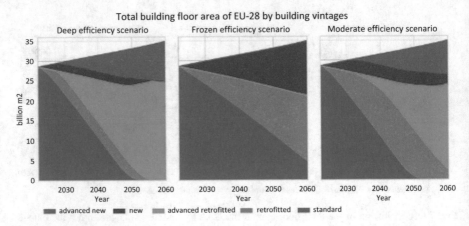

Fig. 7.8 Distribution of the total floor area in EU-27 countries by building vintages across the modelling period

will be crucial to implement advanced efficiency measures for retrofitted and new buildings that will be constructed in 2022–2060. If today's best practices of energy efficiency are applied to all new and retrofitted buildings globally (*deep efficiency* scenario), 43% of the building stock will be classifiable as 'advanced new' buildings and 41% of the building stock as 'advanced retrofitted' buildings in 2060. However, a significant amount of stock will remain less energy-efficient based on the assumption that the construction market cannot adjust immediately to the new practices required to build highly efficient buildings. On the contrary, if the current practice is '*frozen*' and no advanced measures are introduced, 99% of the stock will remain less efficient while having the rest unchanged in 2022 values. It is noteworthy that according to the findings of the HEB model, 66% of the building stock in 2060 does not yet exist in 2022. The *moderate efficiency* scenario assumes that only

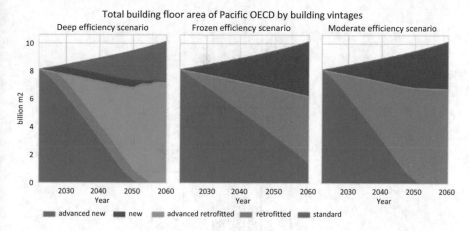

Fig. 7.9 Distribution of the total floor area in Pacific OECD countries by building vintages across the modelling period

present policies will be enforced and there will no further more ambitious goals set throughout the world. Under this scenario, only a minor share (7%) of the floor area will be classifiable as 'advanced' (2% new and 5% retrofitted). This is because most countries with strong policies for energy-efficient buildings (especially the EU) will only play a minor role in constructing a share of new buildings around the globe.

7.2.4   Final Energy Use for Space Heating and Cooling Under the HEB Scenarios

The final energy use for space heating and cooling will largely depend upon the calculated floor areas. After the floor area is calculated for each region, the thermal energy use is calculated. Like the floor area calculations, thermal energy use is also calculated for the four different scenarios.

Among the four scenarios, the final energy use for space heating and cooling under two scenarios clearly shows immense potential for reducing the energy demand of the building sector by 2060. At the global level, if best practices in building construction and retrofitting become standard, the final energy for heating and cooling will decrease from 24 PWh in 2022 to 10 PWh in 2060, which corresponds to a 56% drop, as shown in Table 7.3. However, if existing policies continue in place until 2060, the final energy use will increase by 34% by 2060 relative to the 2022 level. In other words, under the *moderate efficiency* scenario, the global final energy required for space heating and cooling will increase by 34% by 2060 relative to that in 2022. Under the *deep efficiency* scenario, the global final energy demand in 2060 will be 67% less than under the *moderate efficiency* scenario, whereas under the *frozen efficiency* scenario, it will be 37% higher, which corresponds to an 83% increase relative to the 2022 level.

Table 7.3 Results for final energy use for heating and cooling in the key regions and the world

Scenario	Baseline	Moderate efficiency		Deep efficiency			Nearly net zero			Frozen		
	2022 PWh	2060 PWh	Δ% to 2022	2060 PWh	Δ% to 2022	Δ% to Moderate	2060 PWh	Δ% to 2022	Δ% to Moderate	2060 PWh	Δ% to 2022	Δ% to Moderate
China	4.10	6.86	+67%	2.86	−30%	−58%	1.03	−75%	−85%	9.65	+135%	+41%
EU-28	2.82	0.66	−77%	0.37	−87%	−44%	0.00	−100%	−100%	2.97	+5%	+353%
India	2.33	7.50	+222%	2.60	+12%	−65%	1.70	−27%	−77%	8.82	+279%	+18%
Pacific OECD	0.99	0.77	−22%	0.15	−85%	−81%	0.00	−100%	−100%	0.99	+1%	+29%
USA	5.16	4.61	−10%	0.76	−85%	−84%	0.00	−100%	−100%	5.87	+14%	+27%
World	23.59	31.49	+34%	10.49	−56%	−67%	4.42	−81%	−86%	43.05	+83%	+37%

There are two key reasons behind the significant energy savings in the *deep effi-ciency* and *nearly net-zero* scenario compared with the *frozen efficiency* and *moder-ate efficiency* scenarios:

1. Low retrofitting rates
2. Higher proportions of advanced new and retrofitted buildings

More precisely, in the *deep efficiency* and *nearly net-zero* scenarios, the retrofit-ting rate is assumed to be 3% in developed countries and 1.5–1.6% in developing countries after 2027. The same retrofitting rates are assumed in the *moderate effi-ciency* scenario. However, in the *frozen efficiency* scenario, the retrofitting rate is assumed to be no higher than 1.4% across all regions. Similar to the retrofitting rate, under the *deep efficiency* scenario, it is assumed that all new and retrofitted build-ings will have a very low energy demand in the EU, NAM, and PAO after 2030 and in the other parts of the world after 2037. Under the *nearly net-zero* scenario, it is assumed that all new and retrofitted buildings will have a net-zero energy demand in EU, NAM, and PAO after 2030 and in other parts of the world after 2037 because the local onsite solar electric production is included in the definition of the *nearly net-zero* scenario. In the *moderate efficiency* scenario, advanced buildings are only introduced in Western Europe after 2035 for all new buildings, and after 2045, all retrofitted buildings will have a low-energy design. Based on these assumptions, the findings of HEB highlight the importance of ambitious in-act policies.

Key regions, such as China, EU-27, and India, consume most of the global energy, so it is important to know how the building sectors in these regions will perform under different scenarios. Regions such as the USA and EU-27 have much greater potential to reduce space-heating- and space-cooling-related energy use with the help of best practices. Precisely, 73% and 75% of energy consumption related to thermal comfort can be reduced by 2060 in the USA and EU-27, respec-tively, if best practices are followed. The *nearly net-zero* scenario goes one step further than the *deep efficiency* scenario. The results show that the energy consump-tion of buildings for heating and cooling can reach almost zero in the EU, the USA, and Pacific OECD countries by 2055–2057. Although heating- and cooling-related energy consumption in China and India will not reach zero in the modelled period, significant reductions in China and India (85% and 27%, respectively) can be achieved relative to 2022 values. Figures 7.10 and 7.11 show the final energy demands for space heating and cooling in different parts of the world under the dif-ferent scenarios (Fig. 7.12).

Globally, commercial and public buildings in urban areas are the largest consum-ers of space-heating- and space-cooling-related energy. Therefore, best practices should especially focus on commercial and public buildings in urban areas. Commercial and public buildings in urban areas will reduce their consumption by up to 33% by 2060 under the *deep efficiency* scenario. Similarly, urban residential buildings will reduce their consumption by up to 57% globally by 2060 under the *deep efficiency* scenario. Under the *nearly net-zero* scenario, commercial and public buildings still have a significant share of energy consumption in 2060, but the total energy demand is extremely reduced. It is noteworthy that reducing the energy intensity of commercial and public buildings even further will require further

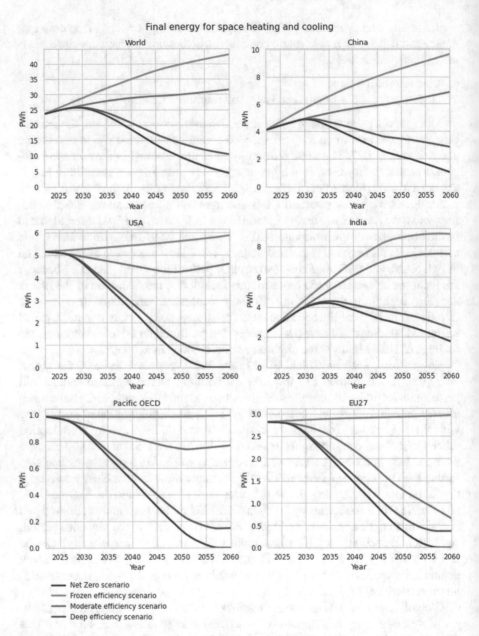

Fig. 7.10 Final energy consumption for space heating and cooling in the world and key regions (in PWh)

investigation of the usage characteristics of different building types. Therefore, even more effort will be required than merely servicing these building with renewable energy. Similar findings are obtained from the analysis of the region-specific final energy demands for the USA, the European Community, India, China, and the OECD Pacific.

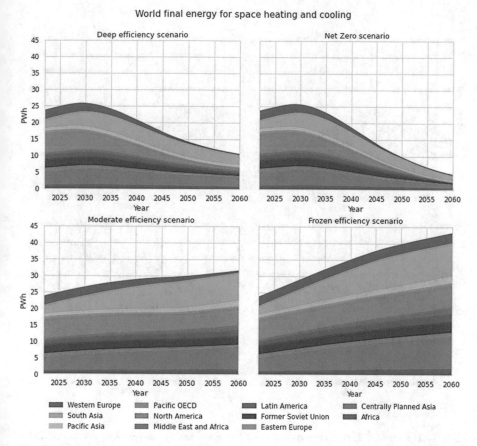

Fig. 7.11 Shares of the total heating and cooling energy consumption attributable to different regions of the world

Commercial buildings are the largest consumer of space-heating- and space-cooling-related final energy in low- to middle-income regions, such as India and China. However, in developed regions, such as the Pacific OECD, EU-28, and the USA, the residential building sector is the largest consumer. The HEB results show that these developed high-income regions can substantially reduce their energy demands in both residential and commercial building sectors if advanced high-efficiency energy measures are standardised over the years. In fact, in these regions, if local energy production is included (i.e. *nearly net-zero* scenario), then the building sector can achieve a net-zero status by 2060. In contrast, the low- to middle-income regions will not be able to achieve a net-zero status by 2060, even if the local production of solar electric energy is added into the calculation. However, regardless of the local energy production, these regions can still achieve a substantial reduction in China, and in India, the rate of increase will be slowed by the introduction of advanced efficiency energy measures, such as new energy-efficient

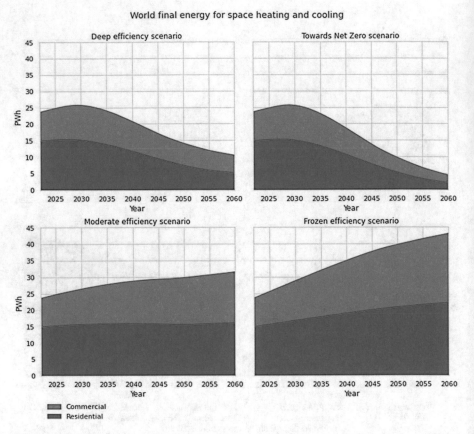

Fig. 7.12 Shares of final energy consumption for space heating and cooling for the world by building category (in PWh)

building codes and the rigorous renovation of existing buildings. More precisely, in India, even with advanced energy-efficiency measures, the final energy demand for heating and cooling will increase by 12% in 2060 relative to that in 2022, which is 65% lower than the final energy demand for 2060 if the existing efficiency measures are followed until 2060.

7.2.5 Key Findings for the HEB Scenarios

The HEB model analysis demonstrates the potential for reducing the energy demand in the building sector if state-of-the-art high-efficiency buildings are implemented worldwide. The findings of the study show that with a higher share of high-efficiency renovations and construction (as assumed in the *deep efficiency* and *nearly net-zero* scenarios), it will be possible to reduce the final thermal energy used globally in the building sector by more than half by 2060. In some regions, such as the EU and Pacific OECD, it will even be possible to achieve net-zero status for the thermal energy demand. However, this pathway towards high-efficiency or net-zero emissions is ambitious in its assumptions and requires strong policy support. On the contrary, if policy support to implement more high-efficiency buildings is not in place (*frozen efficiency* scenario) or even if the present policy scenarios are continued (*moderate efficiency* scenario), the total thermal energy demand of the building sector could increase by 34–83% by 2060 relative to the 2022 level. Furthermore, if the present rate of energy-efficiency measures is continued, 67–80% of the global final thermal energy savings will be locked in by 2060 in the world building infrastructure. The lock-in effect of the building sector also indicates that if the present moderate energy performance levels become the standard in new and/or retrofitted buildings, it will be almost impossible to further reduce the thermal energy consumption in such buildings for many decades to come.

7.3   1.5 °C OECM Pathway for Buildings

Based on the results of the detailed HEB model analysis, the *deep efficiency* scenario was chosen for commercial buildings and the *moderate efficiency* scenario for residential buildings. These scenarios were chosen after stakeholder consultation with representatives of the respective industries, members of the Carbon Risk Real Estate Monitor (CRREM), the Net-Zero Asset Owner Alliance, and academia. To integrate the building sector into the 1.5 °C pathway as part of the OECM, consistent with all other industry and service sectors and the transport sector, the selection of one specific pathway for the building sector as a whole was necessary. The energy demand for the construction sector was also required to calculate the emissions for the Global Industry Classification Standard (GICS) (see Chap. 2). This section documents the calculation process and the results for the residential and commercial building sector and construction.

Table 7.4 shows the assumed development of floor space for residential and commercial buildings, which was taken from the HEB analysis and the projected economic development of the construction sector. The increase in the construction industry is based on the overall global GDP, developed as documented in Chap. 2, and is therefore not directly related to the HEB floor space projections. The direct link between both parameters was beyond the scope of this analysis and is therefore highlighted as a potential source of error.

The global energy intensities for residential and commercial buildings (in kilowatt-hours per square metre (kWh/m^2)) are the second main input for the OECM 1.5 °C building pathway and are taken from the documented HEB analysis. The global values were calculated on the basis of the total HEB results for the global energy demand per year divided by the floor space. The global values are the sum of the values for all 11 regions analysed with HEB. Table 7.5 also shows the reductions

Table 7.4 OECM—global buildings: projected floor space and economic value of construction

Parameter	Units	2019	2025	2030	2035	2040	2045	2050
			Projection					
Residential buildings	[Billion m²]	184	196	211	226	241	255	269
Residential buildings—variation compared with 2019	[%]	0	6%	15%	23%	31%	38%	46%
Commercial buildings	[Billion m²]	101	112	130	146	163	177	192
Commercial buildings—variation compared with 2019	[%]	0	12%	29%	46%	62%	76%	91%
Construction: residential and commercial building—economic value	[bn $GDP]	2149	2699	3186	3753	4321	5607	2149
Variation compared with 2019	[%]	0	26%	48%	75%	101%	131%	161%

Table 7.5 OECM—global buildings: assumed energy intensities

Parameter	Units	2019	2025	2030	2035	2040	2045	2050
			Projections					
Residential buildings: energy intensity	[kWh/m²]	81	78	74	70	66	62	58
Variation compared with 2019	[%]	0%	−4%	−9%	−14%	−19%	−23%	−28%
Commercial buildings: energy intensity	[kWh/m²]	87	81	77	65	53	43	33
Variation compared with 2019	[%]	0%	−7%	−11%	−25%	−39%	−51%	−62%
Construction								
Construction: residential and commercial buildings—energy intensity	[MJ/$GDP]	0.70	0.57	0.56	0.56	0.56	0.55	0.54
Variation compared with 2019	[%]	0%	−19%	−19%	−20%	−20%	−22%	−23%
Energy demand: mining and quarry—sand, stones, clay, gravel	[PJ/yr]	510	526	618	724	829	934	1034
Variation compared with 2019	[%]	0	3%	21%	42%	63%	83%	103%

in the energy intensity for residential and commercial buildings relative to the values in the base year 2019.

The energy intensity of the construction industry was calculated with the total energy demand (in petajoules (PJ)) in 2019, as provided in the IEA World Energy Balances 2019 for Construction and the projected economic values (in $US) for the same year. The energy demand value for construction in the IEA statistics includes the construction of roads and railways, as well as other civil engineering and utility projects, as defined in IEA (2020). Therefore, the shares of the energy demand for residential and commercial buildings must be estimated. The calculated energy intensity for construction work was compared with published values.

Based on the assumptions and input parameters documented in Tables 7.4 and 7.5, the energy demand for all sub-sectors was calculated. Table 7.6 shows the calculated annual energy demand for residential and commercial buildings and for the construction industry. The energy demand consists of the energy required for space heating and cooling ('heating energy') and the electricity demand, which includes all electrical applications in the buildings but excludes electricity for heating and cooling. This separation is necessary to harmonise the input data from the HEB, which do not include electricity for household applications such as washing machines, etc., with the OECM.

The electricity demand for residential buildings is based on the bottom-up analysis of households documented in Sect. 3.1.2. The electricity demand for the service sector is based on a breakdown of electricity and heating in 2019 across all service

Table 7.6 OECM—global buildings: calculated annual energy demand for residential and commercial buildings and construction

Parameter	Units	2019	2025	2030	2035	2040	2045	2050
			Projections					
Residential buildings: total energy demand	[PJ/ yr]	82,565	77,724	77,039	75,274	75,199	66,944	63,147
Variation compared with 2019	[%]	0%	−6%	−7%	−9%	−9%	−19%	−24%
Residential buildings: heat energy demand	[PJ/ yr]	60,417	54,746	56,056	56,739	56,983	56,677	55,989
Variation compared with 2019	[%]	0%	−9%	−7%	−6%	−6%	−6%	−7%
Residential buildings: electricity demand	[PJ/ yr]	22,148	22,979	20,983	18,536	18,216	10,268	7158
Variation compared with 2019	[%]	0%	4%	−5%	−16%	−18%	−54%	−68%
Commercial buildings: total energy demand	[PJ/ yr]	34,567	40,609	44,311	42,549	39,315	35,991	31,676
Variation compared with 2019	[%]	0%	17%	28%	23%	14%	4%	−8%
Commercial buildings: heat energy demand	[PJ/ yr]	28,432	34,736	38,346	36,482	33,137	29,690	25,243
Variation compared with 2019	[%]	0%	22%	35%	28%	17%	4%	−11%
Commercial buildings: electricity demand	[PJ/ yr]	2921	2686	2619	2554	2490	2428	2367
Variation compared with 2019	[%]	0%	−8%	−10%	−13%	−15%	−17%	−19%
Construction of residential and commercial building: energy demand	[PJ/ yr]	1505	1531	1798	2108	2415	2719	3010
Variation compared with 2019	[%]	0%	2%	20%	40%	60%	81%	100%

Table 7.7 OECM–global buildings: energy supply

Parameter	Units	2019	2025	2030	2035	2040	2045	2050
		Projections						
Residential buildings—heating: fossil fuels	[PJ/yr]	38,274	25,484	15,547	10,781	6450	1912	0
Residential buildings—heating: renewable, electric, and synthetic fuels	[PJ/yr]	22,144	29,262	40,509	45,958	50,533	54,765	55,989
Residential buildings—heating: renewable share	[%]	36.7%	53.5%	72.3%	81.0%	88.7%	96.6%	100.0%
Residential buildings—electricity: fossil fuels	[PJ/yr]	16,712	11,017	5412	2117	930	187	0
Residential buildings—electricity: renewables	[PJ/yr]	5436	11,962	15,571	16,418	17,286	10,081	7158
Residential buildings—electricity for heating share	[%]	6.6%	18.2%	20.4%	37.7%	42.0%	48.4%	54.0%
Commercial buildings—heating: fossil fuels	[PJ/yr]	18,011	16,170	10,635	6932	3751	1002	0
Commercial buildings—heating: renewable, electric, and synthetic fuels	[PJ/yr]	10,421	18,567	27,711	29,550	29,386	28,688	25,243
Commercial buildings—heating: renewable share	[%]	36.7%	53.5%	72.3%	81.0%	88.7%	96.6%	100.0%
Commercial buildings—electricity: fossil fuels	[PJ/yr]	2204	1288	676	292	127	44	0
Commercial buildings—electricity: renewables	[PJ/yr]	717	0	−1	−1	2363	−1	−1
Commercial buildings—electricity for heating share	[%]	6.2%	17.0%	19.0%	35.2%	39.2%	45.2%	50.4%
Construction of residential and commercial buildings: fossil fuels	[PJ/yr]	1356	1037	854	809	675	512	273
Residential buildings—heating: renewable, electric, and synthetic fuels	[PJ/yr]	149	494	944	1300	1740	2207	2737
Residential buildings—heating: renewables share	[%]	9.9%	32.3%	52.5%	61.6%	72.0%	81.2%	90.9%

sectors, published in the IEA World Energy Balances. The future values until 2050 are based on the projections for the analysed service and industry sectors documented in Chaps. 5 and 6.

The supply side for the building and construction sectors is based on the 1.5 °C pathway for energy utilities, as documented in Chap. 12 . In contrast to the demand side, the supply values for electricity are provided both for room climatisation (heating and cooling) and for appliances (Table 7.7). The total energy-related CO_2 emissions were calculated based on the energy supply mix for heating and electricity generation (Table 7.8).

Table 7.8 OECM—global buildings: energy-related CO_2 emissions

Parameter	Units	2019	2025	2030	2035	2040	2045	2050
			Projection					
Residential buildings: total emission intensity (heating and electricity)	[kgCO$_2$/ kWh]	0.61	0.35	0.17	0.07	0.04	0.01	0.00
Residential buildings: emission intensity— heating per square metre	[kgCO$_2$/ m2]	7.8	5.0	2.6	1.5	0.8	0.2	0.0
Residential buildings: emission intensity—heat	[kgCO$_2$/ kWh]	0.097	0.064	0.036	0.021	0.013	0.003	0.000
Residential buildings: emission intensity—electricity	[kgCO$_2$/ kWh]	0.509	0.291	0.135	0.052	0.024	0.007	0.000
Residential buildings: total emission intensity— compared with 2019	[%]	0%	−36%	−67%	−81%	−89%	−97%	−100%
Residential buildings: total emissions	[MtCO$_2$/ yr]	4578	2830	1343	605	320	74	0
Residential buildings: emissions—heat	[MtCO$_2$/ yr]	1446	975	553	336	201	54	0
Residential buildings: emissions—electricity	[MtCO$_2$/ yr]	3132	1855	789	269	119	21	0
Residential buildings: total emissions—compared with 2019	[%]	0%	−38%	−71%	−87%	−93%	−98%	−100%
Commercial buildings: total emission intensity (heating and electricity)	[kgCO$_2$/ kWh]	19.88	12.64	7.07	3.75	1.91	0.45	0.00
Commercial buildings: emission intensity—heat per square metre	[kgCO$_2$/ m^2]	19.6	12.5	7.0	3.7	1.9	0.4	0.0
Commercial buildings: emission intensity—heat	[kgCO$_2$/ kWh]	0.227	0.154	0.090	0.057	0.036	0.010	0.000
Commercial buildings: emission intensity—electricity	[kgCO$_2$/ kWh]	0.509	0.291	0.135	0.052	0.024	0.007	0.000
Commercial buildings: total emission intensity— compared with 2019	[%]	0%	−36%	−67%	−81%	−89%	−97%	−100%
Commercial buildings: total emissions	[MtCO$_2$/ yr]	5107	3255	1696	810	425	98	0
Commercial buildings: emissions—heat	[MtCO$_2$/ yr]	1975	1400	907	541	306	78	0
Commercial buildings: emissions—electricity	[MtCO$_2$/ yr]	3132	1855	789	269	119	21	0

(continued)

Table 7.8 (continued)

Parameter	Units	2019	2025	2030	2035	2040	2045	2050
Commercial buildings: total emissions—compared with 2019	[%]	0%	−38%	−71%	−87%	−93%	−98%	−100%
Construction: total emissions	[MtCO₂/ yr]	128	81	54	38	23	12	0
Construction: emissions—fuels	[MtCO₂/ yr]	65	45	34	29	18	11	0
Construction: emissions—electricity	[MtCO₂/ yr]	63.1	36.1	19.8	8.9	4.6	1.6	0.0

The specific energy-related CO_2 emissions are also provided for power and heat generation, as well as per square meter of floor area, for residential and commercial buildings. The specific energy demand and the CO_2 emissions per square meter are key performance indicators for the finance industry for real estate. Moreover, these parameters are used for regulatory frameworks, such as the EU energy performance for building directive (EU, 2010).

Acknowledgements The authors of Sect. 7.2 (Souran Chatterjee, Benedek Kiss, Diana Ürge-Vorsatz) are immensely grateful to the DBH Group for the administrative support provided during the HEB research. This research was funded by the project Sustainable Energy Transitions Laboratory (SENTINEL), which received funding from the European Union's Horizon 2020 research and innovation programme under grant agreement no. 837089. This research was also partly funded by the Energy Demand changes Induced by Technological and Social innovations (EDITS) project, which is part of the initiative co-ordinated by the Research Institute of Innovative Technology for the Earth (RITE) and the International Institute for Applied Systems Analysis (IIASA) (funded by the Ministry of Economy, Trade, and Industry [METI], Japan).

References

Chatterjee, S., Ürge-Vorsatz, D. (2020). *Observed trends and modelling paradigms. Topic: LC-SC3-CC-2-2018 of the Horizon 2020 work program: Modelling in support to the transition to a Low-Carbon Energy System in Europe. Building a low-carbon, climate resilient future: secure, clean and efficient energy.*

Creutzig, F., Roy, J., Lamb, W. F., et al. (2018a). Towards demand-side solutions for mitigating climate change. *Nature Climate Change, 84*(8), 260–263. https://doi.org/10.1038/s41558-018-0121-1

EIA. (2015a). *2015 Residential Energy Consumption Survey (RECS) data, consumption & expenditures (C&E) tables.*

EIA. (2012). *2012 Commercial Buildings Energy Consumption Survey (CBECS) Data.* Energy Information Administration (EIA). https://www.eia.gov/consumption/commercial/data/2012/. Accessed 24 Dec 2021.

ENTRANZE. (2014a). *Welcome to ENTRANZE project page.* https://www.entranze.eu/. Accessed 21 Dec 2021.

EU. (2010). *Directive 2010/31/EU of the European Parliament and of the Council of 19 May 2010 on the energy performance of buildings.* European Parliament.

European Commission. (2021a). *Share of non-residential in total building floor area*. https://ec.europa.eu/energy/eu-buildings-datamapper_en. Accessed 21 Dec 2021.

European Commission. (2021b) *EU buildings database | energy*. European Commission. https://ec.europa.eu/energy/eu-buildings-database_en?redir=1. Accessed 24 Dec 2021.

Eurostat. (2021a). *Distribution of population by degree of urbanisation, dwelling type and income group—EU-SILC survey*. Eurostat. https://ec.europa.eu/eurostat/databrowser/view/ILC_LVHO01__custom_598746/default/table?lang=en. Accessed 21 Dec 2021.

Güneralp, B., Zhou, Y., Ürge-Vorsatz, D., et al. (2017a). Global scenarios of urban density and its impacts on building energy use through 2050. *Proceedings of the National Academy of Sciences of the United States of America, 114*, 8945–8950. https://doi.org/10.1073/pnas.1606035114

Hong, L., Zhou, N., Fridley, D., et al. (2014a) *Modeling China's building floor-area growth and the implications for building materials and energy demand | LBL China Energy Group*. https://china.lbl.gov/publications/modeling-chinas-building-floor-area. Accessed 21 Dec 2021.

Hotmaps. (2021a). *Hotmaps Project—The open source mapping and planning tool for heating and cooling*. https://www.hotmaps-project.eu/. Accessed 21 Dec 2021.

IEA. (2020). *World energy balances 2020*. IEA. https://www.iea.org/data-and-statistics?country=WORLD&fuel=Energy transition indicators&indicator=TFCShareBySector. Accessed 6 Apr 2021.

Mantzos, L., Matei, N. A., Rozsai, M., et al. (2015). *JRC-IDEES: Integrated database of the European energy sector: Methodological note*.

OECD. (2021a). *GDP and spending—Real GDP long-term forecast—OECD Data*. OECD. https://data.oecd.org/gdp/real-gdp-long-term-forecast.htm. Accessed 24 Dec 2021.

Our World in Data. (2021a). *Urban and rural population projected to 2050*. Our World Data. https://ourworldindata.org/grapher/urban-and-rural-population-2050?tab=table&country=~OWID_WRL. Accessed 24 Dec 2021.

Prieto, M. P., de Uribarri, P. M. Á., & Tardioli, G. (2019a). Applying modeling and optimization tools to existing city quarters. In *Urban energy systems for low-carbon cities* (pp. 333–414). Academic. https://doi.org/10.1016/B978-0-12-811553-4.00010-X

Rogelj, J., Popp, A., Calvin, K. V., et al. (2018a). Scenarios towards limiting global mean temperature increase below 1.5 °C. *Nature Climate Change, 8*, 325–332. https://doi.org/10.1038/s41558-018-0091-3

Schnieders, J., Feist, W., & Rongen, L. (2015a). Passive Houses for different climate zones. *Energy and Buildings, 105*, 71–87. https://doi.org/10.1016/J.ENBUILD.2015.07.032

UN-Habitat. (2021a). *Housing, slums and informal settlements | Urban Indicators Database*. UN-Habitat. https://data.unhabitat.org/pages/housing-slums-and-informal-settlements. Accessed 24 Dec 2021.

UN DESA. (2019). *World population prospects 2019*. United Nations Department of Economic and Social Affairs. https://population.un.org/wpp/Download/Standard/Population/. Accessed 25 Oct 2021.

UN DESA. (2018). *World urbanization prospects: The 2018 revision, online edition*.

United Nations Conference on Trade and Development. (2021). *Beyond 20/20 WDS—Table view—Total and urban population, annual*. https://unctadstat.unctad.org/wds/TableViewer/tableView.aspx?ReportId=97. Accessed 21 Dec 2021.

Ürge-Vorsatz, D., Cabeza, L. F., Serrano, S., et al. (2015a). Heating and cooling energy trends and drivers in buildings. *Renewable and Sustainable Energy Reviews, 41*, 85–98. https://doi.org/10.1016/J.RSER.2014.08.039

Urge-Vorsatz, D., Khosla, R., Bernhardt, R., et al. (2020). Advances toward a net-zero global building sector. *Annual Review of Environment and Resources, 45*, 227–269. https://doi.org/10.1146/ANNUREV-ENVIRON-012420-045843

Urge-Vorsatz, D., Petrichenko, K., Antal, M., et al. (2012a). *Building policies for a better world: Best practice policies for low carbon & energy buildings based on scenario analysis*.

Ürge-Vorsatz, D., & Tirado Herrero, S. (2012a). Building synergies between climate change mitigation and energy poverty alleviation. *Energy Policy, 49*, 83–90. https://doi.org/10.1016/J.ENPOL.2011.11.093

World Bank. (2021d). *GDP growth (annual %) | Data.* https://data.worldbank.org/indicator/ NY.GDP.MKTP.KD.ZG. Accessed 21 Dec 2021.

World Bank. (2020). *GDP, PPP (constant 2017 international $) | Data.* World Bank. https://data. worldbank.org/indicator/NY.GDP.MKTP.PP.KD. Accessed 22 Dec 2021.

World Bank. (2021e) *Population, total | Data.* https://data.worldbank.org/indicator/SP.POP. TOTL. Accessed 21 Dec 2021.

Heat Roadmap Europe. (2021). Heat Roadmap. https://heatroadmap.eu/. Accessed 24 Dec 2021.

HEB Scenarios

Chatterjee, S, Ürge-Vorstaz, D. (2020). *D3.1: Observed trends and modelling paradigms. Topic: LC-SC3-CC-2-2018 of the Horizon 2020 work program: Modelling in support to the transition to a low-carbon Energy system in Europe. Building a low-carbon, climate resilient future: Secure, clean and efficient energy.* SENTINEL. https://sentinel.energy/wp-content/ uploads/2021/02/D-3.1-837089-EC.pdf

Creutzig, F., Roy, J., Lamb, W. F., Azevedo, I. M. L., Bruine de Bruin, W., Dalkmann, H., Edelenbosch, O. Y., Geels, F. W., Grubler, A., Hepburn, C., Hertwich, E. G., Khosla, R., Mattauch, L., Minx, J. C., Ramakrishnan, A., Rao, N. D., Steinberger, J. K., Tavoni, M., Ürge-Vorsatz, D., & Weber, E. U. (2018b). Towards demand-side solutions for mitigating climate change. *Nature Climate Change, 8*(4). https://doi.org/10.1038/s41558-018-0121-1

ENTRANZE. (2014b). *Demolition rate and retrofit rate database.* https://www.entranze.eu/

EU Buildings Datamapper (2021). *Share of residential and non-residential in total building floor area.* https://ec.europa.eu/energy/eu-buildings-datamapper_en. Accessed 6 Mar 2021.

EIA. (2015b). *2012 RECS Survey Data.* https://www.eia.gov/consumption/residential/data/2015/ index.php?view=consumption#by%20End%20uses%20by%20fuel. Accessed 12 Mar 2021.

EIA. (2019). *2015 RECS Survey Data.* https://www.eia.gov/consumption/residential/data/2015/ index.php?view=consumption#by%20End%20uses%20by%20fuel. Accessed 12 Mar 2021.

Eurostat. (2021b). *Distribution of population by degree of urbanisation, dwelling type and income group—EU-SILC survey.* https://ec.europa.eu/eurostat/databrowser/view/ILC_LVHO01__custom_598746/default/table?lang=en. Accessed 10 Mar 2021.

Güneralp, B., Zhou, Y., Ürge-Vorsatz, D., Gupta, M., Yu, S., Patel, P. L., Fragkias, M., Li, X., & Seto, K. C. (2017b). Global scenarios of urban density and its impacts on building energy use through 2050. *Proceedings of the National Academy of Sciences of the United States of America, 114*(34). https://doi.org/10.1073/pnas.1606035114

Mantzos, L., Matei, N. A., Mulholland, E. A., Rózsai, M., Tamba, M., & Wiesenthal, T. (2018). *JRC-IDEES 2015.* European Commission, Joint Research Centre (JRC) [Dataset]. https://doi. org/10.2905/JRC-10110-10001. PID. http://data.europa.eu/89h/jrc-10110-10001

Heat RoadMap. (2021). *Energy intensity for heating.* https://heatroadmap.eu/. Accessed 10 Mar 2021.

HotMaps. (2021b). *Energy intensity for heating.* https://www.hotmaps-project.eu/. Accessed 10 Mar 2021.

Hong, L., Zhou, N., Fridley, D., Feng, W., & Khanna, N. (2014b, August). Modeling China's building floor-area growth and the implications for building materials and energy demand. In *2014 ACEEE summer study on energy efficiency in buildings* (pp. 146–157).

IEA. (2019). *2019 Global Status Report for Buildings and Construction: Towards a zero-emissions, efficient and resilient buildings and construction sector.*

Schnieders, J., Feist, W., & Rongen, L. (2015b). Passive Houses for different climate zones. *Energy and Buildings, 105,* 71–87.

OECD. (2021b). *Real GDP long-term forecast (indicator).* https://doi.org/10.1787/d927bc18-en. https://data.oecd.org/gdp/real-gdp-long-term-forecast.htm. Accessed 6 Mar 2021.

Our World in Data. (2021b). *Urban population long-run with 2050 projections (OWID)*. https://ourworldindata.org/grapher/urban-and-rural-population-2050?tab=table&country=~OWID_WRL. Accessed 6 Mar 2021.

Prieto, M. P., Álvarez de Uribarri, P. M., & Tardioli, G. (2019b). Applying modeling and optimization tools to existing city quarters. In *Urban energy systems for low-carbon cities*. Elsevier. https://doi.org/10.1016/B978-0-12-811553-4.00010-X

Rogelj, J., Popp, A., Calvin, K. V., Luderer, G., Emmerling, J., Gernaat, D., … Krey, V. (2018b). Scenarios towards limiting global mean temperature increase below 1.5 °C. *Nature Climate Change, 8*(4), 325.

World Bank. (2021a). *GDP growth (annual %)*. https://data.worldbank.org/indicator/NY.GDP.MKTP.KD.ZG. Accessed 6 Mar 2021.

World Bank. (2021b). *GDP (current US$)*. https://data.worldbank.org/indicator/NY.GDP.MKTP.CD. Accessed 6 Mar 2021.

World Bank. (2021c). *Population, total*. https://data.worldbank.org/indicator/SP.POP.TOTL. Accessed 6 Mar2021.

UN. (2019). *World Population Prospects 2019. File POP/2: Average annual rate of population change by region, subregion and country, 1950–2100 (percentage). Estimates, 1950–2020.* https://population.un.org/wpp/Download/Files/1_Indicators%20(Standard)/EXCEL_FILES/1_Population/WPP2019_POP_F02_POPULATION_GROWTH_RATE.xlsx

United Nations, Department of Economic and Social Affairs, Population Division (2018). *World urbanization prospects: The 2018 revision, online edition*. https://population.un.org/wup/Download/Files/WUP2018-F01-Total_Urban_Rural.xls

UNCTAD. (2021). *Total and urban population, annual*. https://unctadstat.unctad.org/wds/TableViewer/tableView.aspx?ReportId=97. Accessed 5 Mar 2021.

UN-HABITAT. (2021b). *Urban indicators database*. https://data.unhabitat.org/datasets/proportion-of-urban-population-living-in-slums-time-period-between-1990-and-2018/data. Accessed 5 Mar 2021.

Ürge-Vorsatz, D., Cabeza, L. F., Serrano, S., Barreneche, C., & Petrichenko, K. (2015b). Heating and cooling energy trends and drivers in buildings. *Renewable and Sustainable Energy Reviews, 41*. https://doi.org/10.1016/j.rser.2014.08.039

Ürge-Vorsatz, D., Khosla, R., Bernhardt, R., Chan, Y. C., Vérez, D., Hu, S., & Cabeza, L. F. (2020). Advances toward a net-zero global building sector. *Annual Review of Environment and Resources, 45*, 227–269.

Ürge-Vorsatz, D., & Tirado Herrero, S. (2012b). Building synergies between climate change mitigation and energy poverty alleviation. *Energy Policy, 49*, 83–90. https://doi.org/10.1016/j.enpol.2011.11.093

Urge-Vorsatz, D., Petrichenko, K., Antal, M., Labelle, M., Staniec, M., Ozden, E., & Labzina, E. (2012b). *Best practice policies for low energy and carbon buildings. A scenario analysis.*

Chapter 8
Decarbonisation Pathways for Transport

Sven Teske and Sarah Niklas

Abstract An overview of the main drivers of the transport energy demand and the assumed socio-economic development (population and GDP) until 2050 for ten world regions are given. The countries in each world region are tabulated. Detailed documentation of projected shifts in transport modes for all world regions, including technological assumptions and energy intensities, by vehicle type is presented. This section contains the OECM 1.5 °C transport scenarios for aviation, shipping, road, and rail, each broken down into passenger and freight transport. The calculated energy demands and energy-related carbon emissions for all transport modes are provided.

Keywords Global and regional transport demand · Mode shift · Transformative Urban Mobility Initiative (TUMI) · GHG development · GDP · Population · Energy intensities

8.1 Introduction

The transport sector consumed 28% of the final global energy demand in 2019, and its decarbonisation potential is therefore among the most important of all industries. Given its size and diversity, not only with regard to different transport modes and technologies but also regional differences, it is also one of the most challenging sectors. In 2019, transport consumed 65% of the total oil demand globally. Therefore, the transition from oil to electric drives and to synthetic fuels and biofuels is key to achieving the goals of the Paris Climate Agreement. A rapid uptake of electric mobility, combined with a renewable power supply, is the single most important measure to be taken to remain within the carbon budget of the 1.5 °C pathway.

S. Teske (✉) · S. Niklas
Institute for Sustainable Futures, University of Technology Sydney, Sydney, NSW, Australia
e-mail: sven.teske@uts.edu.au; Sarah.Niklas@uts.edu.au

The financial sector *Transport* spans civil aviation, shipping, and road transport, including passenger and freight transport, and all related services. For each transport mode, there are two main sub-sectors:

1. Design, manufacture, and sale of planes, ships, and road vehicles for the transportation of passengers and freight
2. Operation and maintenance of vehicles to provide transport services for passengers and freight

This section is based on multiple closely linked research projects: the One Earth Climate Model (OECM) developed in 2019 (Teske et al., 2019) and 2021 and the TUMI Transport Outlook 1.5 °C (Teske et al., 2021), which was developed within a multi-stakeholder dialogue, including two workshops organised by Deutsche Gesellschaft für Internationale Zusammenarbeit GmbH (GIZ) and the University of Technology Sydney/Institute for Sustainable Futures (UTS/ISF) in June and September 2021. As a result, the OECM methodology described in Chap. 3 has been expanded to achieve higher levels of accuracy and resolution, in both the area of the transport demand projections and the calculation of the regional and global transport energy demands.

The demand projections are based on a bottom-up approach. The actual basis of the passenger transport demand is diverse (e.g. to get groceries, to commute for work, or for leisure and recreation), and the transport demand is expressed in kilometres per person per year. Therefore, the development of this transport demand is dependent upon a number of different factors, among the most important of which are the actual population development and economic situation of a region. Geography and lifestyle also play important roles.

In considering the transport of goods, it is important where the goods are produced, the resources required, and where they are located. Economies with high local production rates have lower transport demands than those with high import/export dependence. However, calculation of the actual transport demand is based on non-energy-related factors. A transport or travel demand does not necessarily lead to an energy demand if a non-energy transport mode, such as walking or cycling, is used—sufficient to satisfy the demand. However, most transport modes require energy, and the amount of energy per kilometre depends upon the energy intensity of the chosen vehicle.

The demand for transport energy does not inevitably lead to CO_2 emissions if the energy is generated from renewable electricity and/or renewable fuels. Therefore, a carbon-neutral global transport sector is possible, while regional and intercontinental travel and global trade are maintained.

The transport demand is dependent upon a huge number of factors—the most important of which are the population size and the economic situation. In general, more people and a higher economic standard entail a higher transport demand. The transport service structure—and therefore the transport mode—also depends on a variety of factors. The actual distance travelled, the travel time required, the availability of certain transport modes, and the costs, among other factors, define the chosen transport mode. Each transport mode includes a variety of vehicles with different energy intensities. The transport mode 'road', for example, has by far the largest number of different vehicle options: buses, a huge variety of car types with different drive trains, motorcycles, bicycles, and even walking.

A global scenario requires the simplification of the transport demand projections. A detailed analysis of the purpose of each of those transport demands in kilometres per day for the entire population is not possible. Therefore, the methodology focuses on the development of regional person–kilometres (pkm) and tonne–kilometres (tkm) per year. The main factors affecting demand changes are population and economic development.

Whereas the industry and service pathways (Chaps. 5 and 6) were developed with accumulated global gross domestic product (GDP) values and bottom-up product-based projections, such as the annual steel production (in million tonnes per year), the demand projections for the buildings and transport sectors have been developed on the basis of specific data from ten world regions, to capture the significant regional differences. The geographic breakdown is based on IEA's ten world regions used in the World Energy Outlook series (see Table 8.1).

8.2   Socio-economic Assumptions

The assumed development of regional populations is based on the projections of the United Nations Department of Economic and Social Affairs, whereas the regional GDP developments are based on World Bank projections. The global values for population and GDP are identical throughout the entire analysis, across all sectors. The regional values are used for the buildings and transport sectors, whereas for all other sectors, the resulting (summed) global values are used (Table 8.2).

8.3   Transport Demand

8.3.1   Global and Regional Transport Demands

The global pandemic began in early 2020 and led to significant travel restrictions across the world. At the time of writing (December 2021), travel restrictions in many countries are still in place.

The global oil demand accounted for 11.5 Gt of energy-related CO_2 in 2019 (IEA, 2020a). The transport sector consumes 65% of total oil demand, which included oil for international bunkers (10.4% of the total oil demand). Road transport consumed more than 40% of the total oil demand in 2019. The sector's growth has been responsible for over half the growth in the total oil demand since 2000 (BloombergNEF, 2020). As a result of the restricted mobility imposed to stop spread of the COVID-19 virus, the global pandemic led to a significant reduction in the oil

Table 8.1 World regions used for the 1.5 °C OECM transport scenario

World region	Countries
OECD Europe	Austria, Belgium, Czech Republic, Denmark, Estonia, Finland, France, Germany, Greece, Hungary, Iceland, Ireland, Italy, Israel, Luxembourg, the Netherlands, Norway, Poland, Portugal, Slovak Republic, Slovenia, Spain, Sweden, Switzerland, Turkey, United Kingdom
OECD North America	Canada, Mexico, United States of America
OECD Pacific	Australia, Japan, Korea (South), New Zealand
Eurasia	Albania, Armenia, Azerbaijan, Belarus, Bosnia-Herzegovina, Bulgaria, Croatia, former Yugoslav Republic of Macedonia, Georgia, Kazakhstan, Kosovo, Kyrgyz Republic, Latvia, Lithuania, Montenegro, Romania, Russia, Serbia, Tajikistan, Turkmenistan, Ukraine, Uzbekistan, Cyprus, Gibraltar, Malta
China	People's Republic of China, including Hong Kong
India	India
Non-OECD Asia	Afghanistan, Bangladesh, Bhutan, Brunei Darussalam, Cambodia, Chinese Taipei, Cook Islands, East Timor, Fiji, French Polynesia, Indonesia, Kiribati, Democratic People's Republic of Korea, Laos, Macao, Malaysia, Maldives, Mongolia, Myanmar, Nepal, New Caledonia, Pakistan, Papua New Guinea, Philippines, Samoa, Singapore, Solomon Islands, Sri Lanka, Thailand, Tonga, Vanuatu, Vietnam
Latin America	Antigua and Barbuda, Argentina, Aruba, Bahamas, Barbados, Belize, Bermuda, Bolivia, Brazil, British Virgin Islands, Cayman Islands, Chile, Colombia, Costa Rica, Cuba, Dominica, Dominican Republic, Ecuador, El Salvador, Falkland Islands, French Guyana, Grenada, Guadeloupe, Guatemala, Guyana, Haiti, Honduras, Jamaica, Martinique, Montserrat, Netherlands Antilles, Nicaragua, Panama, Paraguay, Peru, St. Kitts and Nevis, Saint Lucia, St. Pierre et Miquelon, St. Vincent and Grenadines, Suriname, Trinidad and Tobago, Turks and Caicos Islands, Uruguay, Venezuela
Africa	Algeria, Angola, Benin, Botswana, Burkina Faso, Burundi, Cameroon, Cape Verde, Central African Republic, Chad, Comoros, Congo, Democratic Republic of Congo, Cote d'Ivoire, Djibouti, Egypt, Equatorial Guinea, Eritrea, Ethiopia, Gabon, Gambia, Ghana, Guinea, Guinea-Bissau, Kenya, Lesotho, Liberia, Libya, Madagascar, Malawi, Mali, Mauritania, Mauritius, Morocco, Mozambique, Namibia, Niger, Nigeria, Rwanda, Sao Tome and Principe, Senegal, Seychelles, Sierra Leone, Somalia, South Africa, South Sudan, Sudan, Swaziland, United Republic of Tanzania, Togo, Tunisia, Uganda, Western Sahara, Zambia, Zimbabwe
Middle East	Bahrain, Iran, Iraq, Jordan, Kuwait, Lebanon, Oman, Qatar, Saudi Arabia, Syria, United Arab Emirates, Yemen

demand, especially for road transport and aviation, which are responsible for nearly 60% of oil use (IEA, 2020a). The global oil demand is estimated to have dropped by 8% in 2020. At the time of writing, the global pandemic is still ongoing, although travel restrictions have been relaxed in many countries, increasing in the transport demand relative to that in 2020. In our transport demand projections, we assume that the demand will continue to increase to pre-pandemic levels by 2025.

Table 8.2 Assumed population and GDP developments by region in 2020–2050

Assumed population and GDP developments by region in 2020–2050							
		Units	2019	2025	2030	2040	2050
OECD North America	Population	[Million]	499	524	543	575	599
	GDP	[Billion $]	24,255	27,650	30,513	37,562	45,788
Latin America	Population	[Million]	526	552	571	599	616
	GDP	[Billion $]	7415	8807	10,141	13,761	18,675
OECD Europe	Population	[Million]	579	587	592	598	598
	GDP	[Billion $]	23,433	26,076	28,269	32,807	36,963
Africa	Population	[Million]	1321	1522	1704	2100	2528
	GDP	[Billion $]	6865	9247	11,376	17,498	26,403
Middle East	Population	[Million]	250	276	295	331	363
	GDP	[Billion $]	6120	7230	8857	12,112	17,587
Eurasia	Population	[Million]	346	347	346	343	339
	GDP	[Billion $]	6685	7919	9081	11,853	15,025
NON-OECD Asia	Population	[Million]	1189	1269	1329	1428	1499
	GDP	[Billion $]	11,101	14,577	17,794	25,876	34,234
India	Population	[Million]	1368	1452	1513	1605	1659
	GDP	[Billion $]	10,816	17,084	22,652	37,966	54,074
China	Population	[Million]	1427	1447	1450	1426	1374
	GDP	[Billion $]	26,889	37,997	47,427	64,986	84,825
OECD Pacific	Population	[Million]	208	208	208	204	198
	GDP	[Billion $]	8761	9644	10,407	11,842	13,081
Global	Population	[Million]	7713	8185	8551	9210	9772
	GDP	[Billion $]	132,339	166,230	196,516	266,263	346,656

The pandemic had a dramatic impact on public transport. Fear of being infected with COVID-19 led many people to avoid using public transport and to switch to other transport modes—especially individual transport, such as private cars or (electric) bicycles. *The Future of Public Transport (C40 Cities Climate Leadership Group and International Transport Workers' Federation 2021)*, published in March 2021, reported that as 'public transport ridership has fallen during the COVID-19 pandemic, so has revenue. Public transport agencies across cities worldwide face a critical funding shortfall that threatens jobs and services'.

The energy demand is likely to increase and there is currently no sign that these increases will slow in the near future. The increasing demand for energy for transport has mainly been met by greenhouse gas (GHG)-emitting fossil fuels. Although (battery) electric mobility has recently surged considerably, it has done so from a very low base, which is why, in terms of total numbers, electricity still plays a relatively minor role as an energy carrier in the transport sector.

Apart from their impact on climate, increasing transport levels—especially by car, truck, and aeroplane—also have unwanted side effects: accidents, traffic jams, noise and other pollutants, visual pollution, and the disruption of landscapes by the large-scale build-up of the transport infrastructure. However, road, rail, sea, and air transport are also integral parts of our globalised and interconnected world and guarantee both prosperity and intercultural exchange. Therefore, if we are to cater to people's desire for mobility while keeping the economy running and meeting the Paris climate goals, fundamental technical, operational, and behavioural measures are immediately required.

In this analysis, we discuss potential pathways of transport activity and technological developments by which we can meet the requirement that warming does not exceed pre-industrial levels by more than 1.5 °C—while at the same time maintaining a reasonable standard of mobility. The scenarios in this analysis are based on global and regional scenarios developed by the German Aerospace Centre (DLR), published in February 2019 (Pagenkopf et al., 2019), which have been updated in more detail as part of the Transformative Urban Mobility Initiative (TUMI) research (Teske et al., 2021).

We structured our scenario designs around the following key energy- and emission-reducing measures:

- Powertrain electrification
- Enhancement of energy efficiency through technological developments
- Use of bio-based and synthetically produced fuels only within strict sustainability limits
- Modal shifts (from high- to low-energy-intensity modes) and overall reductions in transport activities in energy-intensive transport modes

The final global energy demand in the transport sector[1] totalled 103 EJ in 2019, according to the IEA Energy Balances (IEA, 2020b). Based on this estimate, the freight and passenger transport demands were estimated from statistical data and energy-efficiency figures.

Figure 8.1 shows that road passenger transport had the largest share of the final transport energy (53%) in 2019. Most of this consisted of individual road passenger modes (mostly cars, but also two- and three-wheel vehicles), which accounted for around 40% of all end energy in the transport sector. In total, road transport (passenger and freight) accounted for around 76% of the total final energy demand for transport.

The majority of all passenger transport—in terms of overall kilometres—is by road. However, international freight transport is more strongly dominated by rail and shipping, which account for 45% of all tonne–kilometres. The high efficiency of rail and shipping means that their share of the global transport energy demand is small relative to the share of global tonnage transported.

Figure 8.2 shows the passenger (pkm) and freight transport (tkm) by transport mode in 2019 (OECD, 2021). Road transport clearly dominates. However, international freight often arrives by ship and is further transported by rail and/or road. OECD America and OECD Europe together make up half the total global energy demand, as shown in Fig. 8.3. China is at nearly the same level as OECD Europe, although it has about twice as many inhabitants as OECD Europe.

---

[1] International aviation and navigation bunkers are not included in this figure.

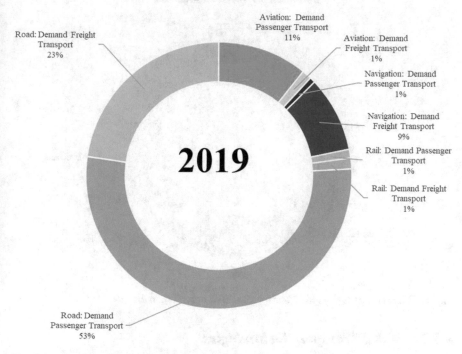

Fig. 8.1 Global final energy use, by transport mode, in 2019 (without international aviation or navigation bunker fuels)

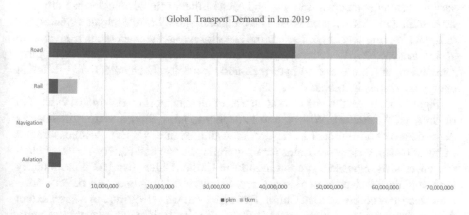

Fig. 8.2 Transport mode performances for road, rail, and aviation

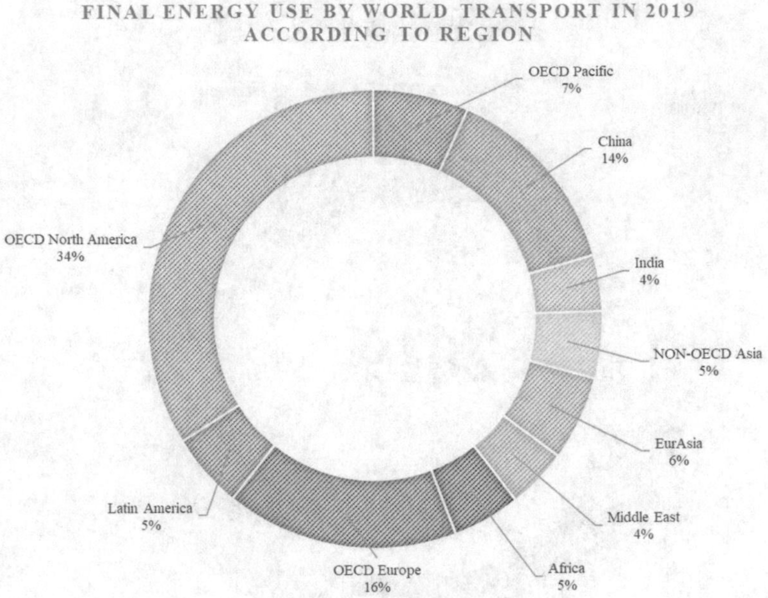

FINAL ENERGY USE BY WORLD TRANSPORT IN 2019
ACCORDING TO REGION

Fig. 8.3 Final energy use by global transport in 2019, according to region

8.3.2 Global Transport Technologies

The energy intensities for different vehicle types and for each of the available drive trains play an important role in the final energy demand. Each transport mode has various different vehicular options, and each of the available vehicles has different drive train and efficiency options. The technical variety of passenger vehicles, for example, is extremely large. The engine sizes for five-seater cars range from around 20 kW to over 200 kW. Moreover, drive trains can use a range of fuels, from gasoline, diesel, and bio-diesel to hydrogen and electricity. Each vehicle has different energy intensities in MJ/pkm.

Figure 8.4 shows the powertrain shares of all transport modes in 2019 (in pkm or tkm) (IEA, 2020b). With a few exceptions, most modes were still heavily dependent on conventional internal combustion engines (ICEs). A small number of buses had electric powertrains (mainly trolley buses) and battery-powered electric buses also increased, predominantly in China. China also has a particularly large number of electric two- and three-wheel vehicles. Almost all battery-powered electric scooters were in China. Passenger rail was electrified to a large extent (e.g. metropolitan and high-speed trains), whereas freight trains were predominantly not electrified.

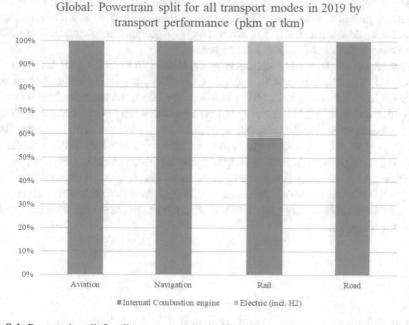

Fig. 8.4 Powertrain split for all transport modes in 2019, by transport performance (pkm or tkm)

8.4   Aviation

The 2020 pandemic led to significant travel restrictions and significantly affected the energy demands of global and domestic aviation (IEA, 2020c). The International Air Transport Association expects flight capacity utilisation to be, on average, 65% below the 2019 level in the second quarter (Q2) of 2020, 40% below in Q3 2020, and 10% below in Q4 2020 (Pearce, 2020). Data show that the global flight numbers were down by 70% at the start of April 2020 relative to those in the previous year. The consumption of kerosene in the whole of 2020 was expected fall by 26% (IEA, 2020c).

8.4.1   Energy Intensity and Emission Factors: Aviation

The energy intensity for aviation freight transport was assumed to be around 30 MJ/tkm in 2019 (Pagenkopf et al., 2019), decreasing by 1% per year until 2025. By 2050, the energy intensity for freight planes is estimated to be 25 MJ/tkm, 17% below today's value. The energy intensity for aviation passenger transport will decrease from 5.8 to 4.2 MJ/pkm between 2020 and 2050. Technical improvements in the aerodynamics, materials, weight, and turbine efficiency for both freight and passenger planes are assumed. The volume of freight (in tkm) and the passenger–kilometres (pkm) are assumed to decrease by 30% globally between 2019 and 2050, an average reduction of around 1% per year.

The emissions factor for kerosene is calculated to be 73.3 g of CO_2 per MJ (gCO_2/MJ) (Jurich, 2016). The specific CO_2 emissions for aviation freight will decrease from 2.3 to 2.0 $kgCO_2$/tkm in 2025. By 2035, the specific emissions will more than halve, to 0.8 $kgCO_2$/tkm, and will be completely decarbonised by 2050.

In passenger aviation transport, specific CO_2 emissions will decrease from 425 gCO_2/pkm in 2019 to 350 gCO_2/pkm in 2025, will halve by 2035, and will be CO_2-free by 2050—analogous to freight transport. Both reduction trajectories will be achieved by the gradual replacement of fossil kerosene with organic kerosene, and after 2040, with synthetic kerosene that is generated with renewable electricity. Because aviation is a truly global sub-sector, the assumptions for aviation are the same for all regions.

8.5   Shipping

Of the global energy demand for shipping, 90% is for freight transport, and only around 10% is for passenger transport (mainly cruise ships and ferries). In 2018, the worldwide cruise ship passenger capacity was 537,000 passengers on 314 ships, and 26 million passengers were transported in 2018 (Cruise Market Watch, 2020). In comparison, around 53,000 merchant ships were registered globally in January 2019: approximately 17,000 cargo ships, 11,500 bulk cargo carriers, 7500 oil tankers, 5700 chemical tankers, and 5150 container ships. The remaining ships included roll-on, roll-off passenger and freight transport ships and liquefied natural gas (LNG) tankers (Statista, 2021).

8.5.1   Energy Intensity and Emission Factors: Shipping

The energy intensity for freight transport by ship was assumed to be 0.19 MJ/tkm in 2019 (Pagenkopf et al., 2019) and will decrease only slightly to 0.18 MJ/tkm in 2030 and 0.17 MJ/tkm in 2050. An equivalent trajectory is assumed for shipping passengers, from 0.056 to 0.054 MJ/pkm in 2030 and to 0.052 MJ/pkm in 2050. Shipping is already by far the most efficient transport mode. However, further technical improvements, especially in ship engines, are required. The volume of freight (in tkm) is assumed to increase by around 0.5% per year globally until 2050, whereas passenger transport volumes will remain at today's levels over the entire modelling period.

The emissions factor for heavy fuel oil is calculated to be 81.3 gCO_2/MJ (Jurich, 2016). The specific CO_2 emissions for shipping freight will decrease from 15 gCO_2/

tkm to 10 kgCO$_2$/tkm by 2030. By 2040, freight shipping will be completely decarbonised. The specific CO$_2$ emissions for passenger shipping transport will decrease from 5 gCO$_2$/pkm in 2019 to 3 gCO$_2$/pkm in 2030, and analogous to freight shipping, passenger transport by ship will be carbon neutral by 2040. Both reduction trajectories will be achieved by the gradual replacement of fossil fuels with biofuels and, after 2040, with renewables-generated synthetic fuels.

8.6 Land Transport

Although the most-efficient transport mode for long distances over land is railways, vehicular road transport for passenger and freight transport dominates by an order of magnitude.

Road transport is the single largest consumer of oil. In 2018, 64% of the global demand was attributed to road transport vehicles, for both freight and passenger transport. The pandemic in 2020 led to a unique development: as a consequence of global lockdown measures, mobility (57% of the global oil demand) declined at an unprecedented rate. The road transport in regions under lockdown decreased by 50–75%, with the global average road transport activity falling to almost 50% of the 2019 level by the end of March 2020 (IEA, 2020c).

Whereas electric-powered planes or ships are still in the early stages of development, there are no technical barriers to the phasing-out of ICEs or the transition to efficient electric vehicles (EVs) for passenger transport and to hydrogen or synthetic biofuels for heavy-duty vehicles. The vehicle technology required is widely available and market shares are rising sharply. In 2012, only 110,000 battery electric vehicles (BEVs) had been sold worldwide. Since then, sales have almost doubled every year, reaching 1.18 million BEVs in 2016, 3.27 million in 2018, and 4.79 million in 2019 (IEA, 2020d).

8.6.1 Energy Intensity and Emission Factors: Land Transport

Individual Transport
Passenger transport by road makes up by far the commonest and most important form of travel. There are numerous technical options to 'move people with vehicles'—bicycles, motorcycles, tricycles, city cars, four-wheel drive SUVs—and each vehicle has very different energy intensity per kilometre. Although this research project aims for high-technology resolution, simplification is required. First and foremost, the data for all existing vehicles for each of the regions and for the global level are neither available nor practical to use. Figure 8.3 shows the energy intensities for the main vehicle types, which form the basis for the energy scenario calculations (Table 8.3).

Table 8.3 Energy intensities for individual transport modes—road transport

Individual transport			Passengers		Vehicle demand	Consumption per passenger	Energy demand
			Average passengers per vehicle	Assumed occupation rate	Average	Average	Assumption for scenario calculation
		Fuel			**Litres/100 km**	**Litres/100 pkm**	**MJ/pkm**
Scooters and motorbikes	2-wheeler	Gasoline	1	1	3.0	3.0	1.21
		Electricity			**kWhel/100 km**	**kWhel/100 pkm**	**MJ/pkm**
E-bikes	2-wheeler	Battery	1	1	1.0	1.0	0.04
Scooters	2-wheeler	Battery	1	1	1.8	1.9	0.06
Motorbikes	2-wheeler	Battery	1	1	4.8	2.4	0.17
Rickshaws	3-wheels	Battery	3	2	8.0	4.0	0.14
		Fuels			**litres/100 km**	**litres/100 pkm**	**MJ/pkm**
Cars	Small	ICE–oil	2	1.8	5.0	2.8	1.12

				kWhel/100 km	kWhel/100 pkm	MJ/pkm
Medium	ICE–oil	4	2	7.5	3.8	1.51
Large	ICE–oil	5	2	10.5	5.3	2.11
Small	ICE–gas	2	1.8	4.5	2.5	0.63
Medium	ICE–gas	4	2	7.0	3.5	1.41
Large	ICE–gas	5	2	10.0	5.0	1.25
Small	ICE–bio	2	1.8	5.0	2.8	0.91
Medium	ICE–bio	4	2	7.5	3.8	1.51
Large	ICE–bio	5	2	10.5	5.3	1.72
Small	Hybrid–oil	2	1.8	4.0	2.2	0.89
Medium	Hybrid–oil	4	2.5	6.0	2.4	0.96
Large	Hybrid–oil	5	2.5	8.5	3.4	1.37
Electricity				kWhel/100 km	kWhel/100 pkm	MJ/pkm
Small	Battery	2	1.8	16.0	8.9	0.32
Medium	Battery	4	2	25.0	12.5	0.45
Large	Battery	5	2	32.5	16.3	0.59
Large	Fuel cell	4	2	37.5	18.8	1.36

Public Transport

There are a wide variety of public transport vehicles, ranging from rickshaws to taxis and from minibuses to long-distance trains. The occupation rates for those vehicles are key to calculating the energy intensity per passenger kilometre. For example, a diesel-powered city bus that transports 75 passengers requires, on average, about 27.5 litres per 100 kilometres. If the bus is operating at full capacity during peak hour, the energy demand per passenger is as low as 400 ml per kilometre—lower than almost all other fossil-fuel-based road transport vehicles. However, if the occupancy drops to 10% (e.g. for a night bus), the energy intensity increases to 3.7 litres, equal to that of a small energy-efficient car. Occupation rates vary significantly and depend upon the time of day, day of the week, and season. There are also significant regional differences, even within a single country, and even more so across larger regions, such as OECD Europe, which is composed of over 30 countries from Iceland to Turkey.

Again, the parameters shown in Table 8.4 are simplified averages and are further condensed for the scenario calculations. Although high technical resolution is possible for the scenario model, it would imply an accuracy that does not exist, because the statistical data required for this are not available on either regional or global levels.

Freight Transport

The energy intensity data for freight transport are not as diverse as those for passenger transport, because the transport vehicle types are more standardised and the fuel demand is well known. However, the utilisation rate of the load capacity varies significantly, and consistent data are not available for the regional and global levels calculated. Therefore, the assumed utilisation rate has a huge influence on the calculated energy intensity per tonne–kilometre. The average energy intensities per tonne–kilometres used in the scenarios are shown in Table 8.5 and are largely consistent with other sources in the scientific literature. The assumed energy intensities for electric and fuel cell/hydrogen freight vehicles are only estimates, because this technology is still in the demonstration phase. Therefore, none of the scenarios calculated factor in large shares of electric freight transport vehicles before 2035.

8.7   Global Transport Demand Projections

A variety of actions will be required for the transport sector to conform to the limit global warming to 1.5 °C. The set of actions described can be clustered into technical and operational measures (e.g. increases in energy efficiency, electrification of drive trains), behavioural measures (e.g. shifts to less-carbon-intensive transport carriers and an overall reduction in transport activity), and accompanying policy measures (e.g. taxation, regulations, urban planning, and the promotion of less-harmful transport modes).

Table 8.4 Energy intensities for public transport—road and rail transport

Public transport			Passengers		Vehicle demand	Consumption per passenger	Energy demand
			Average passengers per vehicle	Assumed occupation rate	Average	Average	Assumption for scenario calculation
		Fuels			**litres/100 km**	**litres/100 pass-km**	**[MJ/pkm]**
Buses	Small	Diesel	12	40%	8.8	1.8	0.73
	Small	Bio	12	40%	8.8	1.8	0.60
	12 m	Diesel	75	40%	27.5	0.9	0.37
	12 m	Bio	75	40%	27.5	0.9	0.30
	Large	Diesel	135	40%	57.5	1.1	0.43
		Electricity	**0**	**0**	**kWhel/100 km**	**kWhel/100 pkm**	**[MJ/pkm]**
	Small	Battery	12	40%	31	6.4	0.23
	Small	Fuel cell	12	40%	77	15.9	0.57
	12 m	Battery	75	40%	143	4.8	0.17
	12 m	Fuel cell	75	40%	358	11.9	0.43
	Large	Overhead lines	135	40%	263	4.9	0.18
Trains		**Fuels**			**Litres/100 km**	**Litres/100 pkm**	**[MJ/pkm]**
	Metros	Diesel	400	40%	150	0.9	0.38
	Metros	Bio	400	40%	150	0.9	0.31
	Commuter trains	Diesel	600	40%	300	1.3	0.50
	Commuter trains	Bio	600	40%	300	1.3	0.41
		Electricity	**0**	**0**	**kWhel/100 km**	**kWhel/100 pkm**	**[MJ/pkm]**
	Trams	Electric	300	40%	495	3.8	0.14
	Metros	Electric	300	40%	1200	4.0	0.14
	Commuter trains	Electric	600	40%	1950	4.8	0.17

Table 8.5 Energy intensities for freight transport—road and rail transport

Freight transport			Maximum load capacity in tonnes	Assumed utilisation rate	Vehicle demand Average	Consumption per tonne Average	Energy demand Assumption for scenario calculation
Trucks		**Fuels**			**Litres/100 km**	**Litres/tkm**	**MJ/tkm**
	3.5 t	Diesel	3.5	40%	11	7.9	3.16
	3.5 t	Bio	3.5	40%	11	7.9	2.57
	7.5 t	Diesel	7.5	40%	20	6.5	2.61
	7.5 t	Bio	7.5	40%	20	6.5	2.13
	12.5 t	Diesel	12.5	40%	25	5.0	2.01
	12.5 t	Bio	12.5	40%	25	5.0	1.64
		Electricity			**kWhel/100 km**	**kWhel/tkm**	**MJ/tkm**
	3.5 t	Battery	3.5	40%	19	13.6	1.34
	3.5 t	Fuel cell	3.5	40%	46	33.2	1.33
	7.5 t	Battery	7.5	40%	41	13.6	0.49
	7.5 t	Fuel cell	7.5	40%	100	33.2	1.19
	12.5 t	Battery	12.5	40%	68	13.6	0.49
	12.5 t	Fuel cell	12.5	40%	166	33.2	1.19
Trains		**Fuels**			**Litre/100 km**	**Litre/tkm**	**MJ/tkm**
	Freight—740 m	Diesel	1000	40%	300	0.8	0.30
	Freight—740 m	Bio	1000	40%	300	0.8	0.25
		Electricity			**kWhel/100 km**	**kWhel/tkm**	**MJ/tykm**
	Freight—740 m	Electric	1000	40%	5840	14.6	0.53

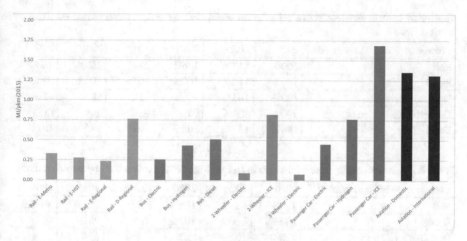

Fig. 8.5 Energy intensities for urban and interurban passenger transport modes in 2019 (world averages). (Source: DLR/ IFFT 2019, Deutsches Zentrum für Luft- und Raumfahrt (DLR), Institut für Fahrzeugkonzepte, Fahrzeugsysteme und Technologiebewertung, Stuttgart, Data from Johannes Pagenkopf et al. 2019)

The key requirements for achieving a reduction of the transport energy demand in the alternative scenarios follow a three-step approach:

- *Reduction* of transport kilometres for passengers and freight with behavioural changes, urban planning, increased local production, and transport logistics
- *Shift* to more-energy-efficient transport modes, e.g. from road to rail for passengers and from aviation to navigation for freight
- *Innovation*—replacing inefficient combustion engines with efficient electric drives

8.7.1 Projection of the Transport Service Demand

The first step in the projection of the global transport demand is calculating the actual service demand in passenger–kilometres travelled and tonnes of goods–kilometres transported. This is essential before the development of the chosen transport mode (road, rail, or ship) is projected.

Under the three scenarios, the global transport demand is the sum of the ten world regions plus bunker fuels. Bunker fuels are all the fuels required for interregional aviation and shipping transport and are therefore not part of any regional demand. The assumed development is based on the population and economic developments in $GDP provided in Table 8.2. The 1.5 °C scenario assumes a reduction in the global pkm of 30% relative to 2020, whereas the global freight demand will increase by 30% based on the assumption of a growing GDP (Tables 8.6 and 8.7).

Table 8.6 Global: development of behavioural changes in passenger travel (based on pkm) by transport mode

Change in % of 2020 demand	2020	2025	2030	2040	2045	2050
Rail	100%	117%	136%	221%	282%	360%
Road	100%	112%	109%	105%	100%	94%
Domestic aviation	100%	101%	96%	71%	58%	47%
Domestic navigation	100%	101%	94%	81%	75%	69%
Total	**100%**	**111%**	**109%**	**108%**	**105%**	**101%**

Table 8.7 Global: development of changes in freight logistics (based on tkm) by transport mode

Change in % of 2020 demand	2020	2025	2030	2040	2045	2050
Rail	100%	133%	186%	238%	269%	305%
Road	100%	110%	107%	103%	98%	92%
Domestic aviation	100%	102%	98%	76%	65%	51%
Domestic navigation	100%	103%	97%	85%	80%	75%
Total	**100%**	**106%**	**103%**	**96%**	**92%**	**89%**

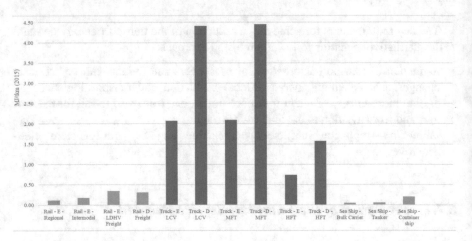

Fig. 8.6 Energy intensities for freight transport modes in 2019 (world averages). (Source: DLR IFFT 2019)

8.7.2 Mode-Specific Technology Efficiency and Improvements Over Time

For passenger transport, trains and buses are much more energy efficient per pkm than passenger cars or airplanes. This situation does not change fundamentally if only electric drive trains are compared (Fig. 8.5). Railways and (especially) ships are clearly more energy efficient than trucks in transporting freight (Fig. 8.6). The efficiency data are based on both literature-reported and on transport-operator documents in this study and on Pagenkopf et al. (2019). Efficiency levels, in terms of pkm or tkm, depend to a large extent on the underlying utilisation of the capacity of the vehicles, which varies across world regions. The numbers presented are average values and differences are evaluated at the regional level.

8.7.3   Powertrain Electrification for Road Transport

Increasing the market penetration of highly efficient (battery and fuel cell) electric vehicles, coupled with the generation of clean electricity, is a powerful lever for decarbonisation and probably the most effective means of moving toward a decarbonised transport system.

All-electric vehicles have the highest efficiency levels of all the drivetrain options. Today, only a few countries have significant proportions of electric vehicles in their fleets. The total number of electric vehicles, particularly for road transport, is insignificant, but because road transport is by far the largest CO_2 emitter of overall transport, it offers a very powerful lever for decarbonisation.

In terms of drivetrain electrification, we cluster the world regions into three groups, according to the diffusion theory (Rogers, 2003):

- *Innovators*: OECD North America (excluding Mexico), OECD Europe, OECD Pacific, and China
- *Moderate*: Mexico, Non-OECD Asia, India, Eurasia, and Latin America
- *Late adopters*: Africa and the Middle East

Although this clustering is rough, it sufficiently mirrors the basic tendencies implemented in our scenarios. The regions differ in the speed with which novel technologies, especially electric drivetrains, will penetrate the market.

In addition to powertrain electrification, there are other potential improvements in energy efficiency, and their implementation will steadily improve these energy intensities over time. Regardless of the type of power train and the fuel used, efficiency improvements on MJ/pkm or MJ/tkm will result from (for example):

- Reductions in powertrain losses through more-efficient motors, gears, power electronics, etc.
- Reductions in aerodynamic drag
- Reductions in vehicle mass through light-weighting
- Use of smaller vehicles
- Operational improvements (e.g. automatic train operation, load factor improvements)

8.7.4   Projection of Global and Regional Modal Shifts

In 2019, road transport predominated over all other transport modes, with almost 95% of all pkm travelled by some form of road vehicle throughout the world. Based on the kilometres travelled, just over 3.5% of journeys were by train and about 2% by plane. Although ship transport is one of the most important means of transport for freight, marine-based passenger transport makes only a very minor contribution at the global level. To implement the 1.5 °C scenario, passenger transport must shift from road to rail. Efficient light rail in cities, commuter trains for short to medium distances, and high-speed trains that offer convenient services are therefore alternatives to individual car journeys.

In the context of urban transport, the use of road transport by cars will be significantly reduced and will move towards public transport by other road vehicles, such as buses or trains. The role of electric bikes and walking must also increase under the 1.5 °C scenario. However, road transport will remain dominant, at well over 80% (Fig. 8.7), until 2050. Therefore, the modal shifts within road transport systems, such as from individual cars to public transport, cycling, or mobility services (such as car sharing), are extremely important.

Maritime shipping is the backbone of world trade. It is estimated that some 80% of all goods are carried by sea. In terms of value, the global maritime container trade is estimated to account for around 60% of all seaborne trade, which was valued at around $US14 trillion in 2019 (Placek, 2021).

In terms of tonnage, aviation plays a comparatively minor role globally. In terms of tonne–kilometres, road transport dominates globally. Every second tonne is transported by road and only 10% by rail (Fig. 8.8). However, the different transport modes cannot be separated because goods delivered by ship are further distributed by road and rail. Therefore, a direct modular shift is often not possible. Ship transport cannot be replaced by trains in most cases, and vice versa. There is competition between road and rail, and modular shifts in favour of rail freight transport will occur. The 1.5 °C pathway assumes that about one-third of the freight transported by trucks will be shifted to rail transport systems.

Fig. 8.7 World passenger transport by mode under the 1.5 °C scenario—shares based on passenger–kilometres

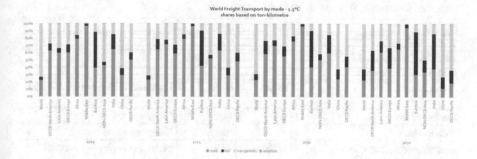

Fig. 8.8 World freight transport by mode under the 1.5 °C scenario—shares based on tonne–kilometres

Compared with passenger transport, freight transport is far more diverse, and regional differences are significant. In Eurasia, a region very similar to the former USSR, rail transport shoulders about half of all freight transport in terms of tonnage. This reflects the significance of the Trans-Siberian Railway line connecting the European part of Russia with Mongolia (Ulan Bator) and China (Beijing).

In Non-OECD Asia, water transport is by far the most important transport mode, which reflects the situations in the island states Indonesia and the Philippines, as well as the vast coastlines of Southeast Asian countries.

8.7.5   Calculation of Transport Energy Demand

The calculation of the transport demand is based on a two-step approach, with all the parameters described in the previous subsections (Sects. 8.7.1, 8.7.2, 8.7.3, and 8.7.4):

1. Calibration of the model with statistics from the past 10–15 years (Table 8.8)
2. Projection of the transport demand based on the changing demand in kilometres and energy intensities by transport mode (Table 8.9)

To calibrate the model, the transport demand of the past decade was recalculated on the basis of the available energy statistics. The International Energy Agencies' (IEA) Advanced World Energy Balances provided the total final energy demands by transport mode—aviation, navigation, rail, and road—by country, region, or globally. However, there is no further specification of the energy usage within each of the transport modes. A further division into passenger and freight transport is therefore calculated using percentage shares. These proportions are determined with a literature research and from the average energy intensity for each of the transport modes for passenger and freight vehicles.

The annual energy demand divided by the average energy intensity by mode generates the annual transport demand in passenger–kilometres per year [pkm/yr]

Table 8.8 Calibration for transport demand calculations

Parameter	Units	Source	
Transport demand			
Aviation, navigation, rail, road—past to present			
Annual demand	[PJ/yr]	Data	Database: IEA Advanced World Energy Balances
Passenger share	[%]	Input	Literature review
Freight share	[%]	Input	Literature review
Average energy intensity—passenger transport	[MJ/pkm]	Data	Literature review—based on the current supply mix
Average energy intensity—freight transport	[MJ/tkm]	Data	Literature review—based on the current supply mix
Passenger–kilometres	[pkm]	Calculation	= Annual demand/energy intensity [compared with OECD statistic]
Tonne–kilometres	[tkm]	Calculation	
Annual growth/decline—passenger–kilometres	[%/yr]	Calculation	= Annual demand previous year/annual demand in the calculated year
Annual growth/decline—tonne–kilometres	[%/yr]	Calculation	
Population—indicator of passenger transport development	[million]	Data	Database: United Nations
GDP/capita—indicator of passenger and freight transport development	[$GDP/capita]	Data	Database: World Bank
GDP—indicator of freight transport development	[$GDP]	Data	Database: World Bank

Table 8.9 Projection of transport demand based on changing demand in kilometres

Parameter: 2020–2050	Units	Process	Comment
= (Passenger–kilometres previous year) × (increase/decline in %/yr)	[pkm]	Aviation, navigation, rail, and road projections	Starting point: base year 2019
= (Tonne–kilometres previous year) × (increase/decline in %/yr)	[tkm]	Calculation	Starting point: base year 2019
INPUT in %/yr	[%/yr]	Calculation	Assumption
INPUT in %/yr	[%/yr]	Input	Assumption
INPUT in %/yr	[million]	Input	Assumption based on UN projections
= $GDP/capita	[$GDP/capita]	Calculation	
INPUT in %/yr	[$GDP]	Calculation	Assumption based on World Bank projections
Time series 2020–2050: passenger–kilometres per year and region	[pkm/yr]	Calculation	Input for the calculation of energy demand
Time series 2020–2050: freight–kilometres per year and region	[tkm/yr]	Result	Input for the calculation of energy demand
		Result	

and tonne–kilometres per year [tkm/yr]. Those results are then compared with the OECD transport statistics, which provide both parameters, pkm/yr and tkm/yr. Calibrating the model with historical data ensures that the basis of the scenario projection for the coming years and decades has been correctly mapped and that the changes can be calculated more realistically.

For the forward projection of the transport demand, the calculation method is reversed. The transport demand for each transport mode is calculated on the basis of the annual change (as a percentage). The calculated total annual passenger–kilometres and tonne–kilometres are the inputs for the energy demand calculations.

8.7.6 Transport Service: Energy Supply Calculation

Like the transport demand calculation, calculation of the transport energy 'supply' begins with the calibration of the model based on historical data, as part of a two-step approach:

1. Calibration of the model with statistics from the past 10–15 years (Table 8.8)
2. Projection of the transport supply based on the transport mode and vehicle-specific parameters (Table 8.9)

As well as the final energy demand for each transport mode, the IEA Advanced World Energy Balances also provide the energy demand by source—soil, gas, biofuels, and electricity. To calculate the exact energy requirement for each transport mode with the corresponding transport requirement (in km), assuming different vehicle technologies, the status quo must be determined. For this purpose, the respective transport energy requirement for each transport mode and fuel type is calculated based on the current vehicle technology market shares and the technology-specific energy intensities per kilometre. The results provide a technology-specific illustration of each sector. Table 8.10 presents an overview of the calculation process for the calibration of the model.

Future energy demands based on the projected pkm and tkm are calculated from market shares and technology-specific energy intensities. In the first step, the overall transport energy demand, e.g. in passenger–kilometres, is distributed to each transport mode. A mode shift from road to rail can be assumed, and the sector-specific demand is further distributed to specific vehicle types—again by the assumption of market shares (Table 8.11).

8.8 Transport: Energy Demand and Supply

In the previous sections, the global energy demand was calculated based on the documented assumptions. However, the transport sector is among the most diverse sectors of all the end-use sectors analysed. A whole range of logistical, technical,

Table 8.10 Calibration for transport demand calculations

Transport supply				
Aviation, navigation, rail, road—past to present				
Annual demand—oil	Data	Database	[PJ/yr]	Data: IEA Advanced World Energy Balances
Proportion of passenger and freight transport	Input	Literature	[%]	Shares of total energy demand from the literature
Average energy intensity—passenger	Input	Literature	[MJ/pkm]	Average energy intensity for the corresponding fuel
Average energy intensity—freight	Input	Literature	[MJ/tkm]	Average energy intensity for the corresponding fuel

The process above is repeated for natural gas, biofuels, synthetic fuels, hydrogen, and electricity

All the energy carriers are summed by transport mode to calculate the total energy demand for aviation, navigation, rail, and road

All energy demands are summed by transport mode to calculate the total energy demand for transport

Table 8.11 Projection of transport supply based on transport mode and vehicle-specific parameters

Transport energy projections			
Aviation, navigation, rail, road			
From calibration	Energy demand: aviation	[PJ/yr]	Based on statistics, databases
From calibration	Energy demand: aviation—share of passenger transport	[%]	
From calibration	Average energy intensity of oil used in vehicles	[MJ/pkm]	
This is repeated for all fuels and all sectors (passenger and freight)			
This process is repeated for the remaining transport modes: navigation, rail, and road			
Projections			
Input	Passenger transport demand—aviation	[pkm/yr]	Input demand projection
Input	Market share of vehicle type 1	[%]	Possible efficiency increases over time
Input	Market share of vehicle type 2	[%]	Possible efficiency increases over time
Input	Market share of vehicle type n	[%]	Possible efficiency increases over time
Input	Energy intensity of vehicle type 1	[MJ/pkm]	Possible efficiency increases over time
Input	Energy intensity of vehicle type 2	[MJ/pkm]	Possible efficiency increases over time
Input	Energy intensity of vehicle type n	[MJ/pkm]	Possible efficiency increases over time
Calculation	pkm per year × market share type 1 × energy intensity type 1 − energy demand for vehicle type 1		
Calculation	Repeated for all vehicle types		
Calculation	Calculated total energy demand vehicle 1 − n		
Result	Energy demand by transport mode		
This is repeated for all transport modes			

and political measures are required to reduce the overall energy demand while maintaining freedom of movement and mobility. The transport sector is closely related to the buildings sector, because urban planning and urban designs go hand in hand with the transport demand—in terms of the distances travelled or goods transported—and the most suitable technical solutions to provide those services. Furthermore, the carbon intensity of the electricity consumed for transport is directly related to the renewable energy share in power generation.

8.8.1 Shipping and Aviation: Dominated by Combustion Engines for Decades to Come

Navigation will probably remain predominantly powered by ICEs in the next few decades. Therefore, we did not model the electrification of freight vessels. However, pilot projects using diesel hybrids, batteries, and fuel cells are in preparation (DNV, 2015). We assumed the same increase in the share of bio- and synthetic fuels over time as in the road and rail sectors.

In aviation, energy efficiency can be improved by measures such as winglets, advanced composite-based lightweight structures, powertrain hybridisation, and enhanced air traffic management systems (Vyas et al., 2013; Madavan, 2016). We project a 1% annual increase in efficiency on a per pkm basis and a 1% annual increase in efficiency on a per tkm basis.

Aviation will probably remain predominantly powered by liquid fossil fuels (kerosene and bio- and synthetic fuel derivatives) in the medium to long term because of the limitations in electrical energy storage. We project a moderate increase in domestic pkm flown in electric aircraft starting in 2030, with larger shares in OECD Europe, because the flight distances are shorter than, for example, in the USA or Australia. Norway has announced plans to perform all short-haul flights electrically by 2040 (Agence France-Presse, 2018).

However, no real electrification breakthrough in aviation is foreseeable unless the attainable energy densities of batteries increase to 800–1000 Wh/kg, which will require fast-charging post-lithium battery chemistries.

That said, it is estimated that over 200 electric aircraft programs are in progress around the world (Downing, 2019). While small electric planes (up to car size) are in the demonstration phase, long-haul flights with electric planes are currently unviable with contemporary battery technology.

From the perspective of technological innovation, electric aviation is an important field of engineering, and investment in this sector must occur now to achieve results in the mid-2030s. Domestic aviation—mainly short-distance flights of up to around 700 km—makes up about 45% of all global flights (Downing, 2019). The electrification of passenger planes for these distances will most likely start in this market segment.

Table 8.12 Aviation—energy demand and supply

Parameter	Unit	2019	2025	2030	2035	2040	2050
				Projection			
Air freight: energy intensity	[MJ/tkm]	32.2	29.1	27.2	26.5	25.8	25.2
Air passenger: energy intensity	[MJ/pkm]	5.8	4.8	4.5	4.4	4.3	4.2
Air freight: energy demand	[PJ/yr]	1445	911	809	712	595	430
Air passenger: energy demand	[PJ/yr]	13,004	8195	7279	6410	5359	3866
Air freight fuel: fossil	[PJ/yr]	580.5	892.4	740.0	284.9	59.5	0.0
Air freight fuel: renewable and synthetic fuels	[PJ/yr]	0.0	18.2	68.7	427.3	535.9	429.5
Air freight fuel: renewables share	[%]	0%	2%	9%	60%	90%	100%
Air freight electricity: fossil	[PJ/yr]	0	0	0	0	0	0
Air freight electricity: renewables	[PJ/yr]	0	0	0	0	0	0
Air freight electricity: renewables share	[%]	0	0	0	0	0	0
Air passenger fuel: fossil	[PJ/yr]	5224	8031	6660	2564	536	0
Air passenger fuel: renewable and synthetic fuels	[PJ/yr]	0	164	619	3846	4823	3866
Air passenger fuel: renewables share	[%]	0%	2%	9%	60%	90%	100%

However, this research has focused on the rapid reduction of CO_2 in the global transport sector, and realistically, electric aviation will not play a role in the reduction of large amounts of carbon before 2040. Nevertheless, the development of this technology is important in the long term (Tables 8.12 and 8.13).

A key target for the global transport sector is the introduction of incentives for people to drive smaller cars and use new, more-efficient vehicle concepts. It is also vital to shift transport use to efficient modes, such as rail, light rail, and buses, especially in large expanding metropolitan areas. Furthermore, the 1.5 °C scenario cannot be implemented without behavioural changes. It is not enough to simply exchange vehicle technologies, but the transport demand must be reduced in terms of the kilometres travelled and by an increase in 'non-energy' travel modes, such as cycling and walking.

With population increases, GDP growth, and higher living standards, the energy demand of the transport sector is expected to increase without technical and behavioural changes. Under the 1.5 °C scenario, efficiency measures, modal shifts, and the behavioural changes mentioned above will reverse the trend in permanent growth (Table 8.14).

The proportion of BEVs among all passenger cars and light commercial vehicles in use is projected to be between 8% and 15% by 2030. This will require a massive build-up of battery production capacity in the coming years. New car sales will already be dominated by battery electric passenger vehicles in 2030 under the 1.5 °C scenario. However, with an assumed average lifetime of 15 years for ICE passenger cars, the existing car fleet will still predominantly use ICEs.

Table 8.13 Shipping—energy demand and supply

Parameter	Unit	2019	2025	2030	2035	2040	2050
					Projection		
Shipping freight: energy intensity	[MJ/tkm]	0.2	0.2	0.2	0.2	0.2	0.2
Shipping-passenger: energy intensity	[MJ/pkm]	0.1	0.1	0.1	0.1	0.1	0.1
Shipping freight: energy demand	[PJ/yr]	11,067	10,659	11,023	11,121	11,233	11,554
Shipping passenger: energy demand	[PJ/yr]	833	802	830	837	846	870
Shipping freight fuel: fossil	[PJ/yr]	2270	10,425	7441	7507	1460	0
Shipping freight fuel: renewable and synthetic fuels	[PJ/yr]	11	235	3582	3614	9773	11,554
Shipping freight fuel: renewables share	[%]	0%	2%	33%	33%	87%	100%
Shipping passenger fuel: fossil	[PJ/yr]	171	785	560	565	110	0
Shipping passenger fuel: renewable and synthetic fuels	[PJ/yr]	1	18	270	272	736	870
Shipping passenger fuel: renewables share	[%]	0%	2%	33%	33%	87%	100%
Shipping passenger electricity: fossil	[PJ/yr]	0	0	0	0	0	0
Shipping passenger electricity: renewables	[PJ/yr]	0	0	0	0	0	0
Shipping passenger electricity: renewables share	[%]	0%	0%	0%	0%	0%	0%

Under the assumption that new ICE passenger cars and buses will not be produced after 2030, BEVs will dominate the passenger vehicle fleet of 2050 under the 1.5 °C scenario. OECD countries and China are assumed to lead the development of BEVs and therefore to have the highest shares, whereas Africa and Latin America are expected to have the lowest BEV shares. Fuel cell-powered passenger vehicles are projected to play a significantly smaller role than BEVs and will only be used for larger vehicles, such as SUVs and buses (Fig. 8.9).

The shares of electric trains and diesel-powered locomotives vary significantly by region (Fig. 8.10). Under the 1.5 °C scenario, all diesel locomotives will be phased out in all regions by 2050. It is assumed that biofuels and synthetic fuels, as well as hydrogen, will play a minor role and that around 90% of all trains—for both passenger and freight transport—will use electric locomotives. The highest utilisation rates of diesel locomotives in 2019 were in the Middle East (98%) and OECD North America (95%), whereas the majority of trains in Europe were electrified.

Highly efficient drives—with a focus on electric mobility—supplied with renewables will result in large efficiency gains. By 2030, electricity will provide 5% of the transport sector's total energy demand under the 1.5 °C scenario, whereas in 2050, the share will be 37%. The majority of electricity consumed in the transport sector

Table 8.14 Road transport—energy demand and supply

Parameter	Unit	2019	2025	2030	2035	2040	2050
					Projection		
Road freight: energy intensity	[MJ/tkm]	1.33	1.17	1.11	0.86	0.79	0.71
Road passenger: energy intensity	[MJ/pkm]	1.47	1.17	1.07	0.73	0.65	0.58
Road freight: energy demand	[PJ/yr]	38,598	28,937	26,027	19,137	16,736	11,058
Road passenger: energy demand	[PJ/yr]	53,302	50,113	39,315	23,445	19,000	13,787
Road freight fuel: fossil	[PJ/yr]	36,898	26,621	23,513	8670	3787	0
Road freight fuel: renewable, electric, and synthetic fuels	[PJ/yr]	1700	2317	2514	10,467	12,949	11,058
Road freight fuel: renewables share	[%]	4.4%	8.0%	9.7%	54.7%	77.4%	100.0%
Road freight electricity: fossil	[PJ/yr]	77	260	166	629	326	0
Road freight electricity: renewables	[PJ/yr]	25	282	478	4883	6081	5928
Road freight electricity share	[%]	0.3%	1.9%	2.5%	28.8%	38.3%	53.6%
Road passenger fuel: fossil	[PJ/yr]	50,954	46,485	34,491	8481	4043	0
Road passenger fuel: renewable, electric, and synthetic fuels	[PJ/yr]	2348	3628	4825	14,963	14,957	13,787
Road passenger fuel: renewables share	[%]	4.4%	7.2%	12.3%	63.8%	78.7%	100.0%
Road passenger electricity: fossil	[PJ/yr]	119	783	1154	8168	8148	6783
Road passenger electricity: renewables	[PJ/yr]	22	88	98	512	477	338
Road passenger electricity share	[%]	0.3%	1.7%	3.2%	37.0%	45.4%	51.6%

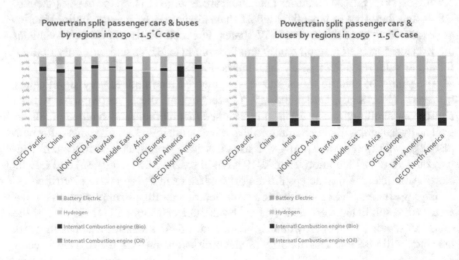

Fig. 8.9 Proportions of powertrains in (fleet) passenger cars and buses by region in 2030 (left) and 2050 (right)

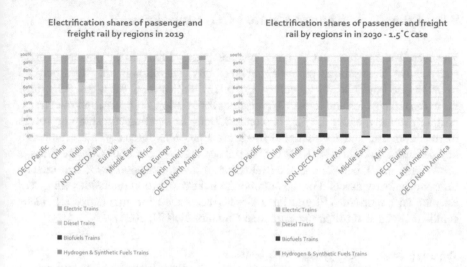

Fig. 8.10 Proportions of electrified passenger and freight rail in 2019 (left) and 2030 (right)—1.5 °C scenario

Table 8.15 Transport sector—final energy demand and supply

Parameter	Units	2019	2025	2030	2035	2040	2050
				Projections			
Total (including pipelines)	**[PJ/yr]**	**104,541**	**102,279**	**88,392**	**64,912**	**57,213**	**45,566**
Fossil fuels	[PJ/yr]	97,777	94,337	74,080	28,304	10,039	0
Biofuels (including biogas)	[PJ/yr]	3810	4499	7939	11,787	17,135	15,299
Synthetic fuels	[PJ/yr]	6	0	296	1310	4023	5517
Natural gas	[PJ/yr]	2149	851	710	629	284	0
Hydrogen	[PJ/yr]	0	541	1877	6534	7837	7700
Electricity	[PJ/yr]	0	2901	4199	16,977	18,179	17,050

will be for land transport—road and rail. Hydrogen and other synthetic fuels generated with renewable electricity will be complementary options to further increase the share of renewable energy in the transport sector, especially for aviation and shipping. In 2050, up to 7700 PJ/yr of hydrogen will be required under the 1.5 °C transport pathway (Table 8.15).

The high reliance on renewable electricity, used either directly in BEVs or to produce synthetic fuels, will require close cooperation between the transport sector and the power sector, not only in terms of the decarbonisation of the power sector itself but also in terms of the increasing electricity demand. In our analysis, the electrification of the transport sector—especially the replacement of ICEs with BEVs—will roughly double the electricity demand of an industrialised country if no further efficiency measures are taken in other sectors, such as the residential and service sectors.

8.9 Transport: Energy-Related CO_2 Emissions

The overall energy-related CO_2 emissions are directly linked to the power sector, as stated above. Under the assumption that electricity generation is fully decarbonised by 2050 (see Power sector trajectory, Chap. 12), Tables 8.16, 8.17, and 8.18 show the carbon intensities and total CO_2 emissions for aviation, shipping, and road transport, respectively, under the 1.5 °C scenario. Both the aviation and shipping values include domestic and international transport. Emissions intensity is an important key performance indicator (KPI) for the finance industry, for both *Climate Change Stress Tests* (see Chap. 2) and the evaluation of investment portfolios that include transport industry assets. For the automobile industry, carbon intensities (in gCO_2/km) are an important KPI and have already been used for mandatory efficiency standards, such as those in the European Community (EU, 2021).

Table 8.16 Aviation—energy-related CO_2 emissions

		2019	2025	2030	2035	2040	2050
Air freight: emission intensity	[g CO_2/tkm]	2360	2092	1822	776	189	0
Air freight: total emission	[million. t CO_2/yr]	94	144	119	46	10	0
Air passenger: emission intensity	[g CO_2/pkm]	426	347	302	129	31	0
Air passenger: total emission (domestic)	[million. t CO_2/yr]	842	1295	1074	413	86	0
Aviation: total	[million. t CO_2/yr]	936	1439	1193	459	96	0

Table 8.17 Shipping—energy-related CO_2 emissions

		2019	2025	2030	2035	2040	2050
Shipping freight: emission intensity	[g CO_2/tkm]	15	15	10	10	2	0
Shipping freight: total emission	[million. t CO_2/yr]	738	3390	2420	2441	475	0
Shipping passenger: emission intensity	[g CO_2/pkm]	5	4	3	3	1	0
Shipping passenger: total emission	[million. t CO_2/yr]	56	255	182	184	36	0
Shipping: total	[million. t CO_2/yr]	794	3645	2602	2625	511	0

Table 8.18 Road transport—energy-related CO_2 emissions

		2019	2025	2030	2035	2040	2050
Road freight: emission intensity	[g CO_2/tkm]	105	90	80	26	14	0
Road freight: total emission	[million. t CO_2/yr]	3034	2189	1933	713	311	0
Road passenger: emission intensity	[g CO_2/pkm]	121	95	85	38	29	23
Road passenger: total emission	[million. t CO_2/yr]	4190	3822	2836	697	332	0
Road transport: total	[million. t CO_2/yr]	7223	6011	4769	1410	644	0

8.10   Transport Equipment

According to the OECD definition, 'Transport equipment (assets) consists of equipment for moving people and objects, other than any such equipment acquired by households for final consumption'(OECD SP, 2021). According to the 2020 edition of the International Energy Agency's World Energy Balances Database Documentation (IEA, 2020b), the energy demand for 'transport equipment' includes industries under Divisions 29 and 30 of the *International Standard Industrial Classification of All Economic* (ISIC) Rev. 4 (ISIC, 2008). Table 8.19 shows the industries that are classified under 'transport equipment'. Based on this classification, the economic values for all sub-sectors were estimated.

Table 8.20 shows the estimated economic breakdown of all sub-sectors of the transport equipment industries. The literature provides various different definitions and economic values for the global automotive industries and for the aviation and shipping industries. However, some of the much higher values (e.g. for the car industry) include the value added for sales and other related services.

Table 8.21 shows the calculated global values for all sub-sectors of the transport equipment industry. The purpose of this analysis is to estimate the energy demand for the manufacture of vehicles, ships, and planes, because the exact statistics for the energy demands of those industries are not available on the global level. To maintain consistency in our methodology, the energy demand for transport equipment provided by the IEA database was used. However, further research is required to determine the industries' exact energy demands.

Table 8.19   Industries classified under 'transport equipment'

Division 29:	Manufacture of motor vehicles, trailers, and semi-trailers
291 2910	Manufacture of motor vehicles
292 2920	Manufacture of bodies (coachwork) for motor vehicles; manufacture of trailers and semi-trailers
293 2930	Manufacture of parts and accessories for motor vehicles
Division 30:	Manufacture of other transport equipment
301	Building of ships and boats
3011	Building of ships and floating structures
3012	Building of pleasure and sporting boats
302 3020	Manufacture of railway locomotives and rolling stock
Detailed structure 51:	Division group class description
303 3030	Manufacture of air and spacecraft and related machinery
304 3040	Manufacture of military fighting vehicles
309	Manufacture of transport equipment n.e.c.
3091	Manufacture of motorcycles
3092	Manufacture of bicycles and invalid carriages
3099	Manufacture of other transport equipment n.e.c.

Table 8.20 Global transport equipment—global GDP shares

Parameter	Units	2019	2025	2030	2035	2040	2045	2050
Global GDP	[bn $GDP]	129,555	142,592	196,715	231,758	266,801	306,519	346,236
Transport equipment—economic value	[bn $GDP]	1257	1612	1902	2241	2580	2964	3348
Transport equipment—Global GDP share	[%]	0.97%	0.97%	0.97%	0.97%	0.97%	0.97%	0.97%
Road (div. 29): manufacture of motor vehicles (four-wheeler) including trailers—weighted share	[%]	75%	75%	75%	75%	75%	75%	75%
Road (div. 30): manufacture of motorcycles (two-wheeler) and other equipment—weighted share	[%]	2%	2%	2%	2%	2%	2%	2%
Road (div. 30): manufacture of bicycles—weighted share	[%]	1%	1%	1%	1%	1%	1%	1%
Rail (div. 30): manufacture of locomotives—weighted share	[%]	5%	5%	5%	5%	5%	5%	5%
Navigation (Div. 30): manufacture of ships, boats, and floating structures—weighted share	[%]	10%	10%	10%	10%	10%	10%	10%
Aviation (div. 30): manufacture of air and spacecraft and related machinery (including military)—weighted share	[%]	7%	7%	7%	7%	7%	7%	7%
Transport equipment—average efficiency gain	[%/yr]	0.00%	0.50%	0.50%	0.50%	0.50%	2.00%	2.00%

Table 8.21 Global transport equipment—estimated GDP values by sub-sector and projection until 2050

Parameter	Units	2019	2025	2030	2035	2040	2045	2050
Transport equipment—economic value	[bn $GDP]	1257	1612	1902	2241	2580	2964	3348
Road vehicle manufacture—economic value	[bn $GDP]	981	1257	1484	1748	2012	2312	2612
Rail vehicle manufacture—economic value	[bn $GDP]	63	81	95	112	129	148	167
Manufacture of navigation ships, yachts, and floating structures—economic value	[bn $GDP]	126	161	190	224	258	296	335
Manufacture of aviation airplanes and spacecraft (including military)—economic value	[bn $GDP]	88	113	133	157	181	207	234
Transport equipment—sector share (global/total GDP)	[%]	1.2%	1.2%	1.2%	1.2%	1.2%	1.2%	1.2%

Table 8.22 Global transport equipment—estimated energy intensities by sub-sector and projection until 2050

Parameter	Units	2019	2025	2030	2035	2040	2045	2050
Transport equipment industry—energy intensities								
Road vehicle manufacture	[MJ/$GDP]	1.87	1.52	1.51	1.50	1.49	1.46	1.43
Rail—locomotive and rail vehicle manufacture	[MJ/$GDP]	1.87	1.52	1.51	1.50	1.49	1.46	1.43
Navigation: ship and yacht manufacture	[MJ/$GDP]	1.87	1.52	1.51	1.50	1.49	1.46	1.43
Aviation: aircraft manufacture	[MJ/$GDP]	1.87	1.52	1.51	1.50	1.49	1.46	1.43
Transport equipment industry—average energy intensity	[MJ/$GDP]	1.87	1.52	1.51	1.50	1.49	1.46	1.43

In the absence of more-detailed information about the energy intensity of the industries analysed, the same values have been assumed for the manufacture of cars, locomotives, ships, and planes. Consistent with this assumption, the same efficiency progress ratio of 0.5% per year has been assumed over the entire scenario period until 2050. More research is required to estimate the energy demand and supply for these industries in the future (Table 8.22).

Based on IEA statistics, the share of electricity in the total energy demand has been calculated as 47%, whereas the remaining 53% is required for heat. The breakdown by temperature level has been estimated as 72% for low-temperature heat (<100 °C) and 10% for medium-temperature heat (100–500 °C), and the remaining demand is for process heat (5% for 500–1000 °C; 13% for >1000 °C). More-detailed assessments of the process heat requirements were not available for this analysis (Table 8.23).

Finally, the calculated energy-related CO_2 emissions for transport equipment are shown in Table 8.24. The emissions are based on the 1.5 °C pathways for electricity

Table 8.23 Global transport equipment—calculated energy demand by sub-sector

Parameter	Units	2019	2025	2030	2035	2040	2045	2050
Energy demand—transport equipment	[PJ/yr]	2351	2442	2869	3363	3852	4337	4801
Energy demand—road vehicle manufacture	[PJ/yr]	1834	1905	2238	2623	3004	3383	3745
Energy demand—locomotive and rail vehicle manufacture	[PJ/yr]	118	122	143	168	193	217	240
Energy demand—ship and yacht manufacture	[PJ/yr]	235	244	287	336	385	434	480
Energy demand—aeroplane manufacture	[PJ/yr]	165	171	201	235	270	304	336
Electricity demand: transport equipment industries	[PJ/yr]	1123	1038	1219	1429	1637	1843	2041
	[TWh/yr]	312	288	339	397	455	512	567
Heat demand	[PJ/yr]	1228	1404	1649	1933	2215	2493	2760
Heat shares:	[%]	52%	57%	57%	57%	57%	57%	57%
Heat demand: < 100 °C	[PJ/yr]	884	1011	1187	1391	1594	1794	1986
Heat demand: 100–500 °C	[PJ/yr]	121	139	163	191	219	246	272
Heat demand: 500–1000 °C	[PJ/yr]	58	66	78	91	105	118	130
Heat demand: > 1000 °C	[PJ/yr]	165	189	222	260	298	335	371

Table 8.24 Global transport equipment—calculated energy-related CO_2 emissions

Parameter	Units	2019	2025	2030	2035	2040	2045	2050
CO_2 emissions								
CO_2 emissions—road vehicle manufacture	[million tCO_2/yr]	183	111	70	44	25	12	0
CO_2 emissions—locomotive and rail vehicle manufacture	[million tCO_2/yr]	12	7	4	3	2	1	0
CO_2 emissions—ship and yacht manufacture	[million tCO_2/yr]	23	14	9	6	3	2	0
CO_2 emissions—aeroplane manufacture	[million tCO_2/yr]	16	10	6	4	2	1	0
Total CO_2 emissions	**[million tCO_2/yr]**	**234.2**	**142.8**	**89.3**	**56.1**	**32.0**	**15.5**	**0.0**

and (process) heat generation (see Chap. 12). The values shown here were used for the Scope 1, 2, and 3 analyses reported in Chap. 12 (Results: industry pathways) and Chap. 13 (Scope 3: industry emissions and future pathways).

References

Agence France-Presse. (2018). Norway aims for all short-haul flights to be 100% electric by 2040 | Air transport | *The Guardian*. https://www.theguardian.com/world/2018/jan/18/norway-aims-for-all-short-haul-flights-to-be-100-electric-by-2040. Accessed 22 Dec 2021.

BloombergNEF. (2020). *Oil demand from road transport: Covid-19 and beyond* | BloombergNEF. https://about.bnef.com/blog/oil-demand-from-road-transport-covid-19-and-beyond/. Accessed 21 Dec 2021.

C40 Cities Climate Leadership Group, International Transport Workers' Federation. (2021). *The future is public transport*. https://www.c40knowledgehub.org/s/article/The-Future-is-Public-Transport?language=en_US. Accessed 21 Dec 2021.

Cruise Market Watch. (2020). *Capacity | Cruise market watch*. https://cruisemarketwatch.com/capacity/. Accessed 22 Dec 2021.

DNV. (2015). *In focus—The future is hybrid by DNV AS—Issuu*. https://issuu.com/dnvgl/docs/in_focus-the_future_is_hybrid. Accessed 22 Dec 2021.

Downing, S. (2019). *6 electric aviation companies to watch*. GreenBiz. https://www.greenbiz.com/article/6-electric-aviation-companies-watch. Accessed 22 Dec 2021.

EU. (2021). *Regulation (EU) 2019/631 of the European Parliament and of the Council of 17 April 2019 setting CO_2 emission performance standards for new passenger cars and for new light commercial vehicles, and repealing Regulations (EC) No 443/2009 and (EU) No 510/201.*

IEA. (2020a). *World energy outlook 2020.*

IEA. (2020b). *World energy balances 2020.*

IEA. (2020c). *Global energy review 2020.*

IEA. (2020d). *Global EV outlook 2020.*

ISIC. (2008). *International Standard Industrial Classification (ISIC) of all economic activities revision 4.* United Nations.

Jurich, K. (2016). *CO_2 emission factors for fossil fuels.*

Madavan, N. (2016). *A NASA perspective on electric propulsion technologies for large commercial aircraft.*

OECD. (2021). *ITF transport outlook 2021.* OECD.

OECD SP. (2021). *The OECD glossary of statistical terms*. https://stats.oecd.org/glossary/. Accessed 22 Dec 2021.

Pagenkopf, J., van den Adel, B., Deniz, Ö., & Schmid, S. (2019). Transport transition concepts. In S. Teske (Ed.), *Achieving the Paris climate agreement goals: Global and regional 100% renewable energy scenarios with non-energy GHG pathways for +1.5°C and +2°C* (pp. 131–159). Springer International Publishing.

Pearce, B. (2020). *Economics updated impact assessment.*

Placek, M. (2021). *Container shipping—Statistics & facts*. Statista. https://www.statista.com/topics/1367/container-shipping/. Accessed 22 Dec 2021.

Rogers, E. M. (2003). *Diffusion of innovations* (5th ed.). Free Press.

Statista. (2021). *Global merchant fleet—Number of ships by type*. https://www.statista.com/statistics/264024/number-of-merchant-ships-worldwide-by-type/. Accessed 22 Dec 2021.

Teske, S., Niklas, S., & Langdon, R. (2021). *TUMI transport outlook 1.5C—A global scenario to decarbonise transport.*

Teske, S., Pregger, T., Simon, S., et al. (2019). *Achieving the Paris climate agreement goals global and regional 100% renewable energy scenarios with non-energy GHG pathways for +1.5°C and +2°C.* Springer International Publishing.

Vyas, A. D., Patel, D. M., & Bertram, K. M. (2013). *Transportation energy futures series: Potential for energy efficiency improvement beyond the light-duty-vehicle sector.* Golden, CO (United States).

Part V
Energy Industry & Engineering

Chapter 9
Renewable Energy for Industry Supply

Sven Teske, Thomas Pregger, Sonja Simon, and Carina Harpprecht ⓘ

Abstract This section focuses on technologies that provide heat, and especially process heat, with renewable energy and electrical systems. All the technologies described, except those that use high-temperature geothermal or concentrated solar heat (CSH) for process heat, are used for the OECM 1.5 °C pathways described in Chaps. 5, 6, 7, and 8. The authors have included geothermal and solar technologies to highlight the further technical options available and to underscore that more research is required in the area of renewable process heat.

Keywords Industry process heat by sector · Renewable process heat · Electric process heat · Solar · Bio energy · Geothermal · Heat pumps · Arc furnace · Hydrogen · Synthetic fuels · Power-to-X)

9.1 Introduction

Heat generation relies currently, to a large extent, on combustion processes. In 2019, 77% of global heating for buildings and industrial process heat came from fossil fuels, whereas only 3.2% was provided by electric heating systems, and 23% was supplied by renewable heating almost entirely from biomass. Only 0.9% derived from solar and geothermal heating systems. To decarbonize the global heat supply is more challenging than to decarbonize the electricity sector, because geographic

S. Teske (✉)
Institute for Sustainable Futures, University of Technology Sydney, Sydney, NSW, Australia
e-mail: sven.teske@uts.edu.au

T. Pregger · S. Simon
Department of Energy Systems Analysis, Institute for Engineering Thermodynamics (TT), German Aerospace Center (DLR), Stuttgart, Germany
e-mail: thomas.pregger@dlr.de; sonja.simon@dlr.de

C. Harpprecht
Institute of Networked Energy Systems, German Aerospace Center (DLR), Stuttgart, Germany

limitations make it difficult to provide high-temperature heat with direct solar or geothermal energy due to their dependency on locally available resources. However, the use of renewable electricity for heating is key to a successful 1.5 °C pathway. This section provides a short overview of the suitable technologies available and the temperature levels that these technologies can generate.

Industry involves a large variety of processes that demand heat. These requirements range, for example, from 40 °C to around 300 °C in the food industry to metal production with furnaces well above 800 °C and cement production with dry kilns at around 1500 °C. Figure 9.1 shows that the metal, chemical, and mineral industries require particularly large amounts of high-temperature process heat.

Decarbonizing process heat for energy-intensive industries, such as the steel, aluminium, cement, and chemical industries, is a major prerequisite to remaining within a 1.5 °C increase in the global temperature. Three main groups of technologies can provide renewable process heat at different temperatures:

1. *Direct heat systems*: geothermal and concentrated solar systems
2. *Electric heat systems*: heat pumps, electromagnetic, di-electric, infrared, induction, resistance, and arc furnace heating
3. *Fuel-based heat systems*: that use bio-energy, hydrogen, and other synthetic fuels

The energy sources for these heat generation technologies are either biomass, geothermal energy, solar energy, or electricity, used either directly or as fuels produced with electricity, such as hydrogen and other synthetic fuels. Whereas the most efficient transformation to renewable energy is the direct application of renewable heat, many industrial processes require higher temperatures or fuels, which cannot be provided directly by renewables. Therefore, as the next best option in terms of efficiency, the direct electrification of processes is preferable. However, some processes

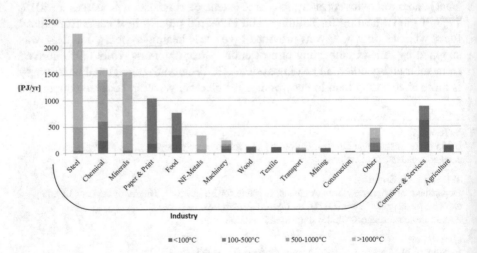

Fig. 9.1 Distribution of process heat demand across all branches of industry in Europe. (Naegler et al., 2015)

will still rely on fuel input in the future. In this case, power-based synthetic fuels will be required, with increasing efficiency losses along the chain from hydrogen to synthetic gas to synthetic liquid fuels. To comply with the 1.5 °C carbon budget, all the electricity used for heat or fuel production must be produced from renewable energy.

In Table 9.1, we compare the technology options used in the OECM for generating different levels of heat. Their descriptions are provided in the following subchapters. It can be seen that the use of fuels is most suitable for high-temperature process heat, although direct electrification is also suitable to some extent. Geothermal energy is particularly suitable for lower temperatures, whereas direct solar energy can only be used to generate high-temperature heat via central receivers. This illustration does not consider the opportunities for or barriers to renewable heat integration that may arise from technical or local structural constraints.

Table 9.2 shows the average breakdown between heat and fuels and electricity in percentages of the industry sectors analysed, which are included in the OECM. The data demonstrate the large share of high-temperature heat in the primary industries. The average energy demand for the steel industry in 2019, for example, was mainly for the generation of process heat (86%), with 14% for electricity. The value for process heat includes fuels.

Table 9.1 Technology options to generate renewable heat by temperature level

	0 °C	50 °C	100 °C	200 °C	300 °C	400 °C	500 °C	600 °C	700 °C	800 °C	900 °C	1000 °C	1100 °C	1200 °C	1300 °C	1400 °C	> 1500 °C
OneEarth Climate Model (OECM) temperature levels used in the 1.5 °C pathway	Low temperature			Medium-temperature process heat				Industrial process heat					High-level process heat				
Bioenergy																	
Fixed-bed boiler																	
Fluidized-bed boiler																	
Gasification																	
Gasification for syngas production																	
Anaerobic digestion																	
Fermentation—ethanol																	
Fermentation—methanol																	
Geothermal																	
Dry steam plants																	
Flash plants																	
Binary cycle																	
Combined cycle																	
Solar thermal																	
Flat-panel solar collectors																	
Vacuum tubes																	
Parabolic trough																	
Linear Fresnel lens system																	
Parabolic dish																	
Central receiver / solar tower																	
Electric heating																	
Residential heat pumps																	
Industry heat pumps																	
Electromagnetic heating																	
Non-thermal electromagnetic																	
Electric resistance heating																	
Electric arc furnace																	

Table 9.2 Average electricity and heat shares by industry in 2019 (heat includes electricity for heating and fuels)

Industry sector	Electricity (non-heat related)	Process heat from fuels and electricity	Shares of required heat levels (Naegler et. al 2015)			
	[%]	[%]	< 100 °C	100–500 °C	500–1000 °C	> 1000 °C
Iron and steel	14%	86%	5%	2%	19%	75%
Chemicals and petrochemicals	25%	75%	18%	22%	48%	12%
Non-ferrous metals	52%	48%	10%	4%	20%	66%
Aluminium	60%	40%	8%	2%	18%	72%
Non-metallic minerals	17%	83%	5%	2%	30%	63%
Cement	19%	81%	5%	2%	30%	63%
Transport equipment	47%	53%	72%	10%	5%	13%
Machinery	34%	66%	57%	15%	9%	20%
Mining and quarrying	41%	59%	13%	2%	28%	57%
Food and tobacco	30%	70%	54%	46%	0%	0%
Paper, pulp, and print	32%	68%	20%	80%	0%	0%
Wood and wood products	29%	71%	37%	63%	0%	0%
Construction	35%	65%	48%	18%	11%	23%
Textiles and leather	42%	58%	100%	0%	0%	0%
Unspecified (industry)	40%	60%	43%	19%	12%	25%

However, whereas the share of the actual heat demand will remain stable for each sector in the OECM until the end of the scenario period in 2050, the electricity used to produce heat will increase. A more detailed bottom-up analysis broken down into primary and secondary steel and aluminium and new manufacturing processes has been undertaken. The assumptions for the process heat calculation for each industry sector are presented in Chap. 5.

In the following section, these different technologies are outlined, and their respective areas of application are explained.

9.2 Direct Renewable Process Heat

9.2.1 Bioenergy and Biofuels

'Biomass' is a broad term used to describe materials of recent biological origin that can be used as a source of energy. It includes wood, crops, algae, and other plants, as well as agricultural and forest residues (Teske & Pregger, 2015). Biomass is used to generate electricity, heat, and fuels. The following section focuses on heat generation.

The majority—around 90%—of bioenergy is used in direct combustion processes to generate heat and/or electricity, mostly for domestic and low-temperature applications. However, many studies and scenarios that have considered the potential of biomass have envisaged a shift in its currently limited potential to allow the generation of high-temperature industrial process heat, in the transition towards a renewable energy system (Lenz et al., 2020).

In principle, two biomass conversion routes are available for the production of heat for industry, using several biomass technologies:

(a) Thermochemical processes:

- Direct combustion
- Gasification
- Pyrolysis

(b) Biochemical conversion processes:

- Anaerobic digestion
- Fermentation

9.2.1.1   Thermochemical Processes

Direct Combustion

The direct combustion technologies relevant to the generation of process heat can be differentiated according to the state in which the biomass is fed into and burned in the process.

In fixed-bed combustion applications, the air is first passed through a fixed bed for drying, gasification, and charcoal combustion. In the second step, the combustible gases produced are burned with air, usually in a zone separated from the fuel bed. Fixed-bed combustion is adaptable to a variety of fuels such as wood, straw chips, and pellets. Therefore, the range of capacities is large, ranging from 10 kW to 60 MW.

The fluidized-bed technology involves the combustion of particulate solid fuel in an inert material bed (usually sand), which is fluidized by the flow of a gas. This type of flow allows efficient gas–solid contact, so it is widely used in covering particles, drying, granulation, blending, combustion, and gasification processes (Philippsen et al., 2015). This technology provides almost complete combustion, with very stable temperatures and low emissions. The prerequisites are fuels with particle sizes <100 mm and ash melting temperatures >1000 °C (Kaltschmitt et al., 2009). Entrained-flow combustion is suitable for fuels that are available as small particles, such as sawdust or fine shavings, which are pneumatically injected into the furnace. Fluidized-bed combustion is generally used in larger systems (> 20 MW), because it is expensive (Teske & Pregger, 2015; ARENA, 2019).

Gasification

Biomass gasification is a method for upgrading solid biomass and is especially valuable in processing biomass of low caloric value or moist biomass, e.g. many residues. The partial oxidation of the biomass fuel provides a combustible gas mixture mainly consisting of carbon monoxide (CO). Gasification provides a homogeneous fuel and controlled combustion, which can increase the efficiency along the whole biomass chain, although at the expense of additional investments in the more sophisticated technology, or the efficient use of low-quality biomass. During the first step, the volatile components of the fuel are vaporized in a complex set of reactions at temperatures <600 °C. Gasification is an intermediate step between pyrolysis and combustion. It is a two-step, endothermic process (IEA BioEnergy Agreement Task 33, 2020).

Biomass gasification is increasingly used to generate high temperature levels. The most commonly available gasifiers use wood or woody biomass, whereas especially designed gasifiers can convert non-woody biomass materials (Norfadhilah et al., 2017). Gasification is more efficient than combustion, providing better-controlled heating, higher efficiencies in power production, and the possibility for co-producing chemicals and fuels (Kirkels & Verbong, 2011). Gasification can also reduce emission levels better than power production with direct combustion and a steam cycle. Finally, gasification can also be the first step in the production of synthetic fuels (Malico et al., 2019) (see next section).

Pyrolysis

Pyrolysis is a technology that 'upgrades' biomass, providing products of high caloric value for combustion. It has been long used in the production of charcoal (Malico et al., 2019). Technically, thermal decomposition occurs in the absence of oxygen. It is also always the first step in combustion and gasification processes, where it is followed by the total or partial oxidation of the primary products (IEA BioEnergy Agreement Task 34, 2021). Pyrolysis produces a solid (charcoal), liquid (pyrolysis oil or bio-oil), and gas product. The relative amounts of the three products are determined by the operating temperature and the residence time used in the process. Lower temperatures produce more solid and liquid products, and higher temperatures, more biogas. All the products are then available for the production of industrial process heat.

9.2.1.2 Biochemical Conversion Processes

Anaerobic digestion and fermentation are the two main biochemical processes that provide energy from biomass with high moisture content, such as food waste or agricultural residues, including liquid manure.

Anaerobic Digestion

In a biogas plant, organic waste is broken down by bacteria in an oxygen-free (= anaerobic) environment in about two-thirds methane (CH_4) and one-third CO_2. This gas is used either directly in power, heating, or cogeneration plants, or purified gas is fed into renewable gas pipelines. For its direct injection into natural gas pipelines, the CH_4 content must be increased to approximately 95% (Wall et al., 2018). The quality of the renewable gas produced depends on the energy content of the

feedstock. Possible feedstocks include food waste, livestock manure, process effluent, sewage sludge, and domestic biowaste.

Alcoholic Fermentation

The alcoholic fermentation of sugar and starch is a 'state-of-the-art' technology.
Plants with high sugar and starch contents, such as sugar cane, are broken down into
ethanol and methanol by microorganisms. Because the use of sugar and starch
plants for this purpose is in direct competition with human nutrition, one direction
of research focuses on the fermentation of lignocellulose, e.g. from straw or grass.
Although lignocellulosic processes are more complex than the fermentation of car-
bohydrates, the first production plants have been developed in Germany (DBFZ,
2015). The products can be used as combustible fuels for power, heat, or cogeneration plants and as a vehicle fuel. However, in the future, these products will become
more important as low-emission feedstocks in a circular economy, with increased
competition for the limited biomass potential (Table 9.3).

9.2.1.3   Bioenergy and Reduction of Greenhouse Gas (GHG) Emissions

Bioenergy is not necessarily carbon neutral. Depending on the feedstock, which can
be agricultural or forestry waste, other biogenic residues, or energy crops, bioenergy production has different upstream burdens in terms of the consumption of
materials and energy, land-use changes, and emissions that have a significant impact
on GHG emissions. Given the environmental effects of the production of energy
crops, the global use of biomass in the 1.5 °C pathway is limited to 100 EJ per year,

Table 9.3 Bioenergy for process heat—overview

Process	Technology	Heat level	Remarks
Thermochemical	Fixed-bed boiler	800–1000 °C	
	Fluidized-bed boiler	750–850 °C	
	Gasification	750–900 °C	
	Gasification for syngas production	400–900 °C	Low H_2 content
		1200–1700 °C	High H_2 content
Biochemical systems	Anaerobic digestion	550–900 °C	High H_2 content increases temperature level
	Ethanol/methanol production via fermentation	–	Combustion temperature depends on application

Sources: IEA, ARENA, UTS/ISF, and DLR (own research)

which is the estimated threshold of carbon-neutral sustainable biomass based on residuals and organic waste (Thrän et al., 2011).

9.2.2 *Geothermal*

Geothermal resources consist of thermal energy from the Earth's interior stored in both rocks and trapped steam or liquid water (IPCC-SRREN CH_4, 2011). Although geothermal resources are available in all countries, their utilization is concentrated in regions where geothermal heat is available close to the Earth's surface. Geothermal 'hotspots' with high temperature levels occur in the western part of the USA, west and central Eastern Europe, Turkey, Iceland, and 'the ring of fire' around the Pacific, from Japan, the Philippines, South-East Asia, and Indonesia to New Zealand.

Geothermal energy resources are classified by temperature level (Huddlestone-Holmes, 2014). Each temperature level involves different technologies and applications. The global average thermal gradient is around 25–30 °C per km depth (Beardsmore & Cull, 2001), which results in an average crustal temperature of around 150 °C at a depth of 5000 m. Higher temperatures can be achieved by drilling deeper or by focusing on areas with favourable conditions. In such areas, the following temperatures are usually possible:

- Low temperature (<90 °C)—direct heat used near the surface and from boreholes drilled to <2000 m depth
- Medium temperature (90–150 °C)—direct heat used near the surface and from boreholes usually drilled to >2000 m depth
- High temperature (<150–250 °C)—from boreholes drilled to depths up to 5000 m

To date, high-temperature geothermal systems are almost exclusively used for power generation. However, high-temperature geothermal systems, around 200 °C, can also be utilized to provide direct process heat (ARENA, 2019).

Geothermal systems predominantly provide low-temperature process heat, which can be used, for example, in the food-processing industry (see Fig. 9.1).

In high-temperature hydrothermal reservoirs, water occurs naturally underground in its liquid form under pressure. As it is extracted, the pressure drops and the water is converted to steam. The residual salty water is sent back to the reservoir through injection wells, sometimes via another system that uses the remaining heat (Teske & Pregger, 2015). The hot water produced from intermediate-temperature hydrothermal or enhanced geothermal system (EGS) reservoirs can be used in heat exchangers, to generate power in a binary cycle, or directly in heat applications. The recovered fluids are also injected back into the reservoir (Younger, 2015).

The key technologies for EGS are:

- *Exploration and drilling* involve the localization and analysis of geothermal resources, including the depth required and the dimensions for drilling. The maximum depth is currently around 5 km, using methods similar to those of the oil and gas industry. Exploration and drilling are among the most cost-intensive parts of a geothermal project and include technical risks (IRENA-Geo, 2017).
- *Reservoir engineering* focuses on determining the volume of the geothermal resource and the optimal plant size. Ideally, the sustainable use of the resources and the safety and efficiency of the operation are considered.
- *Geothermal power and heat plants* use the steam created from heating water by natural underground sources to power the turbines that produce electricity and/or process heat. Three main technologies are used:
    - Dry steam plants
    - Flash plants (single, double, or triple)
    - Binary combined-cycle plants or hybrid plants
- Dry steam plants operating at sites with intermediate- or high-temperature resources ($\geq$150 °C), with unit sizes ranges between 20 $MW_{electric}$ and 110 $MW_{electric}$ (IRENA-Geo, 2017).
- Flash plants—similar to dry steam plants but the steam is obtained from a separation process, called 'flashing'; this is currently the commonest type of operational geothermal electricity plant (IRENA-Geo, 2017).
- Binary-cycle plants operate with low- and medium-temperature heat (100–170 °C) and heat exchangers that transfer the heat into a closed loop (IEA Geo, 2011). For process heat, fluid ammonia/water mixtures are used in Kalina cycles or hydrocarbons in organic Rankine cycles,[1] which have boiling and condensation points that must match the geothermal resource temperature (IRENA-Geo, 2017). The typical plant sizes are in the 10–50 MW range.
- Combined-cycle/hybrid plants use two different heat cycles—one for heat and one for power generation. This increases the overall electric efficiency. Other heat, such as that from solar thermal power plants, can feed into the heat cycle to increase the temperature and output.

9.2.3 Concentrated Solar

Concentrating solar technologies generate high-temperature heat that can be used for industry processes (CSH) or to produce electricity via steam turbines (concentrated solar power—CSP).

---

[1] A Rankine cycle power system is a heat engine that converts thermal energy into work. Similar to the vapor compression heat pump, it comprises four main components: a boiler (sometimes called an 'evaporator'), a turbine, a condenser, and a pump (Fig. 9.2). The working fluid, in a low-pressure slightly subcooled liquid state, is brought to high pressure by the pump. The pump consumes power (ARENA2019).

Like high-temperature geothermal plants, concentrating solar technologies are currently predominantly utilized to generate power. However, the process heat temperature required for direct use in industrial processes (to 400 °C) is technically possible (DLR-ISR, 2021).

Direct normal irradiation—sunlight not dispersed by clouds, fumes, or dust in the atmosphere—is concentrated by mirrors to a single point or line to heat a liquid, solid, or gas to a temperature between 400 °C and >> 1000 °C, depending on the technology used. Concentrating solar plants require direct sunlight, which limits the areas of application to regions with more than 2000 h of direct sunlight per year.

There are several different CSP/CSH system types, but all require four main elements: a concentrator, a receiver, some form of transfer medium or storage, and a power conversion system or a connection that directs process heat to the site of its applications. An overview of the commonest concentrating solar systems is given by Pitz-Paal (2016).

Parabolic trough plants use rows of parabolic trough collectors, each of which reflects solar radiation into an absorber tube. The troughs track the sun around one axis, which is typically oriented north–south. Synthetic oil circulates through the tubes and is heated to approximately 400 °C. The hot oil from numerous rows of troughs is passed through a heat exchanger. The direct evaporation of water in the parabolic troughs, which has been developed to operational maturity for years, will allow the realization of decentralized plants with relatively small solar fields, because heat exchangers and (possibly) toxic synthetic heat transfer fluids will no longer be required. Increasingly, CSP plants use thermal storage systems, such as molten salt, to store high-temperate heat (up to 400 °C) to allow their operation without sunlight or at night. The land requirements are around 2 km^2 for a 100 MW plant, depending on the collector technology and assuming that no storage is available.

Linear Fresnel systems use a series of long, narrow, flat Fresnel mirrors instead of a parabolic trough to concentrate solar radiation to a linear absorber positioned above the lenses. All the other parts of the system correspond to those of parabolic trough plants.

Central receivers or solar towers focus solar radiation to a single point and achieve higher temperatures than parabolic troughs or Fresnel lenses. This technology uses a circular array of mirrors (heliostats) in which each mirror tracks the sun, reflecting the light onto a fixed receiver on top of a tower. Temperatures exceeding 1000 °C can be achieved. A heat-transfer medium absorbs the highly concentrated radiation reflected by the heliostats and converts it into thermal energy to be used for the subsequent generation of super-heated steam for turbine operation or as industrial process heat. The heat transfer medium is currently either water/steam, molten salts, liquid sodium, or air and possibly also pressurized gas or air at very high temperatures. The unit sizes range from 20 to 200 MW.

Parabolic dishes use a shaped reflector to concentrate sunlight onto a receiver located at their focal points. The receiver moves with the dish. The concentrated beam radiation is absorbed into the receiver to heat a fluid or gas to approximately 750 °C. This is then used to generate electricity via Stirling engines or a micro-turbine attached to the receiver. Dishes have been used to power Stirling engines up to 900 °C and also to generate steam. The largest solar dishes have a 485 m^2 aperture and are in research facilities or demonstration plants. Individual unit sizes are in the double-digit kilowatt range and can be combined in modular systems to form utility-scale plants. The generation of process heat is possible but is not yet commercially available.

Concentrated solar heat (CSH) system applications: In addition to its use in different types of solar reflector systems, a CSH system can be used to directly feed into industrial processes, or to desalinate water. The significant cost reduction with solar photovoltaic systems has led to a focus on the application of concentrated solar for to heat generation rather than to the generation of electricity.

Thermal storage, when integrated into a system, is an additional and increasingly important asset in concentrated solar plants, providing heat outside the hours of sunshine and even during the night. Additional concentrator area can be added to produce heat for storage purposes, increasing the capacity factor. There are three categories of storage medium that can be used in CSP plants (Pitz-Paal, 2020):

- *Advanced sensible heat-storage* systems use two tanks with molten salts at different temperatures. These temperatures are high, at around 300–600 °C. Other materials, such as ceramic particles, have also been used and evaluated in research projects as sensitive heat-storage materials.
- *Latent heat-storage* systems transfer heat as a phase change, which occurs in a specific narrow temperature range in the relevant material. The phase-change materials most frequently used for this purpose are molten salt, paraffin wax, and water/ice (Jouhara et al., 2020).
- *Thermochemical energy storage* is achieved via a reversible chemical reaction, resulting in the highest energy density of all thermal storage options, but with a reaction efficiency that decreases with time. For example, different thermochemically active redox materials can be used for the thermochemical storage of CSP (Buck et al., 2021).

The storage capacity currently installed is, on average, around 8 full-load hours.

Concentrated solar power plants have been developed to generate electricity, but the technology has significant potential to provide high-temperature process heat in sunny regions, such as Australia, Chile, North Africa and the Sahara, parts of Central Asia, India, and China, as well as the Middle East. Research is targeting CSP as a source of high-temperature process heat that can directly feed reactors for endothermic chemical reactions. Currently, solar metal-oxide redox cycles and sulphur cycle processes have been developed that rely on temperatures of 1000–1500 °C (Roeb et al., 2020). The first applications of this technology, for hydrogen production, have achieved technology readiness levels of 5–6, for example, in the SUN-to-LIQUID project (Koepf et al., 2019). Newer projects go beyond hydrogen and integrate the direct air capture of CO_2 for the production of chemical feedstocks, such as methanol (Prats-Salvado et al., 2021).

9.3 Electric Process Heat

9.3.1 Heat Pump Technology

Heat pumps are largely known as electric heating (and cooling) systems that supply buildings with space heat and hot water. However, in general, heat pumps are devices that transfer heat from one medium at a lower temperature to another medium at a higher temperature. Therefore, they allow the efficient recycling of low-temperature heat, such as waste heat.

Heat pumps use a refrigeration cycle to provide heat or cold. They use renewable energy from the ground, water, or air to move heat from a relatively low-temperature reservoir (the 'source') to a temperature level required for a specific thermal application (the 'output'). Heat pumps commonly use two types of refrigeration cycles:

- Compression heat pumps use mechanical energy, most commonly electric motors or combustion engines, to drive the compressor in the unit. Consequently, electricity, gas, or oil is used as auxiliary energy.
- Thermally driven heat pumps use thermal energy to drive the sorption process—either adsorption or absorption—to make ambient heat useful. Different energy sources can be used as auxiliary energy: waste energy, biomass, solar thermal energy, or conventional fuels.

Compression heat pumps are most commonly used, but thermally driven units are considered a promising future technology. The efficiency of a heat pump is described by the coefficient of performance (COP), the ratio between the annual useful heat output and the annual auxiliary energy consumption of the unit. In the residential market, heat pumps work best for relatively warm heat sources and low-temperature applications, such as space heating and sanitary hot water. They are less efficient in providing higher-temperature heat and cannot be used for heat over 90 °C. For industrial applications, different refrigerants can be used to efficiently provide heat of 80–90 °C, so they are only suitable for part of the energy requirements of industry.

Heat pumps are generally distinguished by the heat source they exploit:

- Ground-source heat pumps use the energy stored in the ground at depths from around 100 m up to the surface. They are used for deep borehole heat exchangers (300–3000 m), shallow borehole heat exchangers (50–250 m), and horizontal borehole heat exchangers (a few meters deep).
- Water-source heat pumps are coupled to a (relatively warm) water reservoir typically at around 10 °C, such as wells, ponds, rivers, and the sea.
- Aero-thermal heat pumps use the outside air as a heat source. Because the outside temperatures during the heating period are generally lower than soil or water temperatures, ground-source and water-source heat pumps are typically more efficient than aero-thermal heat pumps.

Heat pumps require additional energy apart from the environmental heat extracted from the heat source, so the environmental benefit of heat pumps depends upon both their efficiency and their emissions associated with the production of working energy. When a heat pump has a low COP and a high share of electricity from coal power plants, for example, the CO_2 emissions relative to the useful heat produced might be higher than for conventional gas condensing boilers. However, efficient heat pumps powered with renewable electricity are emission-free.

Reversible heat pumps can be operated in both heating and cooling modes. When they operate in cooling mode, heat is extracted from, for example, a building, and 'pumped' into either a reservoir or the open environment, without storage. When a reservoir is used, the heat can be reused. Alternatively, renewable cooling can be provided by circulating a cooling fluid through the relatively cool ground before it is distributed in a building's heating/cooling system ('free cooling'). However, in a GHG-emission-free system, this cooling fluid must not be based on hydrofluorocarbons (HFCs) or chlorofluorocarbons (CFCs) but on ammonia, water, or air (ARENA, 2019).

In principle, high-enthalpy geothermal heat can provide the energy required to drive an absorption chiller. However, only a very limited number of geothermal absorption chillers are in operation throughout the world. Heat pumps have become increasingly important in buildings but can also be used for industrial process heat. Industrial heat pumps offer various opportunities for all types of manufacturing processes and operations and use waste process heat as their heat source. They deliver heat at medium temperatures for use in industrial processes, heating, or pre-heating or for space heating and cooling in industry. Heat pumps with operating temperatures below 100 °C are state-of-the-art technologies, and high-temperature industrial heat pumps in the range of 160–200 °C are beginning to enter the market. Essential aspects of the future use of heat pumps are efficient system integration and flexibility via heat storage.

9.3.2   Electric Heating Systems

There are four main technological types of electric heating systems, which use different physical methods. Each of them has different temperature levels and applications.

1. Electromagnetic heating

    (a) Dielectric heating
    (b) Infrared heating
    (c) Induction

2. Non-thermal electromagnetic heating

    (a) Ultraviolet
    (b) Pulsed electric field

Table 9.4 Electromagnetic process heating technologies

Technology	Induction	Radio	Microwave	Infrared	Visible light	Ultraviolet
Frequency	50–500 kHz	10–100 MHz	200–3000 MHz	30–400 THz		1–30 PHz
Maximum temperature	3000 °C	2000 °C	2000 °C	2200 °C		–
Power density (kW/m²)	50,000	100	500	300		100
Efficiency	50–90%	80%	80%	60–90%		
Application	Rapid internal heating of metals	Rapid internal heating of large volumes	Rapid internal heating of large volumes	Very rapid heating of surfaces and thin material		Non-thermal curing of paints and coating

Source: ARENA, 2019

    (c) Microwave[2].

3. Electric resistance heating
4. Electric arc furnaces

Electromagnetic heating systems are used to transfer energy to a target material or process without the need for a heat transfer medium. The main advantage of this technology is that heat can be generated and delivered to the point of need, which makes this an energy-efficient technology (ARENA, 2019).

Non-thermal electrical systems generate heat directly on the target object, and no additional medium is required to transfer the heat (Xiong, 2021). Both technology groups use different frequencies to generate heat (Table 9.4).

9.3.2.1 Electric Resistance Heating

Materials conduct electricity to different degrees. The lower the electrical conductivity of a material, the higher the heat developed within that material. This physical law—*ohmic resistance*—is used in electric resistance heating devices. There are two types of electrical resistance heating:

- Direct resistance: the targeted material is heated by an electricity current.
- Indirect resistance heating: a resistive heating element transfers heat to the target material by radiation and convection.

---

[2] Microwaves can generate significantly higher temperatures over time. Objects continue to heat while microwaves are emitted.

This technology is among the oldest electrical heating systems and has been used for room heat, industrial ovens, furnaces, and kilns for decades. Different configurations of indirect resistance heating are:

- Electric furnaces: use high-temperature heating elements, usually made of silicon carbide (SiC), molybdenum disilicide ($MoSi_2$), or nichrome (NiCr), that can reach temperatures in the range of 1000–2000 °C.
- Electric ovens: the ohmic heating elements mounted in the oven heat the products through convection and radiation, achieving temperatures up to 1000 °C.
- Electric boilers: unit sizes are from kilowatts to megawatts, with possible heat generation to temperatures up to 220 °C.

9.3.2.2 Electric Arc Furnaces

An electric arc occurs when an electric current jump between electrodes. As the current passes through air (or another gas), it produces a plasma discharge, generating heat and light. Lightning is a natural form of electric arc (ARENA, 2019). Electric arc furnaces are predominately used in steel recycling, to melt scrap steel. However, they are also used in other industries that require temperatures up to 1500 °C, such as the processing of copper and other metals.

9.4 Synthetic Fuels and Hydrogen

When the direct use of renewable heat sources (first choice) or electrification (second choice) is not applicable, industrial processes will still rely on the input of fuels based on renewable electricity. For efficiency reasons, hydrogen is the next best choice. However, as a last fuel option, synthetic hydrocarbons can provide the necessary energy.

9.4.1 Hydrogen: The Basics

Hydrogen can be used as a feedstock, a fuel, an energy carrier, and for energy storage and has many possible applications across the industry, transport, energy, and buildings sectors. Molecular hydrogen does not occur in nature but can be produced using any primary source of energy, such as gas, oil, or coal. It can be produced by electrolysis, which requires electricity, or by directly splitting water with a solar high-temperature process. Therefore, hydrogen is not an energy source—it is a secondary energy carrier and an energy storage medium. The combustion of hydrogen gas only generates water and no further GHG emissions are produced.

The chemical formula for this process—the scientific term is 'oxidation'—is

$$2H_2 + O_2 \, ? \, 2H_2O.$$

Today, most of the world's hydrogen is still produced in CO_2-intensive processes: steam–methane reformation (SMR) (gas, approximately 50%), oil product reformation (30%), and coal gasification (18%). In SMR, carbon (CO_2) is separated from hydrogen by the steam reformation of natural gas. This method involves the conversion of hydrocarbons and steam into hydrogen and CO (known as 'syngas').

According to the London-based *Committee on Climate Change (CCC)*, SMR has an emissions factor of around 285 g of CO_2 per kilowatt-hour (kWh) (9.5 kg of CO_2 per kg of hydrogen), and coal gasification has an emission factor of around 675 g of CO_2 per kilowatt of hydrogen, accounting only for energy use and process emissions. Therefore, arguments for the early establishment of an energetic use of hydrogen based on fossil energies are usually combined with arguments for the implementation of carbon capture and storage (CCS) technologies. Counterarguments point to the lock-in effect of investments in high-carbon infrastructure, which comes at the expense of financial resources for the expansion of renewable energies.

If hydrogen is to contribute to climate neutrality, it must achieve a far larger scale and its production must become fully decarbonized. According to the International Energy Agency (IEA), the current production of hydrogen—mainly based on natural gas—is responsible for CO_2 emissions of around 830 million tonnes per year. By comparison, Germany's total CO_2 emissions in 2020 are estimated to have been 722 million tonnes. It is estimated that 6% of global natural gas and 2% of global coal production are used for hydrogen production, whereas only about 0.1% of global dedicated hydrogen production is produced with water electrolysis.

9.4.1.1 Status Quo: Global Demand for Hydrogen in Industry

There are various applications for hydrogen in industry, as shown in Table 9.5, but only two main areas consume most of the global hydrogen produced today: ammonia production and refining processes. About 90% of ammonia is used for the production of fertilizers and the majority of the remaining 10% for cleaning products. In 2018, 43% of the global hydrogen demand was used for ammonia production and 52% for refining processes. Refineries use hydrogen to lower the sulphur content of diesel fuel. The remaining 6% of global hydrogen production is distributed across the other applications shown in Table 9.5.

The demand for hydrogen has grown continuously over the past decades, and the market shares for ammonia production and refinery processes have remained similar. The use of hydrogen for energy storage does not yet show in the global energy statistics.

Table 9.5   Current areas of hydrogen use in industry

Industry sector	Key applications
Chemical	Ammonia, polymers, resins
Refining	Hydrocracking, hydrotreatment
Iron and steel	Annealing, blanketing gas, forming gas
General industry	Semiconductors, propellant fuel, glass production, hydrogenation of fats (liquid vegetable oils made creamy), cooling of generators

Source: Pregger et al. 2019

9.4.1.2   Possible Applications of Hydrogen in Decarbonization Pathways

Although there is a huge diversity of market projections and possible future applications for hydrogen, there is a broad consensus among all market analysts that the market for hydrogen will grow significantly in the coming decade. As well as its current application as a feedstock in chemical industries, hydrogen is expected to expand in the energy sector. Once electricity has been generated and used to produce hydrogen, this hydrogen can store energy in the form of a gas or (pressurized) liquid and replace fossil and/or biofuels in power plants (including fuel cells), cogenerating or heating plants to generate electricity, as heat, or as a transport fuel for vehicles. Figure 9.2 provides an overview of the possible future applications of hydrogen.

An important new industry sector for hydrogen is primary steel production. Based on current knowledge, the use of hydrogen for steel production is among the most promising processes to decarbonize the steel industry (Recharge 2020).

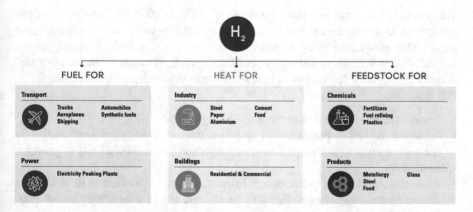

Fig. 9.2   Areas of hydrogen application. (BNEF 2020)

9.4.1.3 New Processes to Produce Hydrogen

In the public debate, colours are often used to refer to different processes for hydrogen production (IRENA 2020–1):

- 'Green' is the term applied to the production of hydrogen using water and electricity from renewable sources.
- 'Black', 'grey', and 'brown' refer to the production of hydrogen from coal, natural gas, and lignite, respectively. This process transforms the fossil fuel into hydrogen and CO_2. The life cycles of GHG emissions in the fossil-fuel-based production of hydrogen are very high.
- 'Blue' is grey hydrogen, except that during its production, CO_2 emissions are reduced by the use of CCS technology. It is also referred to as 'low-carbon hydrogen', because the full-life cycle GHG emissions are lower during its production than when hydrogen is produced with fossil fuels alone.
- 'Turquoise' (aqua, turquois) refers to hydrogen produced from methane in a thermal process (methane pyrolysis). Instead of CO_2, the process generates solid carbon (fixed carbon).

However, the only desirable production route is renewable hydrogen ('green')—only in this case is it a zero-emissions technology. In all other-coloured methods, hydrogen production still demands fossil fuels, which are the greatest cause of climate change. Consequently, our report focuses on renewable ('green') hydrogen as a key element of climate neutrality.

9.4.1.4 Hydrogen and 'Power-to-X'

When hydrogen is used as a fuel, 'power-to-X' (PtX) is often used as a term for the conversion processes and technologies involved. Power or 'P' is the electricity or input on the production side. 'X' can stand for any resulting fuel, chemical, power, or heat. PtX has received increasing public attention, because these technologies allow the indirect electrification of sectors that are (as yet) dependent on fossil fuels. PtX includes:

- Power-to-heat (PtH): transforming electricity to heat
- Power-to-gas (PtG): generation of hydrogen from electricity and (optionally) its use with a carbon source to synthesize methane (via methanation) or produce ammonium.
- Power-to-liquid (PtL): generation of hydrogen from electricity and its use with a carbon source to synthesize liquid hydrocarbons as a fuel or energy carrier (e.g. Fischer–Tropsch or methanol route)

9.4.2   Synthetic Fuels

Some (chemical) processes require either a liquid fuel or a carbon source, and will also do so in the future. Therefore, synthetic fuels are a prerequisite for carbon-neutral industry. On the one hand, these synthetic fuels are based on renewable power. On the other hand, the production of synthetic liquid and gaseous hydrocarbons and methanol requires carbon sources. In a fossil-fuel-free circular carbon economy, only a few carbon sources will be available: carbon from biomass and CO_2 emissions—from either waste incineration or flue gases, such as the process-related emissions from cement production. Therefore, possible CO_2 sources are not only industrial plants but also biogas plants or the direct air capture of CO_2. Depending upon the carbon source and the output, the different PtX processes for the generation of synthetic fuels are defined as:

- PtG: power-to-gas
- PtL: power-to-liquid
- BtL: biomass-to-liquid.
- PBtL: power-and-biomass-to-liquid.

Hydrogen, methane, and liquid hydrocarbons are studied in numerous research projects for their possible use in long-term electricity storage, a balancing option for variable wind and photovoltaic power, and fuels for transportation (see, e.g. Pregger et al., 2019). Liebich et al. (2021) give an overview of the main production routes for synfuels from a variety of energy sources, locations, and transport options, as well as their ecological and economic advantages, disadvantages, and future prospects. A variety of technical concepts and test facilities are also available, ranging from PBtL based on biomass and hydropower in Sweden to PtL based on CO_2 from cement production in Germany and PV power imports from Saudi Arabia. Because the costs are currently relatively high, the potential generation of synfuels and their use in the short term are not economically feasible, but they are primarily considered from the perspective of political expediency for the extensive decarbonization of the entire energy system.

The sustainable production of biofuels, even BtL, is limited by the availability of biomass feedstocks, e.g. residues and solid biomass. Research into and the development of the most-efficient generation routes for synthetic fuels are therefore very important, both for the decarbonization of the transport sector and for the security of future fossil-free fuel supply. Industry co-benefits can arise when emission reduction targets (e.g. for transport) lead to the accelerated development of synfuel production capacities. Synthetic fuel production processes can also provide the necessary feedstocks, such as methanol, for the chemical industries. The future availability of appropriate carbon sources from biomass or process-related emissions for industry is currently unclear, especially for biomass. The direct use of (solid) biomass in industry or the building sector might significantly reduce the remaining potential for biomass, and transport sectors (such as aviation and heavy-duty traffic) might also compete for BtL.

Here, the specific advantages of liquid synthetic hydrocarbons will also play a role. They require no special storage or transport containers, and losses during storage are negligible. The energy density is 100 times higher than today's batteries and 10 times higher than hydrogen at a pressure of 200 bar. Therefore, their handling, transport, and storage are much easier and safer, making the transport sector a major competitor for limited synfuel production. Significant improvements in development will be necessary along the complete process chain, which is as yet far from optimized.

References

ARENA. (2019, November). *Renewable energy options for industrial process heat.* Lovegrove, K., Alexander, D., Bader, R., Edwards, S., Lord, M., Mojiri, A., Rutovitz, J., Saddler, H., Stanley, C., Urkalan, K., Watt, M. https://arena.gov.au/assets/2019/11/renewable-energy-options-for-industrial-process-heat.pdf

Beardsmore, G. R., & Cull, J. P. (2001). *Crustal heat flow; a guide to measurement and modelling* (p. 324). Cambridge University Press.

BNEF. (2020). BNEF - Hydrogen Economy Outlook, Key messages, March 30, 2020. https://data.bloomberglp.com/professional/sites/24/BNEF-Hydrogen-Economy-Outlook-Key-Messages-30-Mar-2020.pdf

Buck, R., Agrafiotis, C., Tescari, S., Neumann, N., & Schmücker, M. (2021). Techno-economic analysis of candidate oxide materials for thermochemical storage in concentrating solar power systems. *Frontiers in Energy Research, 9,* 694248. https://doi.org/10.3389/fenrg.2021.694248

DBFZ. (2015). *Focus on Bioenergie-Technologien/Bioenergy-Technologies* (Fokusheft Energetische Biomassenutzung). Deutsches BiomasseForschungsZentrum.

DLR-ISR. (2021, May). *Solar thermal power plants. Heat, electricity and fuels from concentrated solar power*. German Aerospace Center (DLR) Institute of Solar Research, Pitz-Paal et al. https://www.dlr.de/content/en/downloads/publications/brochures/2021/brochure-study-solar-thermal-power-plants.pdf;jsessionid=DE4A92846D5767F2DBF2E3CB1BF69AA3.delivery-replication2?__blob=publicationFile & v=8

Huddlestone-Holmes, C. (2014, July). *Geothermal energy in Australia.* Prepared for the ARENA International Geothermal Energy Group, CSIRO. https://arena.gov.au/assets/2017/02/Geothermal-Energy-in-Australia.pdf

IEA BioEnergy Agreement Task 33. (2020). *Project website, IEA Task 33—Gasification of biomass and waste; November 2021.* http://www.task33.ieabioenergy.com/content/thermal_gasification

IEA Geo. (2011). *Technology roadmap—Geothermal heat and power.* International Energy Agency. https://iea.blob.core.windows.net/assets/f108d75f-302d-42ca-9542-458eea569f5d/Geothermal_Roadmap.pdf

IEA Task 34. (2021, November). *Technology Collaboration Programme, IEA, Task 34: Direct thermochemical liquefaction.* Website https://task34.ieabioenergy.com/pyrolysis-principles/

IPCC-SRREN CH4. (2011). Goldstein, B., Hiriart, G., Bertani, R., Bromley, C., Gutiérrez-Negrín, I., Huenges, E., Muraoka, H., Ragnarsson, A., Tester, J., Zui, V. Geothermal energy. In IPCC special report on renewable energy sources and climate change mitigation; Edenhofer, O., Pichs-Madruga, R., Sokona, Y., Seyboth, K., Matschoss, P., Kadner, S., Zwickel, T., Eickemeier, P., Hansen, G., Schlömer, S., von Stechow, C. (eds). Cambridge University Press.

IRENA. (2020). *Green Hydrogen: A guide to policy making*. International Renewable Energy Agency, Abu Dhabi, ISBN: 978-92-9260-286-4, https://www.irena.org/-/media/Files/IRENA/Agency/Publication/2020/Nov/IRENA_Green_hydrogen_policy_2020.pdf

IRENA-Geo. (2017). *Geothermal power: Technology brief.* International Renewable Energy Agency. https://www.irena.org/publications/2017/Aug/Geothermal-power-Technology-brief

Jouhara, H., Żabnieńska-Góra, A., Khordehgah, N., Ahmad, D., & Lipinski, T. (2020). Latent thermal energy storage technologies and applications: A review. *International Journal of Thermofluids, 5–6*, 100039. ISSN 2666-2027. https://doi.org/10.1016/j.ijft.2020.100039

Kaltschmitt, M., Hartmann, H., & Hofbauer, H. (Eds.). (2009). *Energie aus Biomasse— Grundlagen, Techniken und Verfahren.* Heidelberg, Springer.

Kirkels, A. F., & Verbong, G. P.J. (2011, January). Biomass gasification: Still promising? A 30-year global overview. *Renewable and Sustainable Energy Reviews, 15*(1), 471–481. https://ideas.repec.org/a/eee/rensus/v15y2011i1p471-481.html

Koepf, E., Zoller, S., Luque, S., Thelen, M., Brendelberger, S., González-Aguilar, J., Romero, M., & Steinfeld, A. (2019). Liquid fuels from concentrated sunlight: An overview on development and integration of a 50 kW solar thermochemical reactor and high concentration solar field for the SUN-to-LIQUID project. In *AIP conference proceedings.* AIP Publishing LLC.

Lenz, V., Szarka, N., Jordan, M., & Thrän, D. (2020). Status and perspectives of biomass use for industrial process heat for industrialized countries. *Chemical Engineering & Technology, 43*(8), 1469–1484.

Liebich, A., Fröhlich, T., Münter, D., Fehrenbach, H., Giegrich, J., Köppen, S., Dünnebeil, F., Knörr, W., Biemann, K., Simon, S., Maier, S., Albrecht, F., Pregger, T., Schillings, C., Moser, M., Reissner, R., Hosseiny, S., Jungmeier, G., Beermann, M., … Bird, N. (2021). *System comparison of storable energy carriers from renewable energies.* Umweltbundesamt, ifeu—Institut für Energie- und Umweltforschung, Deutsches Zentrum für Luft- und Raumfahrt, Joanneum Research Forschungsgesellschaft mbH.

Malico, I., Nepomuceno Pereira, R., Gonçalves, A. C., & Sousa, A. M. O. (2019). Current status and future perspectives for energy production from solid biomass in the European industry. *Renewable and Sustainable Energy Reviews, 112*, 960–977.

Naegler, T., Simon, S., Klein, M., & Gils, H. C. (2015). Quantification of the European industrial heat demand by branch and temperature level. *International Journal of Energy Research, 39*(39), 2019–2030.

Norfadhilah, H., Tokimatsu, K., & Yoshikawa, K. (2017). Prospective for power generation of solid fuel from hydrothermal treatment of biomass and waste in Malaysia. *Energy Procedia, 142*, 369–373. ISSN 1876-6102. https://doi.org/10.1016/j.egypro.2017.12.058

Philippsen, C. G., Vilela, A. C. F., & Zen, L. D. (2015). Fluidized bed modeling applied to the analysis of processes: Review and state of the art. *Journal of Materials Research and Technology, 4*(2), 208–216. ISSN 2238-7854. https://doi.org/10.1016/j.jmrt.2014.10.018

Pitz-Paal, R. (2020). Chapter 19: Concentrating solar power. In T. M. Letcher (Ed.), *Future energy* (3rd ed., pp. 413–430). Elsevier.

Prats-Salvado, E., Monnerie, N., & Sattler, C. (2021). Synergies between direct air capture technologies and solar thermochemical cycles in the production of methanol. *Energies, 14*(16), 4818.

Pregger, T., Schiller, G, Cebulla, F., Dietrich, R.-U., Maier, S., Thess, A., Lischke, A., Monnerie, N., Sattler, C., Le Clercq, P., Rauch, B., Köhler, M., Severin, M., Kutne, P., Voigt, C., Schlager, H., Ehrenberger, S., Feinauer, M., Werling, L., Zhukov, V. P., Kirchberger, C., Ciezki, H. K, Linke, F., Methling, T., Riedel, U., Aigner, M. (2019) Future fuels—Analyses of the future prospects of renewable synthetic fuels. Energies, 13(1), 138.

Recharge. (2020). *'World first' as hydrogen used to power commercial steel production*, 28 April 2020, Leigh Collins. https://www.rechargenews.com/transition/-worldfirst-as-hydrogen-used-to-power-commercial-steel-production/2-1-799308

Roeb, M., Brendelberger, S., Rosenstiel, A., Agrafiotis, C., Monnerie, N., Budama, V., & Jacobs, N. (2020). *Wasserstoff als ein Fundament der Energiewende Teil 1: Technologien und Perspektiven für eine nachhaltige und ökonomische Wasserstoffversorgung.* Deutsches Zentrum für Luft- und Raumfahrt. https://elib.dlr.de/137796/1/DLR_Wasserstoffstudie_Teil_1_final.pdf

Teske, S., & Pregger, T. (2015, September). *Energy[R]evolution—A sustainable world energy outlook.* DLR, Instituteof Engineering Thermodynamics, Systems Analysis and Technology Assessment, Thomas Pregger, Sonja Simon, Tobias Naegler; Sven Teske, Greenpeace International.

Thrän, D., Bunzel, K., Seyfert, U., Zeller, V., Buchhorn, M., Müller, K., Matzdorf, B., Gaasch, N., Klöckner, K., Möller, I., Starick, A., Brandes, J., Günther, K., Thum, M., Zeddies, J., Schönleber, N., Gamer, W., Schweinle, J., Weimar, H. (2011). *Global and regional spatial distribution of biomass potentials—status Quo and options for specification.* DBFZ Report No. 7. Deutsches BiomasseForschungsZentrum (DBFZ).

Wall, D., Dumont, M., & Murphy, J. D. (2018). *Green gas: Facilitating a future green gas grid through the production of renewable gas.* International Energy Agency (IEA) Bioenergy Task 37: 2 2018. http://task37.ieabioenergy.com/files/daten-redaktion/download/Technical%20 Brochures/green_gas_web_end.pdf

Xiong, G., Jia, J., Zhao, L., Liu, X., Zhang, X., Liu, H., Zhou, W. (2021). Nonthermal radiation heating synthesis of nanomaterials. *Science Bulletin, 66*(4), 386–406. ISSN 2095–9273, https://doi.org/10.1016/j.scib.2020.08.037

Younger, P. L. (2015). Geothermal energy: Delivering on the global potential. *Energies, 8*(10), 11737–11754. https://doi.org/10.3390/en81011737

Chapter 10
Transition of the Energy Industry to (Net)-Zero Emissions

Sven Teske

Abstract The status quo in the global oil, gas, and coal industries in terms of their economic value, geographic distribution, and company structures is given. The current fossil fuel production volumes and decline rates required under 1.5 °C-compatible pathways for coal, oil, and natural gas are discussed. The assumptions made when calculating *scope 1* and *2* emissions and current and future energy intensities are defined. The role of power and gas utilities under the OECM 1.5 °C scenario is discussed, together with the projected trajectories for renewable power- and heat-generating plants and those for hydrogen and synthetic fuel. Future structures of the global primary and secondary energy industries are suggested.

Keywords Global oil, gas, and coal industries · International production trajectory · Fossil fuel decline rates · 1.5 °C fossil fuel trajectories · Utilities

10.1 Introduction

The Paris Agreement achieved consensus among all member states to maintain global warming well below 2 °C above pre-industrial levels while pursuing efforts to limit the increase to 1.5 ° C, because this will significantly reduce the risk and impacts of climate change (UNFCCC, 2015). The role of governments in implementing national climate targets and endeavours to reduce emissions at the country level is crucial for achieving global success. Considering the current concentration of CO_2 in the atmosphere (416.2 parts per million) (US DC, 2021), a global effort is required to reduce emissions to as close as possible to zero while removing atmospheric CO_2 by restoring ecosystems.

To reduce emissions to zero in line with a 1.5 °C increase, the use of coal, oil, and gas must be phased out by at least 56% by 2030. However, current climate debates have not involved an open discussion of the orderly withdrawal from the coal, oil, and gas industries. Instead, the political debate about coal, oil, and gas has

S. Teske (✉)
Institute for Sustainable Futures, University of Technology Sydney, Sydney, NSW, Australia
e-mail: sven.teske@uts.edu.au

© The Author(s) 2022
S. Teske (ed.), *Achieving the Paris Climate Agreement Goals*,
https://doi.org/10.1007/978-3-030-99177-7_10

247

continued to focus on supply and price security, neglecting the fact that mitigating climate change is only possible when fossil fuels are phased out.

The finance industry set various 'net-zero' targets in the run up to the Climate Conference COP26 in Glasgow in November 2021. One of these target setting organizations is the Net-Zero Asset Owner Alliance (NZAOA) (see Chap. 2). But what does this mean for the primary energy industry?

This section focuses on the fossil fuel trajectory of the OECM 1.5 °C pathways presented in this book and what it means for the primary energy industry and electricity and (natural) gas utilities to supply end users—customers from industry, services, or private households.

10.2   The Energy Industry: Overview

Oil, gas, and coal are all hydrocarbons—combinations of hydrogen and carbon—that originate in decomposed organic materials. Different combinations of heat and pressure—depending on geological conditions—create different forms of hydrocarbons: oil, gas, and brown or hard coal (NG, 2021).

Oil and gas often occur together, and with the proximity of both fossil resources, primary energy companies are active in oil *and* gas extraction. Geographically, the largest extraction companies for oil and gas are based in the USA, Saudi Arabia, and Russia, which were responsible for 43% of global production in 2020 (IEA OIL, 2021). By far the largest producer of coal is China, which contributed 53% of global production in 2018 (Statista, 2021c).

The geographic distribution of fossil fuels is also reflected in the structure of the industry. In 2020, the top five oil and gas companies were as follows: (1) China Petroleum & Chemical Corp. (SNP); (2) PetroChina Co. Ltd. (PTR); (3) Saudi Arabian Oil Co. (Saudi Aramco); (4) Royal Dutch Shell PLC (RDS.A); and (5) BP PLC (BP). Only Shell and BP are involved in some coal mining, whereas the top three companies focus on oil, gas, and related products for the chemical industry (IN, 2020).

The largest coal companies are BHP and Rio Tinto, both based in Australia, and China Shenhua Energy and have no or only relatively minor involvement in oil and gas extraction.

The Global Industry Classification System (GICS) category *10 Energy* includes all steps in the value chain for the production of primary energy from fossil fuels (oil, gas, and coal), from exploration and extraction to the refinement and processing of fuels as commodity for industry clients, such as the chemical industry and utilities.

Oil, gas, and coal are among the most fundamental commodities of the current global economy. Oil is not only used as a fuel for cars, planes, and ships but also as a commodity to produce, for example, asphalt, plastics, and a variety of other products. In response to the COVID-19 pandemic, the global market size for oil and gas exploration and production in 2020 was at a 10-year low, at US$1.8 trillion, compared with US$2.9 trillion in 2019. The market size a decade earlier (2011) was estimated to be US$5.3 trillion (IBIS, 2021), more than twice as high as 2020. In comparison, the global market value of coal mining companies was US$0.66 trillion in 2020 and US$0.79 trillion in 2021 (Statista, 2021a), about half the value of US$1.27 trillion a decade earlier (2011).

10.2.1   1.5 °C Pathway for the Primary Energy Industry

The primary energy demand analysis—and therefore the projections for the primary energy industry and possible future operation strategies—is the product of the energy demand projections for all end-use sectors, as presented in previous chapters, and the energy supply concept. The challenge for the primary energy industry is to supply energy services for sustained economic development and a growing global population while remaining within the global carbon budget to limit the global temperature rise to 1.5 °C.

The trajectory for oil, gas, and coal depends on how quickly an alternative energy supply can be built up and how energy consumption can be reduced technically and/ or by behavioural changes. The OECM 1.5 °C pathway represents such a trajectory and is based on a detailed bottom-up sectorial demand and supply analysis, as documented in previous chapters.

However, for the primary energy industry, it is important to assess whether or not new oil, gas, or coal extraction projects are required to meet the demand, even under an ambitious fossil-fuel phase-out scenario.

A specific analysis was undertaken in parallel with the development of the OECM 1.5 °C pathway and with scenario data from a previous version of the OneEarth Climate Model (OECM), published in 2019 (Teske et al., 2019). On the basis of publicly available oil, gas, and coal extraction data, future production volumes were calculated and compared with the 1.5 °C trajectory (Teske & Niklas, 2021).

The calculation was based on the assumptions that no new fossil fuel extraction projects would be developed from 2021 onwards and that all existing projects will see a production decline at standard industry rates. These assumptions are supported by the IEA Net-Zero by 2050 report, which concludes that there can be 'no new oil, gas or coal development if world is to reach net zero by 2050' (IEA NZ, 2021b).

A scenario designated the *Existing International Production Trajectory* ('no expansion') was developed and modelled, specifically to understand what global fossil fuel production will look like under the following assumptions:

- No new fossil fuel projects are developed.
- Existing fossil fuel production projects stop producing once the resource at the existing site is exhausted, and no new mines are dug or wells are drilled in the surrounding field.
- Production at existing projects declines at standard industry rates:

- Coal: −2% per year
- Oil: −4% per year onshore and 6% per year offshore
- Gas: −4% per year on- and offshore

The *no expansion* scenario was compared with the OECM 1.5 °C pathway for coal, oil, and gas to understand whether security of supply is possible under an immediate implementation of a 'stop exploration' policy.

The decline rates for oil, gas, and coal that would result from the implementation of the 1.5 °C pathway and the assumed annual production decline rates for oil, gas, and coal are compared in Table 10.1.

- **Coal production** will must decline by 9.5% per year between 2021 and 2030, and then by at least 5% per year beyond 2030. By 2025, global coal production must fall to 3.7 billion tonnes, equivalent to China's production in 2017.
- **Oil production** must fall by 8.5% per year until 2030, and by 6% thereafter. By 2040, global oil production must fall to the equivalent of the production volume of just one of the three largest oil producers (the USA, Saudi Arabia, or Russia). The oil demand for non-energy use, such as the petrochemical industry, is not included in this analysis.
- **Gas production** must decline by 3.5% per year between 2021 and 2030, and decline even further to 9% per year beyond 2030. The 1.5 °C scenario also projects that the existing gas infrastructure, including gas pipelines and power plants, will be retrofitted after the gas phase-out to accommodate hydrogen and/or renewable methane produced with electricity from renewable sources.

Table 10.2 shows the modelled trajectories for global coal, oil, and gas production under the 1.5 °C scenario and the *no expansion* scenario. Projections beyond 2025 are extrapolated based on the fossil fuel production values for 2018 and 2019, taken from the BP Statistical Review and IEA World Energy Balances (IEA WEB, 2021).

The highest rates of overproduction are for hard coal and brown coal (lignite). On a global average, even existing mines cannot remain in operation until their resources are depleted, when calculations are made under the assumed production

Table 10.1 Decline rates required to remain within the 1.5 °C carbon budget versus the production decline rates under 'no expansion'

	Average annual decline rate required to remain within the 1.5 °C carbon budget (67%)		Typical industry production decline rates (global average)
	2021–2030	2030–2050	2021–
Coal	−9.5%	−5%	−2%
Gas: onshore and offshore	−3.5%	−9%	−4%
Oil: onshore	−8.5%	−6%	−4%
Oil: offshore			−6%

Table 10.2 Comparison of the 1.5 °C scenario and the *no expansion* scenario (excluding non-energy use)

Total fossil fuel production (PJ/yr)							
Total	[PJ/yr]	2019	2025	2030	2035	2040	2050
No expansion	[PJ/yr]	418,757	396,466	333,153	280,973	237,843	172,340
1.5 °C phase-out pathway	[PJ/yr]			330,140	235,409	136,281	72,225	0
Non-energy use		39,304	37,760	39,008	38,599	39,129	40,468
Coal							
No expansion	[Mt/yr]	5867	5493	4972	4501	4074	3338
1.5 °C phase-out pathway	[Mt/yr]		3447	1413	655	227	0
Production delta—thermal use	[Mt/yr]		2046	3559	3846	3847	3338
Lignite							
No expansion	[Mt/yr]	2206	1785	1606	1446	1301	1054
1.5 °C phase-out pathway	[Mt/yr]		345	322	73	0	0
Production delta—thermal use	[Mt/yr]		1440	1284	1373	1301	1054
Gas							
No expansion	[Billion cubic meters]	3693	3387	2762	2252	1836	1221
1.5 °C phase-out pathway	[Billion cubic meters]		3558	3177	2606	1792	238
Production delta—thermal use	[Billion cubic meters]		−171	−415	−354	44	983
Oil							
No expansion	[Thousand barrels per day]	74,491	76,778	60,671	47,943	37,885	23,656
1.5 °C phase-out pathway	[Thousand barrels per day]		69,088	55,087	28,852	18,885	14,369
Production delta—thermal use	[Thousand barrels per day]		7690	5584	19,091	19,000	9287

decline rates. No new mines need be opened to supply the remaining demand for coal.

The results for natural gas are less clear, and the production decline rates vary significantly. Shale gas production wells, in particular, have significantly higher production decline rates than conventional onshore or offshore natural gas extraction wells. The demand and supply values under a 1.5 °C scenario are similar, and a large overproduction of gas under the defined scenarios seems unlikely. However, a more detailed and production-side-specific analysis is required. The demand and supply for oil on a global level are similar—meaning that the assumed average production decline rates for oil wells and the reduction in demand are in the same order of magnitude.

Our analysis shows that even with no expansion of fossil fuel production, the current productions levels—especially for coal—will exhaust the carbon budget associated with the 1.5 °C target before 2030. Without the active phase-out of fossil-fuel production, production will significantly surpass what can be produced under a 1.5 °C scenario by 2025 onwards, for all fossil-fuel types.

The following section provides an overview of the breakdown of gross production, the losses during fuel processing, refinement, or the production of other fossil fuel products for hard coal, brown coal (lignite), gas, and oil. These parameters are required to calculate the *scope 1* and 2 emissions of the primary energy industry and are therefore documented. All parameters for the base year (2019) in Tables 10.3, 10.5, 10.6, and 10.8 are based on IEA World Energy Balances Statistics and projections under the OECM 1.5 °C pathway. Losses are calculated with statistical data from previous years and remain stable over the entire modelling period until 2050.

10.2.1.1   1.5 °C Trajectory: Hard Coal

The gross production of hard coal is the second highest for any fossil fuel, after oil. Around 35,000 PJ/yr of all the coal consumed globally is imported from other countries, or in other words, the coal consumed is not a regional energy resource. The main coal producers are China, Indonesia, and Australia. Interestingly, the largest importer of coal in 2019 was China, followed by India and the European Community (IEA Coal, 2020) (Table 10.3).

Table 10.4 shows the assumed losses in the coal industry and 'own energy uses', which are required for secondary projects, such as coking coal and coal liquification. The current shares of coal export and import for hard coal are also shown. All

Table 10.3 Global coal trajectory—OECM 1.5 °C

Parameter	Unit	2019	2025	2030	2035	2040	2050
Hard coal: gross production, including non-energy use	[Mt/ year]	5867	3447	1413	655	227	0
	[PJ/yr]	140,820	82,716	33,904	15,711	5453	0
Compared with 2019	[%]		−41%	−76%	−89%	−96%	−100%
Hard coal: total global imports	[PJ/yr]	35,473	20,837	8541	3958	1374	0
Hard coal: total global exports	[PJ/yr]	−29,316	−17,220	−7058	−3271	−1135	0
Hard coal: mining—own energy use	[PJ/yr]	−2586	−1519	−623	−289	−100	0
Hard coal secondary products: coking coal, coal liquification, etc.	[PJ/yr]	−2783	−1635	−670	−310	−108	0
Hard coal (primary): Own consumption—electricity	[PJ/yr]	2180	1281	525	243	84	0
Hard coal: non-energy use	[PJ/yr]	2205	2429	2445	2483	2453	0

Table 10.4 Global coal production—assumptions for transport shares and technical losses in percent

Parameter	Units	2019	2025	2030	2035	2040	2050
Hard coal mining—own energy use	[%]	−2%	−2%	−2%	−2%	−2%	−2%
Hard coal mining—share of secondary coal products from gross production	[%]	−2%	−2%	−2%	−2%	−2%	−2%
Hard coal import	[%]	25%	25%	25%	25%	25%	25%
Hard coal export	[%]	−21%	−21%	−21%	−21%	−21%	−21%

Table 10.5 Global lignite trajectory—OECM 1.5 °C

Parameter	Unit	2019	2025	2030	2035	2040	2050
Lignite: gross production (including non-energy use)	[Mt/year]	2206	345	322	73	0	0
	[PJ/yr]	20,955	3276	3062	695	0	0
Compared with 2019	[%]		−84%	−85%	−97%	−100%	−100%
Lignite: mining	[PJ/yr]	−14	−2	−2	0	0	0
Lignite secondary production (BKB plants)	[PJ/yr]	−203	−32	−30	−7	0	0
Lignite: own consumption—electricity	[PJ/yr]	519	81	76	17	0	0

parameters are calculated on the basis of 2019 values and remain at the same level for the entire modelling period.

10.2.1.2 1.5 °C Trajectory: Brown Coal

Brown coal (or lignite) mines are in direct proximity to power plants, so the fuel is on-site and not exported. The use of brown coal is limited to fewer countries than that of hard coal (Table 10.5).

10.2.1.3 1.5 °C Trajectory: Oil

Crude oil is the largest single energy source globally. Its production is regionally concentrated, and more than 60% of all oil produced crosses borders between its production and consumption. In 2019, about 0.3% of the oil produced was consumed by the extraction process itself—generating part of the *scope 1* emissions of the oil industry—and another 1.7% was losses in refineries and other prediction-related processes (Table 10.6).

Table 10.7 provides an overview of assumptions for transport shares and technical losses in percent as well as the specific emissions that are assumed for the calculation of the *scope 1* emissions for oil production.

Table 10.6 Global oil trajectory—OECM 1.5 °C

Parameter	Units	2019	2025	2030	2035	2040	2050
Oil: gross production (including non-energy use)	[Million barrels/ day]	74	69	55	29	19	14
	[PJ/yr]	166,397	154,328	123,053	64,450	42,185	32,097
Compared with 2019	[%]		−7%	−26%	−61%	−75%	−81%
Oil: global imported oil	[PJ/yr]	102,009	94,610	75,437	39,511	25,861	19,677
Oil: global exported oil	[PJ/yr]	−98,475	−91,333	−72,824	−38,142	−24,965	−18,995
Oil: extraction	[PJ/yr]	−496	−460	−367	−192	−126	−96
Oil: refineries	[PJ/yr]	−584	−542	−432	−226	−148	−113
Oil: own consumption— electricity	[PJ/yr]	−2207	−2047	−1632	−855	−560	−426

Table 10.7 Global oil production— assumptions for transport shares and technical losses in percent

Parameter	Units	2019	2025	2030	2035	2040	2050
Oil: global imported oil	[%]	61.3%	61.3%	61.3%	61.3%	61.3%	61.3%
Oil: global exported oil	[%]	−59.2%	−59.2%	−59.2%	−59.2%	−59.2%	−59.2%
Oil: extraction	[%]	−0.3%	−0.3%	−0.3%	−0.3%	−0.3%	−0.3%
Oil: refineries	[%]	−0.4%	−0.4%	−0.4%	−0.4%	−0.4%	−0.4%
Oil: own consumption— electricity	[%]	1.3%	1.3%	1.3%	1.3%	1.3%	1.3%
Oil transport: share—pipeline transport	[%]	70.0%	70.0%	70.0%	70.0%	70.0%	70.0%
Oil transport: share— marine transport	[%]	23.0%	23.0%	23.0%	23.0%	23.0%	23.0%
Oil transport: share— land transport	[%]	7.0%	7.0%	7.0%	7.0%	7.0%	7.0%
Oil: power and cogeneration—own consumption and grid losses	[%]	7.3%	5.8%	4.3%	2.5%	1.8%	1.1%
Global oil: methane emissions—upstream	[CH_4/Mt]	37.8	33.7	26.9	14.1	9.2	7.0
Global oil: methane emissions— downstream	[CH_4/Mt]	0.22	0.20	0.16	0.08	0.05	0.04
Black carbon	[Mt BC/yr]	9.07	7.72	6.80	6.00	5.35	4.22
Carbon monoxide	[Mt CO/yr]	900	796	727	657	598	466
Nitrous oxide	[ktN_2O/yr]	0.36	0.37	0.36	0.23	0.16	0.00
Nitrous oxide	[$MtCO_2$eq/ yr]	96.07	97.97	94.21	60.24	42.60	0.00

The assumed methane emissions are based on the IEA Methane Tracker (IEA MT, 2021). It is assumed that methane emissions will be reduced by 30% according to the *Global Methane Pledge* (EU-US, 2021), as announced at the Climate Conference COP26 in Glasgow, which has been supported by 44 countries (GMI, 2021) at the time of writing (December 2021).

10.2.1.4   1.5 °C Trajectory: Natural Gas

About one-third of all-natural gas produced crosses a national border between extraction and consumption (Table 10.8). The vast majority is transported via pipelines, which leads to a fractured world market with different prices, roughly broken down into the Americas, Europe, and the Middle East and Russia, as well as the Asia Pacific Region, which is more focused on liquified natural gas (LNG) transported by ships.

The share of gas flaring in the total production is part of the *scope 1* emissions of the gas industry and is assumed to decrease from 4% currently to 2% in 2025 and to end by 2030, according to *Zero Routine Flaring by 2030* by the World Bank (ZRF, 2030).

Finally, the assumed shares for import and export and various losses, as well as the transport modes, for natural gas are shown in Table 10.9.

Table 10.8 Global gas trajectory—OECM 1.5 °C

Parameter	Units	2019	2025	2030	2035	2040	2050
Gas: gross production (for regional demand—	[BCM/ yr]	3693	3558	3177	2606	1792	238
including import and non-energy use)	[PJ/yr]	129,888	125,132	111,739	91,657	63,022	8371
Compared with 2019	[%]		−4%	−14%	−29%	−51%	−94%
Gas: export	[PJ/yr]	−43,723	−42,122	−37,613	−30,853	−21,214	−2818
Gas: extraction, including gas works	[PJ/yr]	−7430	−7158	−6391	−5243	−3605	−479
Gas: processing— blending, gas-to- liquefaction (GTL) plants, LNG regasification	[PJ/yr]	227	219	195	160	110	15
Gas (primary): own consumption—electricity	[PJ/yr]	1083	1044	932	764	526	70
Gas: flaring	[PJ/yr]	4806	2503	0	0	0	0
Gas flaring share of total production	[%]	4%	2%	0%	0%	0%	0%

Table 10.9 Global gas production—assumptions for transport shares and technical losses in percent

	Units	2019	2025	2030	2035	2040	2050
Gas flaring share of global production	[%]	3.7%	2.0%	0.0%	0.0%	0.0%	0.0%
Gas: import	[%]	32.5%	32.5%	32.5%	32.5%	32.5%	32.5%
Gas: export	[%]	−33.7%	−33.7%	−33.7%	−33.7%	−33.7%	−33.7%
Gas: extraction, including gas works	[%]	−5.7%	−5.7%	−5.7%	−5.7%	−5.7%	−5.7%
Gas: processing— blending, gas-to-liquefaction (GTL) plants, LNG regasification	[%]	0.2%	0.2%	0.2%	0.2%	0.2%	0.2%
Gas (primary): own consumption—electricity	[%]	0.8%	0.8%	0.8%	0.8%	0.8%	0.8%
Gas transport: share— pipeline transport	[%]	98.00%	98.00%	98.00%	98.00%	98.00%	98.00%
Gas transport: share— marine transport	[%]	1.00%	1.00%	1.00%	1.00%	1.00%	1.00%
Gas transport: share— land transport	[%]	1.00%	1.00%	1.00%	1.00%	1.00%	1.00%
Losses during gas transport	[%]	2.41%	2.00%	2.00%	2.00%	2.00%	2.00%
Gas: power and cogeneration—own consumption and grid losses (for *scope* 2)	[%]	7.30%	5.80%	4.30%	2.50%	1.80%	1.10%
Emission factor: primary energy gas to CO_2	[ktCO_2/PJ]	56	56	56	56	56	56
Hydrogen fuel efficiency: electrolysis	[PJ/PJ]	0.677	0.68	0.71	0.71	0.73	0.73
Synthetic: efficiency of synfuel production	[PJ/PJ]	0.374	0.379	0.396	0.396	0.407	0.427

10.2.1.5 Global Renewables Trajectory (Power, Heat, and Fuels) Under OECM 1.5 °C

The primary energy industry—oil and gas companies and coal companies—is at the crossroads. The fossil fuel demand and therefore its extraction must decline sharply to remain within the carbon budget. However, both the global population and the global economy are projected to increase over the next three decades. Therefore, the energy demand will remain high. Even under the ambitious energy efficiency assumptions of the OECM 1.5 °C pathway, the global final energy demand will decrease by less than 10%. Therefore, the (primary) energy industry has an important role to play.

Table 10.10   Global renewables trajectory—power, thermal and fuels—under OECM 1.5 °C

Renewables and other fuels/technologies (RE-FT)	Units	2019	2025	2030	2035	2040	2050
Renewable power	[PJ/yr]	23,818	42,997	71,757	125,450	163,909	218,706
Nuclear power	[PJ/yr]	9255	7605	5466	3306	998	0
Renewable (thermal) heat	[PJ/yr]	3334	8318	14,214	19,485	23,750	31,951
Renewable (electric) heat pumps	[PJ/yr]	966	13,436	18,987	43,982	55,333	68,709
Renewable fuels	[PJ/yr]	0	720	3562	12,998	22,603	34,850
For comparison (*): deviation natural gas production	[PJ/yr]	0	1773	13,393	20,082	28,635	22,911
Total *final energy* RE-FT production	[PJ/yr]	37,374	73,076	113,985	205,221	266,594	354,216
Compared with 2019	[%]	0%	96%	205%	449%	613%	848%
RE-FT intensity of the economy	[PJ/billion $GDP]	0.3	0.4	0.6	0.9	1.0	1.0
Compared with 2019	[%]		52%	101%	207%	246%	255%
For comparison: coal intensity of the economy	[PJ/billion $GDP]	1.25	0.5	0.2	0.1	0.0	0.0
Compared with 2019	[%]		−59%	−85%	−94%	−98%	−100%

(*) Data shows the natural gas reduction trajectory under OECM 1.5 °C in PJ/year

However, the way energy is produced must change and if primary energy companies transition to renewable energy—not just electricity but also heat and fuels—the energy industry must move towards new business models that are closer to those of utilities, renewable project developers, and energy technology companies. Large-scale renewable energy projects, such as offshore wind farms, are in regard to investment needs within the same order of magnitude as offshore oil and gas projects. The skill sets of the offshore oil and gas workforce can also be accommodated well within the offshore wind industry (see Box 10.2).

Table 10.10 shows the global renewable power, heat, and fuel generation requirements under the OECM 1.5 °C trajectory. The overall renewable energy intensity—in petajoules (PJ) per billion $GDP—is compared with the overall current coal energy intensity. Renewables will take over the role of coal in supplying the global economy with energy by around 2030. The overall renewable energy required to supply the needs of industry, services, transport, and buildings will reach the levels of oil, gas, and coal, at around 150,000 PJ/yr, between 2030 and 2035.

Only 5 years later, renewables will provide energy equal to the current contributions of oil and gas combined. Therefore, the potential new market opportunities for both the 'traditional' primary energy industry and utilities are significant, whereas the borders between the primary and secondary energy industries (= utilities) will start to blur.

10.3   Global Utilities Sector

Power and gas utilities are a secondary energy industry. Until now, utilities have purchased (fossil) fuels from the primary energy industry and converted them to electricity in power plants or distributed the fuels—mainly gas—directly to customers to meet their demand for power and heat. Therefore, utilities are positioned between the primary energy industry and the end-use sector. Electricity generation is among the core businesses of utilities. Therefore, the significant increase in the electricity demand due to the electrification of transport and heat under the OECM 1.5 °C pathway can be seen as a business opportunity.

The global market for the generation, transmission, and distribution of electric power was estimated to be US$3.2 trillion in 2020 (PRN, 2021). The 20 largest electric utilities had a cumulative market value of US$686 million (Statista, 2021b). Market analysts expect a significant increase in the electricity demand (IEA EMR, 2021a; IRENA & JRC, 2021) (Fig. 10.1).

In a comparison of 14 global and regional energy scenarios, the International Renewable Energy Agency (IRENA) found that all projections agree that the demand for electricity will increase sharply:

> Total global electricity generated in 2040 ranges from around 40,000 terawatt hours (TWh) in the IEA Sustainable Development Scenario (SDS) to nearly 70,000 TWh in the Bloomberg New Energy Outlook 2021 (BNEF NCS) where electricity generation grows two-and-a-half times from 2019 to 2040. This is due to electric vehicle uptake, electrification in industry and buildings and green hydrogen production. (IRENA & JRC, 2021)

The OECM 1.5 °C pathway will lead to an annual increase in electricity generation from about 26,000 TWh in 2019 to 76,000 TWh. Although there is clear agreement that the global electricity demand will increase, the predictions on how this electricity will be generated are very different. Despite the significant growth in renewable power generation during the last decade, short-term projections still expect that fossil-fuel-based power generation will continue to grow.

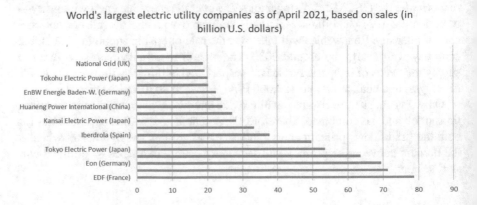

Fig. 10.1 World's largest electric utility companies. (Statista, 2021c)

The Electricity Market Report of the International Energy Agency (IEA) expects that fossil-fuel-based electricity will provide 40% of the additional electricity demand in 2022 and that coal-fired power generation will jump back to 2019 levels after a 4.6% decline in 2020 (IEA EMR, 2021a). Therefore, the lead of renewable power generation is fragile.

10.3.1 Global Power and Natural Gas Utilities: Infrastructural Changes Under the 1.5 °C Scenario

The assumed development of new manufacturing technologies, vehicle technologies, and building standards to achieve lower energy intensities for products and services has been presented in Part IV—*Sector-Specific Pathways* (Chaps. 5, 6, 7, and 8). Power and gas utilities will be significantly affected by the suggested changes. Therefore, the business model must be adapted, as well as the operational organization, to supply secure electricity to all customers.

Throughout the description of the OECM 1.5 °C pathway in this book, the increased electrification of the transport and heating sectors is the overarching scenario narrative and runs across all sectors. Increased electrification will lead to 'sector coupling', i.e. the interconnection of the heating and transport sectors with the electricity sector. The sectors are still largely separate at the time of writing. However, the interconnection of these sectors offers significant advantages in terms of the management of the energy demand and the utilization of generation management with storage technologies. The synergies of sector coupling in terms of the infrastructural changes required to transition to 100% renewable energy systems are well-documented in the literature (e.g. Brown et al., 2018; Bogdanov et al., 2021; Bermúdez et al., 2021; Jacobson, 2020).

10.3.1.1 Power Utilities

Power utilities undertake three main tasks: power generation, the transmission and distribution of electricity, and electricity services. In countries in which the electricity market is liberalized, these tasks are separated and are performed by three independent (unbundeled) companies, for generation, transmission, and distribution. All three areas of responsibility will change significantly under the 1.5 °C scenario.

Power Generation

Fossil-based and nuclear power generation with average sizes of 500–1000 MW per production site require only a small number of power plants at few locations. Widely distributed solar photovoltaic generation, with an average size of 3–5 kW per system, will often be located at customers' premises or private homes, leading to thousands or even millions of decentralized power plants. Utilities and/or power-plant operators have access to a coal power plant for maintenance, for example, but decentralized power generation is different. Solar photovoltaic generators are usually neither owned by utilities, nor are they serviced by them in terms of technical maintenance, and utilities therefore have little influence on the quantity of electricity generated or the time of generation. Electricity is also consumed partly locally and may not even reach the public power grid.

Offshore wind farms, in contrast, are centralized power plants with installed capacities within the range of an average conventional coal power plant and are usually not in direct proximity to the electricity demand. In contrast to oil and gas companies, utilities usually have no experience of working offshore, so the skills of the workforce must change.

Transmission

Under the 1.5 °C scenario, the power grid will change significantly over the next decade in response to three major changes: in the volume, load, and the location of generation.

First, the amount of electricity that must be transported will increase significantly. Electricity will replace fuels for heating and mobility, and the additional energy previously transported by other energy infrastructure, such as pipelines, will flow at least partly via power grids. End users—both private households that use heat and charge vehicles with electricity and industry clients—will not just increase the amount of electricity they use in kilowatt hours per outlet but also the loads required in kilowatts or even megawatts.

A home charger for Tesla vehicles, for example, operates at 230 V and 8–32 A, depending on the location and model (Tesla, 2021), resulting in a load of 1.8–7 kW. Therefore, the load of an average household will approximately double. The replacement of a coal-fired process (such as replacing a heating oven for steel production with an electric arc furnace) can increase the load by ≥ 300 MW.

Higher loads at the customer connection point and increased on-site generation will require a significantly stronger power grid. Furthermore, on days of higher wind and/or solar electricity production, electricity can 'reverse the flow'. With centralized power plants, the electricity is fed into the system—the transmission grid—at high- or medium-voltage levels and is taken out at medium-voltage levels by industry customers and at low-voltage levels (from the distribution grid) by residential or commercial customers. Solar rooftop systems feed electricity in at a low-voltage level. During times of high production, solar electricity can flow from a low-voltage level to a medium-voltage level, although this requires special transformer stations.

Electricity System Services

Like the sites of electricity in- and output, also the time of generation is not centrally managed by a power-plant operator (who would ramp up and down the power plant) but by a 'swarm' of solar electricity generators and onshore wind turbines, whose operation depends on the availability of sunshine and wind. Weather forecasts, the related power generated, and demand projections will be increasingly important for utilities and grid operators. The operation and management of decentralized storage systems, often operated by private households, must also be considered. Power grid operators are among the most important enablers of the energy transition, because an efficient and safe power grid is the backbone of the decarbonized energy industry.

Distribution

Widely distributed generation and storage capacities, in increasing proximity to the electricity demand, will change the relationship between the utility and the customer. The customer is no longer just a consumer but a 'prosumer'—a producer and consumer of electricity. Therefore, the business concept must change significantly, and utilities may see themselves competing with the electricity produced their own customers. A utility must increase its services and integrate local electricity generation (Table 10.11).

10.3.1.2   Gas Utilities

The changes in gas utilities under the OECM 1.5 °C scenario are more profound than those for power utilities, because the main product—natural gas—will be phased out globally by 2050. Tables 10.8 and 10.9 show the projected trajectories. However, the OECM acknowledges the significant value of the existing gas infrastructure and recommends that the gas distribution network be repurposed to utilize it for the future decarbonized energy supply. According to the Global Energy Monitor, 900,757 km of natural gas transmission pipelines were in operation globally at the end of 2020. Research has shown that there are no fundamental technical barriers to the conversion of natural gas pipelines for the transport of pure hydrogen.

Box 10.1 summarizes the key results of the comprehensive research project 'Repurposing Existing Gas Infrastructure: Overview of existing studies and reflections on the conditions for repurposing' by the European Union Agency for the Cooperation of Energy Regulators (ACER), published in July 2021.

Table 10.11 Renewable power, heat capacities, and energy demand for hydrogen and synthetic fuel production under the 1.5 °C scenario

Parameter	Units	2019	2025	2030	2035	2040	2050
Solar photovoltaic (rooftop + utility scale)	[GW$_{electric}$]	537	4197	8212	14,093	15,658	16,950
Solar photovoltaic (utility-scale share, 25% of total capacity)	[GW$_{electric}$]	134	1049	2053	3523	3915	4238
Concentrated solar power	[GW$_{electric}$]	5	113	657	1979	2770	3603
Solar thermal and solar district heating plants	[GW$_{thermal}$]	388	2463	4087	5402	6173	8154
Onshore wind	[GW$_{electric}$]	617	1350	2528	4393	5733	7620
Offshore wind	[GW$_{electric}$]	0	233	451	934	1293	2024
Hydropower plants	[GW$_{electric}$]	1569	1419	1576	1726	1830	1980
Ocean energy	[GW$_{electric}$]	1	44	91	262	379	701
Bioenergy power plants	[GW$_{electric}$]	77	198	174	200	200	231
Bioenergy cogen plants	[GW$_{electric}$]	49	111	175	304	520	668
Bio-district heating plants	[GW$_{thermal}$]	5221	6586	8924	6262	5476	3817
Geo energy power plants	[GW$_{electric}$]	12	37	92	165	267	441
Geo energy cogen plants	[GW$_{electric}$]	1	1	6	8	10	17
Gas power plant for H$_2$ conversion	[GW$_{electric}$]	0	9	56	243	375	650
Gas power cogen for H$_2$ conversion	[GW$_{electric}$]	0	0	0	32	70	199
Fuel cell and synthetic fuel cogen plants	[GW$_{electric}$]	0	0	0	32	70	199
Nuclear power plants	[GW$_{electric}$]	429	322	232	141	43	0
Industrial/district heat pumps + electrical process heat	[GW$_{thermal}$]	157	2223	3302	7461	8909	11,060
Hydrogen fuel production—electricity demand	[TWh$_{electric}$/yr]	0	294	1278	4577	7088	10,784
Hydrogen fuel production—as above, but in PJ/yr	[PJ/yr]	0	1059	4601	16,478	25,517	38,822
Synthetic fuel production—electricity demand	[TWh$_{electric}$/yr]	0	0	82	364	1118	1533
Synthetic fuel production—as above, but in PJ/yr	[PJ/yr]	0	0	296	1310	4023	5517

Therefore, the OECM assumes the conversion of natural gas pipelines to transport hydrogen, either for direct use as a replacement for natural gas in (process) heating systems, as feedstock for chemical processes, or for energy storage purposes. Therefore, the calculation of *scope 1* and *2* emissions (Chap. 13) factors in a transition to hydrogen and synthetic fuels, with the provided factors repurposed for conversion losses (Table 10.9).

Box 10.1 Conversion of Existing Natural Gas Pipelines to Transport Hydrogen
Key results from the research of the European Union Agency for the Cooperation of Energy Regulators (ACER, 2021)
Pipeline transport capacity: natural gas vs pure hydrogen—technical aspects

- In a gaseous state, the energy density of hydrogen is only slightly lower (10–20%) than that of natural gas, under the same pressure and temperature conditions.
- A pure hydrogen pipeline can have up to 80% of the maximum energy flow capacity of a natural gas pipeline, depending on the operating conditions.
- The capacity for compression power must be increased by a factor of three to transport pure hydrogen in a natural gas pipeline to achieve a similar transport capacity.
- If the compressor power capacity is not increased, the transport capacity will decrease.... This could be an option when the volumes of hydrogen to be transported are low, during the early stages of the hydrogen market uptake.
- Hydrogen can accelerate the degradation of steel pipelines, which occurs primarily in the form of embrittlement, which causes cracks and may eventually result in pipeline failure. However, technical remedies to prevent embrittlement are readily available:

  (i)   Inner coating to chemically protect the steel layer
  (ii) Intelligent pigging (monitoring)
  (iii) Operational pressure management (avoiding large pressure changes)
  (iv) Admixing degradation inhibitors (e.g. 1000 ppm oxygen)

Transmission pipeline conversion
The main advantages of repurposing pipelines are:

- Natural gas pipeline networks are already available and socially accepted (routes, including rights of way and use).
- Natural gas networks can be converted to carry hydrogen less expensively than the building new, dedicated hydrogen pipes. Such conversion can also be done gradually, depending on the development of the hydrogen supply demand. This will confer new uses on parts of the existing natural gas network, which has extensive geographic coverage throughout the EU.
- Technologies for converting the natural gas infrastructure to hydrogen operation are already largely available and tested.

10.3.2   1.5 °°C Trajectory for Power and Gas Utilities

Table 10.12 shows the development of the demand and supply of natural gas and electricity for the global utilities sector—including combined heat and power (CHP)—under the OECM 1.5 °C pathway. Figure 10.2 shows the significant increase—by a factor of 10—in the global generation of renewable electricity. The projected transition of gas utilities to the distribution of hydrogen and synthetic

Table 10.12 Global utilities sector—electricity and gas distribution under the OECM 1.5 °C

Sub-sector	Units	2019	2025	2030	2035	2040	2050
Power							
Total public power generation (incl. CHP, excluding auto producers, losses)	[TWh/yr]	25816.8	29139.5	36610.6	54467.7	63982.1	75243.9
Compared with 2019	[%]		13%	42%	111%	148%	191%
Coal: public power generation (incl. CHP, excluding auto producers)	[TWh/yr]	8337.8	5134.1	1876.5	496.8	193.2	0.0
Compared with 2019	[%]		−38%	−77%	−94%	−98%	−100%
Lignite: public power generation (incl. CHP, excluding auto producers)	[TWh/yr]	1871.0	389.8	286.7	291.1	83.2	0.0
Compared with 2019	[%]		−79%	−85%	−84%	−96%	−100%
Gas: public power generation (incl. CHP, excluding auto producers)	[TWh/yr]	6127.1	5610.8	4997.4	3962.7	2575.3	0.0
Compared with 2019	[%]		−8%	−18%	−35%	−58%	−100%
Nuclear: power generation	[TWh/yr]	2764	2113	1518	918	277	0
Renewables: public power generation (incl. CHP, excluding auto producers)	[TWh/yr]	6716.5	15892.1	27931.4	48798.7	60853.1	75243.9
Compared with 2019	[%]		137%	316%	627%	806%	1020%
Electricity carbon intensity	[gCO$_2$/kWh]	509.0	290.6	135.4	52.2	23.6	0.0
Electricity intensity: variation compared with 2019	[%]		−43%	−73%	−90%	−95%	−100%
Gas							
Gas: transport AND distribution	[BCM/year]	3693	3558	3177	2606	1792	238
	[PJ/yr]	129,888	125,132	111,739	91,657	63,022	8371
Compared with 2019	[%]	0%	−4%	−14%	−29%	−51%	−94%
Synthetic and hydrogen fuels	[PJ/yr]	0	720	3562	12,998	22,603	34,850
Total energy transport and distribution (gas, synthetic fuels, and hydrogen)	[PJ/yr]	129,888	125,851	115,301	104,655	85,625	43,221

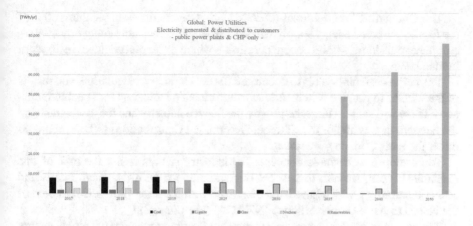

Fig. 10.2 Global power utilities sector—electricity under the OECM 1.5 °C scenario

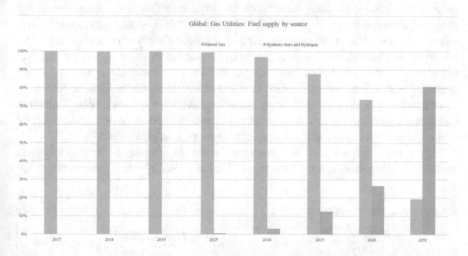

Fig. 10.3 Global gas utilities sector—gaseous fuel supply under the OECM 1.5 °C scenario

fuels will represent 50% of their sales by 2045. Therefore, the transition is assumed to have a lead time of about 10 years for the implementation of the required technical and regulatory changes (Fig. 10.3).

10.4   Energy and Utilities Sectors: A Possible Structure

Of all the industries analysed, the energy industry—often referred to in this book as the primary energy sector—classified as GICS *10 Energy*, will experience the most drastic changes. The decarbonization of the global *energy* sector requires the complete phase-out of fossil fuels in combustion processes to generate energy —the very core business of the energy industry.

   The OneEarth Climate Model (OECM) 1.5 °C scenario assumes that 100% of the fossil-fuel-based energy supply will be replaced by renewable energy by 2050—complete transition within one generation—which is unprecedented in modern human history.

   The purpose of this book is to document the development, calculation, and results of the OECM to provide benchmark key performance indicators for specific industries. These will support target setting by the finance industry and those who develop the net-zero targets and/or the National Determined Contributions (NDCs) required under the Paris Climate Agreement.

   To develop new business concepts for industry sectors is not the task of this research. Instead, we aim to support our assumptions with technical details and scenario narratives, which have been discussed with the scientific advisory board of the Net-Zero Asset Owner Alliance (NZAOA) (see Chap. 2).

   The *energy* and *utility* sectors must grow together to implement the global energy transition within only three decades. Utility-scale solar power plants and onshore and offshore wind farms are large infrastructural projects that require investments in the range of several hundred millions to billions of dollars. The operation and maintenance of offshore wind farms are very similar to those of offshore oil and gas rigs. The transport and distribution of natural gas from the point of its extraction to the end user is the core business activity of (natural) gas utilities. Power utilities oversee the entire gamut of production, from generation to distribution. Based on the OECM decarbonization pathway described in this book, we propose a horizontal integration of all three sub-sectors, which integrates the core areas of expertise and avoids stranded assets by repurposing the existing fossil-fuel infrastructure, such as pipelines.

   Figure 10.4 shows a possible structure for the decarbonized *energy* and *utility* sectors. The (primary) energy industry will focus on utility-scale power generation

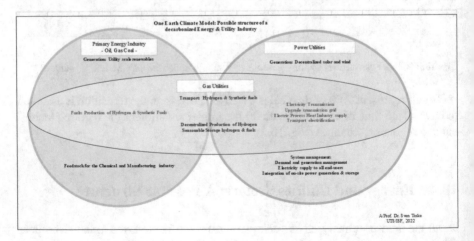

Fig. 10.4 One Earth Climate Model: possible structure of a decarbonized *energy* and *utilities* industries

and the production of hydrogen and synthetic fuels for the supply of energy and chemical feedstock. Gas utilities will focus on the transport of hydrogen and fuels and offer decentralized hydrogen production and storage services to the power sector. Power utilities will concentrate on the power grid, the management of the electricity system, and the integration of decentralized renewable power generation and storage systems, including those from 'prosumers'.

Box 10.2 Occupational Match between Offshore Oil and Gas, and Offshore Wind energy (Briggs et al., 2021)

Briggs et al. (2021) found that the main occupational pathways into offshore wind with are from other technically related sectors (such as offshore industries and the energy sector), as new entrant apprentices or graduates, and from a workforce with skills that cut across sectors (e.g. business/commercial, IT and data analytics, drone and underwater remotely operated vehicle (ROV) operators, etc.).

Consequently, the development of offshore wind energy could be an important source of alternative employment for the offshore oil and gas workforce.

A major UK study of offshore wind found that there are three main pathways for workers into the industry:

- Movement from other, technically related industries (offshore industries, energy sector)
- Apprenticeships and graduates
- Movement of workers with cross-sector skills (e.g. business/commercial, IT and data analytics, drone/ROV operators, etc.)

A range of studies have found significant movement has occurred from the offshore oil and gas workforce into the offshore wind industry, because these workers often have the foundation skills required to work on offshore installation vessels and offshore platforms and the specialized knowledge of the environmental challenges associated with operating and maintaining offshore infrastructure (IRENA, 2018).

A Scottish study found that only 15% of jobs in these industries have no skills match. For around two-thirds of jobs, there is a 'good' or 'some' skills match, including in many professional jobs, construction and installation, electrical and mechanical trades, technicians, and subsea pipelines. For a range of administrative, quality control, logistic, and project management jobs, there are 'partial' skills overlaps, which suggests that these workers could be transitioned with training (Fig. 10.5).

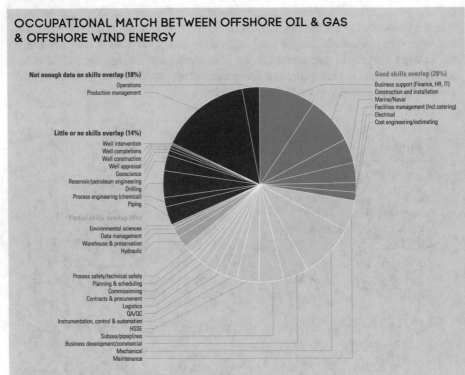

Fig. 10.5 Occupational match between offshore oil and gas and offshore wind energy. (Friends of the Earth; Global Witness and Greener Jobs Alliance, 2019)

References

ACER. (2021, July 16). *Transporting pure hydrogen by repurposing existing gas infrastructure: Overview of existing studies and reflections on the conditions for repurposing.* European Union Agency for the Cooperation of Energy Regulators, Trg republike 3, 1000 Ljubljana, Slovenia. https://extranet.acer.europa.eu/official_documents/acts_of_the_agency/publication/transporting%20pure%20hydrogen%20by%20repurposing%20existing%20gas%20infrastructure_overview%20of%20studies.pdf

Bermúdez, J., Jensen, I., Münster, M., Koivisto, M., Kirkerud, J. G., Chen, Y.-K., & Ravn, H. (2021). The role of sector coupling in the green transition: A least-cost energy system development in northern-central Europe towards 2050. *Applied Energy, 289*, 116685. https://doi.org/10.1016/j.apenergy.2021.116685

Bogdanov, D., Gulagi, A., Fasihi, M., & Breyer, C. (2021). Full energy sector transition towards 100% renewable energy supply: Integrating power, heat, transport and industry sectors including desalination. *Applied Energy, 283*, 116273. ISSN 0306-2619. https://doi.org/10.1016/j.apenergy.2020.116273

Briggs, C., Hemer, M., Howard, P., Langdon, R., Marsh, P., Teske, S., & Carrascosa, D. (2021). *Offshore wind energy in Australia* (92 pp.). Blue Economy Cooperative Research Centre.

Brown, T., Schlachtberger, D., Kies, A., Schramm, S., & Greiner, M. (2018). Synergies of sector coupling and transmission reinforcement in a cost-optimised, highly renewable European energy system. *Energy, 160*, 720–739. ISSN 0360-5442. https://doi.org/10.1016/j.energy.2018.06.222

EU-US. (2021, September 18). *Joint EU-US press release on the global methane pledge*. https://ec.europa.eu/commission/presscorner/detail/en/IP_21_4785

Friends of the Earth (FoE), Global Witness and Greener Jobs Alliance. (2019). *Sea change: Climate emergency, jobs and managing the phase-out of UK oil and gas extraction* (May Issue).

GMI. (2021). *Global Methane Initiative (GMI)*. https://www.globalmethane.org/

IBIS. (2021, February 18). *Industry Statistics—Global; Global oil & gas exploration & production—Market size 2005–2027*. Online https://www.ibisworld.com/global/market-size/global-oil-gas-exploration-production/

IEA. (2021a). *Electricity market report, July 2021*. IEA. https://www.iea.org/reports/electricity-market-report-july-2021

IEA. (2021b). *Net zero by 2050*. IEA. https://www.iea.org/reports/net-zero-by-2050

IEA Coal. (2020) *Coal 2020*. IEA. https://www.iea.org/reports/coal-2020

IEA MT. (2021). *Methane tracker 2021*. IEA. https://www.iea.org/reports/methane-tracker-2021

IEA OIL. (2021). *International Energy Agency, Oil information overview, Supply and demand*. https://www.iea.org/reports/oil-information-overview/supply-and-demand

IEA WEB. (2021). *IEA World Energy Balances*. https://www.iea.org/data-and-statistics

IN. (2020, September 10). *Investopedia, 10 biggest oil companies*, Nathan Reiff. Online https://www.investopedia.com/articles/personal-finance/010715/worlds-top-10-oil-companies.asp

IRENA. (2018). *Offshore wind investment, policies and job creation—Review of key findings for G7 ministerial meetings* (September Issue).

IRENA and JRC. (2021). *Benchmarking scenario comparisons: Key indicators for the clean energy transition*. International Renewable Energy Agency, European Commission's Joint Research Centre.

Jacobson, M. (2020). *100% clean, renewable energy and storage for everything*. Cambridge University Press. https://doi.org/10.1017/9781108786713

NG. (2021). National Geographic, Resource Library, Petroleum, website viewed December 2021 https://education.nationalgeographic.org/resource/petroleum

PRN. (2021, March). *Online news Cision PR Newswire, Global electric power generation, transmission, and distribution market report (2021 to 2030)—COVID-19 impact and recovery*.   https://www.prnewswire.com/news-releases/global-electric-power-generation-transmission-and-distribution-market-report-2021-to-2030%2D%2D-covid-19-impact-and-recovery-301248676.html

Statista. (2021a). *Leading hard coal producing countries worldwide in 2018*. https://www.statista.com/statistics/264775/top-10-countries-based-on-hard-coal-production/

Statista.   (2021b).   https://www.statista.com/statistics/274670/biggest-electric-utilities-in-the-world-based-on-sales/

Statista. (2021c). *Market value of coal mining worldwide from 2010 to 2021 (in billion U.S. dollars)*. https://www.statista.com/statistics/1137437/coal-mining-market-size-worldwide/

Teske, S., & Niklas, S. (2021). *Fossil fuel exit strategy: An orderly wind down of coal, oil and gas to meet the Paris Agreement, June 2021*. UTS, February 2021. https://adobeindd.com/view/publications/e0092323-3e91-4e5c-95e0-098ee42f9dd1/z7xq/publication-web-resources/pdf/Fossil_Fuel_Exit_Strategy.pdf

Teske, S., Pregger, T., Naegler, T., et al (2019). Energy Scenario Results. In: Teske, S. (ed.) Achieving the Paris Climate Agreement Goals. In: Teske, S. (ed.) Global and Regional 100% Renewable Energy Scenarios with Non-energy GHG Pathways for +1.5°C and +2°C. Springer Open https://link.springer.com/book/10.1007/978-3-030-05843-2

Tesla. (2021). *GEN 2 mobile connector owner's manual, Australia*. https://www.tesla.com/sites/default/files/pdfs/charging_docs/gen_2_umc/gen2_mobile_connector_en_au.pdf

UNFCCC. (2015). *Paris Agreement to the United Nations Framework Convention on Climate Change*.

US DC. (2021). US Department of Commerce, "ESRL Global Monitoring Laboratory—Research Areas." https://gml.noaa.gov/about/theme1.html

ZRF. (2030). *Zero Routine Flaring by 2030*. World Bank. Website https://www.worldbank.org/en/programs/zero-routine-flaring-by-2030#5

Part VI
Non-energy GHG and Aerosol Emissions

Chapter 11
Climate Sensitivity Analysis: All Greenhouse Gases and Aerosols

Sven Teske

Abstract This section provides an overview of all greenhouse gases (GHGs) and aerosols, the sources, their contributions to overall emissions, and their likely cumulative effects on global temperature increases. The non-energy GHG modelling in this chapter is an update of the probabilistic assessment of the global mean temperature published in the first part of *Achieving the Paris Climate Agreements*, Chap. 12 (Meinshausen 2019). The 1.5 °C energy and non-energy pathways were assessed by Climate Resource—specialists in assessing the warming implications of emissions scenarios. The analysis focuses on the derivation of the trajectories of non-CO_2 emissions that match the trajectories of energy and industrial CO_2 emissions and evaluates the multi-gas pathways against various temperature thresholds and carbon budgets until 2100. (120).

Section 7.2 is based on the following: 'Documentation of 'UTS scenarios – Probabilistic assessment of global-mean temperatures' by Climate Resource Malte Meinshausen, Zebedee Nicholls, October 2021.

Keywords Non-energy GHG modelling · Agriculture, forestry, and other land use (AFOLU) emissions · N2O · CH4 · Global warming potential (GWP), Temperature projections and exceedance probabilities

11.1 Introduction

In previous chapters, we focused on the *energy* sector and the role of land use in certain industry sectors. This section provides an overview of all greenhouse gases (GHGs) and aerosols, the cause of their emission, their contribution to overall

S. Teske (✉)
University of Technology Sydney – Institute for Sustainable Futures (UTS-ISF),
Sydney, NSW, Australia
e-mail: sven.teske@uts.edu.au

© The Author(s) 2022
S. Teske (ed.), *Achieving the Paris Climate Agreement Goals*,
https://doi.org/10.1007/978-3-030-99177-7_11

emissions, and the likely cumulative effect of global temperature increases. The non-energy GHG modelling in this chapter is an update of Meinshausen (2019). The major sources of non-energy-related emissions—process emission from cement, steel, and aluminium production—have been quantified as part of the *industry* demand analysis (Chap. 5).

11.2 Overview: Greenhouse Gases and Aerosols (Substances, Origins, and Projected Development)

11.2.1 Energy-Related CO$_2$ Emissions

Energy-related CO$_2$ emissions all derive from oil, gas, or coal and are defined as '1A' emissions according to the Intergovernmental Panel on Climate Change (IPCC) 2006 guidelines (IPCC 2006) for the National Greenhouse Gas Inventory, shown in Fig. 11.1. These emissions are caused by combustion processes, such as those in power or heating plants, engines of cars, truck, planes, and ships, and any other use of fossil fuels that involves a combustion process. All the pre-2019 values are historical statistical data, whereas all the data points from 2020 onwards are the results of the 1.5 °C energy scenario documented in previous chapters.

11.2.1.1 Fugitive CO$_2$ Emissions

According to the IPCC (2019), fugitive CO$_2$ emissions can be broken down into energy- and industry-related emissions and are categorized as 'non-1A' emissions. Energy-related fugitive emissions are further subdivided into fugitive coal emissions from underground or surface mines, including the CO$_2$ from methane (CH$_4$) utilization or flaring from underground coal mines.

Fugitive emissions from oil and natural gas include the products of unconventional oil and gas exploration, such as tar sand and fracking gas, and emissions from abandoned wells. Fugitive emissions also arise from fuel transformation processes, such as in oil refineries, charcoal and coking coal production, or gasification processes. Fugitive emissions constitute only a fraction of the emissions from energy-related combustion processes. In our analysis of industry-specific emissions, they are included in the *Scope 1, 2,* and *3* emissions.

11.2.1.2 Industrial Process Emissions

The second category according to the IPCC guidelines (IPCC 2019) is emissions from industrial processes and product use. The main emissions in this group are non-energy-related CO$_2$ from steel and cement manufacturing and include chemical

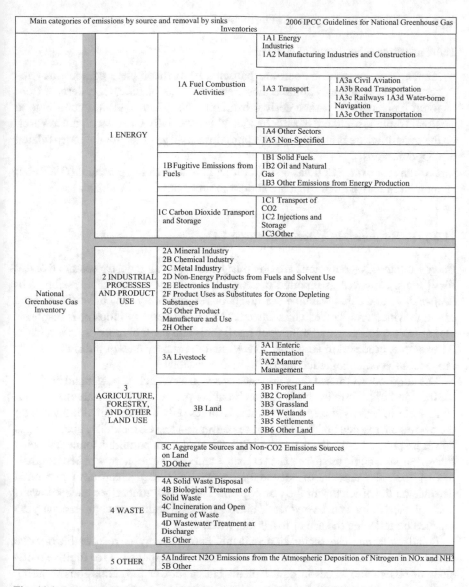

Main categories of emissions by source and removal by sinks Inventories			2006 IPCC Guidelines for National Greenhouse Gas	
National Greenhouse Gas Inventory	1 ENERGY	1A Fuel Combustion Activities	1A1 Energy Industries 1A2 Manufacturing Industries and Construction	
			1A3 Transport	1A3a Civil Aviation 1A3b Road Transportation 1A3c Railways 1A3d Water-borne Navigation 1A3e Other Transportation
			1A4 Other Sectors 1A5 Non-Specified	
		1B Fugitive Emissions from Fuels	1B1 Solid Fuels 1B2 Oil and Natural Gas 1B3 Other Emissions from Energy Production	
		1C Carbon Dioxide Transport and Storage	1C1 Transport of CO2 1C2 Injections and Storage 1C3 Other	
	2 INDUSTRIAL PROCESSES AND PRODUCT USE	2A Mineral Industry 2B Chemical Industry 2C Metal Industry 2D Non-Energy Products from Fuels and Solvent Use 2E Electronics Industry 2F Product Uses as Substitutes for Ozone Depleting Substances 2G Other Product Manufacture and Use 2H Other		
	3 AGRICULTURE, FORESTRY, AND OTHER LAND USE	3A Livestock	3A1 Enteric Fermentation 3A2 Manure Management	
		3B Land	3B1 Forest Land 3B2 Cropland 3B3 Grassland 3B4 Wetlands 3B5 Settlements 3B6 Other Land	
		3C Aggregate Sources and Non-CO2 Emissions Sources on Land 3D Other		
	4 WASTE	4A Solid Waste Disposal 4B Biological Treatment of Solid Waste 4C Incineration and Open Burning of Waste 4D Wastewater Treatment ar Discharge 4E Other		
	5 OTHER	5A Indirect N2O Emissions from the Atmospheric Deposition of Nitrogen in NOx and NH3 5B Other		

Fig. 11.1 Main categories of emissions by source and their removal by sinks, as used by the IPCC. (Source: 2006 IPCC Guidelines for National Greenhouse Gas Inventories, p. 6, (IPCC 2006))

substances used in the chemical industry, in aluminium production, or as technical gases for refrigeration. Although the volume of these chemical substances is small, their global warming effect is often significant. Details are provided in Sects. 11.2.5 and 11.2.6.

11.2.1.3 Black and Organic Carbon and Carbon Monoxide

There are three other forms of carbon:

- 'Black' and 'organic' carbon are particles from incomplete combustion processes that accumulate in the atmosphere. Whereas black carbon derives from fossil fuels, organic carbon derives from biofuels. These particles contribute to cloud formation and are hazardous to health, especially when inhaled. However, the quantities are relatively small, approximately 1% of the total energy-related CO_2 emissions.
- *Carbon monoxide* (CO) has a small direct global warming potential (GWP) but indirect radiative effects that are similar to those of CH_4 (IPCC 2001).

11.2.1.4 Responsible Industry Sectors: CO_2

Energy-related CO_2 emissions are obviously caused by all industry sectors that use fossil fuels. However, as reported in Chap. 4, the categorization of emissions into *Scopes 1, 2,* and *3* helps define the levels of responsibility for emissions and the extent to which emission can be reduced. The primary energy industry is responsible for the exploitation and mining of fossil fuels and for fuel transformation from, for example, crude oil to kerosene. Therefore, the primary energy industry directly influences the potential reduction of fugitive emissions.

The conversion of fossil fuels into secondary energy, such as power and heat, and the transport of fuels to industrial, commercial, or private consumers are the responsibility of power, gas and energy utilities. Utilities only have a limited influence on the overall energy demand but can reduce conversion losses, including in coal or gas power plants. Although the amount of CO_2 released from burning a tonne of coal, a litre of oil, or a cubic metre of gas is constant and only varies across different qualities of fuel, the amount of secondary energy units (e.g. electricity) generated depends on the efficiency of the power plant. The GHGs emitted for each kilowatt-hour of electricity can be reduced, although the overall emissions can only be reduced by reducing the use of fossil fuel itself.

Finally, the end-use sector of fossil-fuel-based energy is responsible for the actual total demand for fossil fuels. End users are not responsible for fugitive emissions or conversion losses in power plants but can lower CO_2 emissions by using more-efficient end-use applications, such as energy-efficient cars, and by driving less. However, a total phase-out of energy-related CO_2 emission is possible with the use of carbon-free renewable energy sources.

11.2.2   Agriculture, Forestry, and Other Land Use (AFOLU) Emissions

In the climate science context, emissions from agriculture, forestry, and other land-uses are referred to as *AFOLU* emissions. The AFOLU sector contributes to the emission of multiple GHGs and aerosol species, including CO_2, CH_4, and nitrous oxide (N_2O). More details about AFOLU emissions and industry responsibilities are provided in Sect. 6.1 (Overview of the Global Agriculture and Food Sector) and Chap. 14. In the 1.5 °C pathway, the phase-out of AFOLU emissions by 2030—mainly by the cessation of deforestation and the introduction of negative emissions by the creation of carbon sinks with nature-based solutions, such as are-forestation and soil management—is vital. AFOLU emissions must decline sharply until 2030, in concert with the introduction of negative emissions and the absorption of CO_2, between 2035 and 2100 (Fig. 11.9).

11.2.3 N_2O Emissions

The long-lived GHG N_2O is emitted by human activities, such as fertilizer use, burning fossil and biofuels, and wastewater treatment (IPCC 2007 AR4). However, natural processes in soils and oceans also release N_2O. More than one-third of all N_2O emissions are anthropogenic and primarily derive from agriculture (IPCC 2007 AR4). In this analysis, we focus on human sources of N_2O.

11.2.3.1   Responsible Industry Sectors: N_2O

Of all GHG emissions, 6% are N_2O. About 71% of all N_2O emissions are caused by the use of synthetic and organic fertilizers in the agricultural sector. Of all N_2O emissions, 15% are related to the chemical production of fertilizers, fibres, and synthetic products; around 10% are the by-products of combustion processes; and 4% arise from wastewater treatment plants (IPCC AR4).

11.2.4 CH_4 Emissions

Methane is a GHG with an estimated lifetime of 12 years. About 17% of all GHG emissions are CH_4. Anthropogenic CH_4 is predominantly emitted from manure and as gastroenteric releases from livestock; from rice paddies; as fugitive emissions from the mining of coal, oil, and gas; and in gas transport leakages. There are also natural sources CH_4, such as gas hydrates, freshwater bodies, oceans, termites, and wetlands, and other sources such as wildfires. Globally, wetlands are the largest natural source of CH_4, with emissions estimated to be 102–200 Mt./year on average in 2008–2017, which constituted approximately one-quarter of global CH_4 emissions (UNEP 2021).

11.2.4.1 Responsible Industry Sectors: CH_4

Anthropogenic sources contribute to about 60% of total global CH_4 emissions, 90% of which come from only three sectors: 40% from the fossil-fuel industry, approximately 35% from the agriculture sector, and approximately 20% from waste and landfill utilities (UNEP 2021).

- *Primary energy sector*: CH_4 released during oil and gas extraction, or the pumping and transport of fossil fuels. About 23% of all CH_4 emissions are anthropogenic, of which 12% originate in coal mining.
- *Agriculture*: Methane emissions from enteric fermentation and manure management represent roughly 32% of global anthropogenic emissions. Rice cultivation adds another 8% to anthropogenic emissions. Agricultural waste burning contributes close to 1%.
- *Waste*: The third largest amount of anthropogenic CH_4 emissions are from landfills and waste management, which contribute 20% of global anthropogenic CH_4 emissions.

The remaining CH_4 emissions are mainly from wastewater treatment (UNEP 2021).

11.2.5   Other GHGs

Although CO_2, CH_4, and N_2O are the main GHG gases, representing approximately 90% of all GHG emissions, a large number of other GHGs and aerosol precursors are emitted, including substances used as feedstock for the chemical industry, such as ammonia, or chemical substances used for technical processes. The largest group of these chemical substances is controlled by the Montreal Protocol, which phases down the consumption and production of different ozone-depleting substances, including halons, chlorofluorocarbons (CFCs), and hydrofluorocarbons (HFCs) (UNEP MP 2021).

11.2.6   Global Warming Potential (GWP)

Greenhouse gases warm the earth by trapping energy and reducing the rate at which energy escapes the atmosphere. These gases differ in their ability to trap heat and have various radiative efficiencies. They also differ in their atmospheric residence

Table 11.1 Main greenhouse gases and their global warming potential (GWP)

Greenhouse gas	Formula	100-year GWP (AR4)
Carbon dioxide	CO_2	1
Methane	CH_4	25
Nitrous oxide	N_2O	298
Sulphur hexafluoride	SF_6	22,800
Hydrofluorocarbon-23	CHF_3	14,800
Hydrofluorocarbon-32	CH_2F_2	675
Perfluoromethane	CF_4	7390
Perfluoroethane	C_2F_6	12,200
Perfluoropropane	C_3F_8	8830
Perfluorobutane	C_4F_{10}	8860
Perfluorocyclobutane	$c\text{-}C_4F_8$	10,300
Perfluoropentane	C_5F_{12}	13,300
Perfluorohexane	C_6F_{14}	9300

Source: IPCC AR4, compilation by the Climate Change Connection, Manitoba/Canada
Note: GWP values were changed in 2007. The values published in the 2007 IPCC Fourth Assessment Report (*AR4*) were refined from the IPCC Second Assessment Report (*SAR*) values. However, both values (AR2 and AR4) can be found throughout the literature

times. Each gas has a specific global warming potential (GWP), which allows comparisons of the amount of energy the emission of 1 tonne of a gas will absorb over a given time period, usually a 100-year average time, compared with the emissions of 1 tonne of CO_2 (Vallero 2019). Table 11.1 shows the main GHGs and their GWPs. Although the quantities of substances considered under the Montreal Protocol are small, their GWPs are significantly higher than those of the main GHGs.

11.3 Assessment of the 1.5 °C Energy and Non-Energy Pathways

This section is based on the analysis of *Climate Resource* under contract to the University of Technology Sydney (UTS) as part of the Net-Zero Sectorial Industry Pathways Project (UTS/ISF 2021). The study is an update of the previous OneEarth Climate Model (OECM) publication (Teske et al. 2019). However, the Generalized Quantile Walk (GQW) methodology used (Meinshausen & Dooley 2019) has been developed further.

The energy and industrial CO_2 emissions pathways are based on the OECM 1.5 °C energy scenario described in previous chapters, whereas the non-CO_2 GHG emission time series have been described with the advanced GQW methodology.

The probabilistic global mean temperature, radiative forcing, and concentration implications of the scenarios are also examined with the reduced complexity model MAGICC, in the same set-up used by the IPCC's Sixth Assessment Report (IPCC

AR6 2021). The emissions pathways developed are analysed in terms of their 1.5 °C, 2 °C, and 2.5 °C exceedance probabilities over time until 2050 and 2100. The climate projections are performed with a probabilistic modelling set-up that includes additional feedbacks, such as permafrost-related CH_4 and CO_2 emissions.

11.3.1 Accounting for Non-Energy Sectors

The IPCC Assessment Report 6 (IPCC AR 2021), published in August 2021, contains five scenarios, each of which represents a different emissions pathway. These scenarios are called the *Shared Socioeconomic Pathway* (*SSP*) scenarios. The most optimistic scenario, in which global CO_2 emissions are cut to net zero around 2050, is the SSP1-1.9 scenario. The number at the end (1.9) stands for the approximate end-of-century radiative forcing, a measure of how hard human activities are pushing the climate system away from its pre-industrial equilibrium. The most pessimistic is SSP5-8.5. The SSP1-1.9 scenario, described in detail by Rogelj et al. (2018), assumes that the global community takes strong mitigation action consistent with the sustainable development goals. As a result, this scenario sees strong reductions in GHG emissions.

In this analysis, the energy-related CO_2 emissions data are the results of the OECM 1.5 °C pathway documented in previous chapters. These sectorial energy scenarios include key fossil-fuel combustion activities, as defined under category 1A emissions of the IPCC 1996 guideline definitions (IPCC 2006), shown in Fig. 11.1.

Climate Resource has added CO_2 emissions that fall under other fossil fuel and industrial activities, such as fugitive emissions, cement production, and waste disposal and management, from the SSP1-1.9 scenario, a scenario in which there is strong mitigation action. The SSP1-1.9 scenario has been chosen, because it has similar reductions of CO_2 fossil-fuel emissions as the OECM 1.5 °C scenario. The combination of both time series creates an emission pathway that is likely to include all fossil-fuel and industrial uses.

Within the energy sector category, the non-1A category emissions are those that derive from fugitive emissions and fossil-fuel fires. In adding these emissions, we assume that they will remain constant into the future, and we derive their magnitude based on the detailed sectorial breakdown provided by Hoesly et al. (2018). The data categorization follows the latest scientific standards (Nicholls et al. 2021; Gidden et al. 2019).

The assumption that the non-1A emissions—industrial and fugitive emissions— will remain constant is an oversimplification, given the likelihood that changes (such as flaring during gas production) will vary into the future. With a complete fossil-fuel phase-out, there will be no further natural gas extraction and therefore no emissions from gas flaring. However, these emissions represent <1% of total CO_2 emissions, so the effect of this simplification will be of the order of hundredths of a degree, even in a baseline scenario. We chose not to assume that these emissions will continue to represent a fixed fraction of the total energy sector emissions, because the energy sector emissions will become negative in the twenty-first century under the SSP1-1.9 scenario and negative emissions from fossil-fuel fires seem highly unlikely.

In the industry sector, the non-1A category emissions mainly derive from cement and metal production. We assume that these emissions will represent a fixed fraction of the *industry* sector emissions, with the fixed fraction varying by region. We derive the fixed fraction from the ratio of non-1A category emissions in the *industry* sector to the total emissions in the *industry* sector in 2014 in the data of Hoesly et al. (2018). This assumption is once again a simplification. However, in the absence of other data sources, it is a simple and justifiable choice. Moreover, given that these emissions represent approximately 6% of the total emissions and that the fixed fraction assumption captures at least some of the underlying scenario dynamics, we expect the effect of this assumption to be limited to the order of a few hundredths of a degree centigrade.

We combined the 1A CO_2 emissions of the OECM with the estimate of non-1A emissions to create a complete time series of fossil CO_2 emissions (see Fig. 11.2). Whereas the 1A emissions from the OECM will reach zero in 2050, the non-1A sector emissions are generally considered harder to mitigate, so we assume that they will not reach zero in 2050.

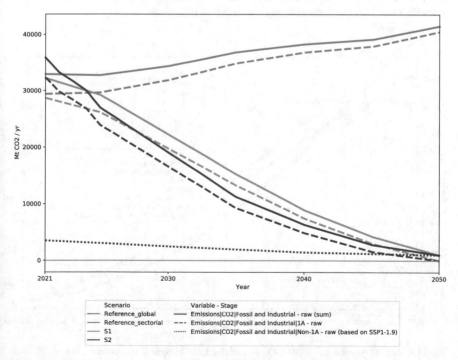

Fig. 11.2 Three non-1A fossil CO_2 emissions (Reference, 2.0 °C) and the OECM 1.5 °C pathway

The analysis performed above suggests there will be a small non-zero amount of emissions from these non-1A sectors in 2050, even under an ambitious mitigation scenario. As a result, the total fossil CO_2 emissions will generally follow the trajectory provided by the OECM but will be slightly higher, because the non-1A sector emissions are included, and in 2050, the total fossil CO_2 emissions will be close to, but not equal to, zero. Creating a scenario in which they reach exactly zero by 2050 would require further analysis of these non-1A sectors. Figure 11.2 shows the inclusion of non-1A fossil CO_2 emissions. The data for the additional scenario represent the reference case and the 2.0 °C scenario published by Teske et al. (2019).

11.3.2 Harmonization

In a second step, the projected emissions are harmonized to historical emissions estimates of the Global Carbon Project 2020 (GCP 2021a). To estimate the rebound after the COVID-related reduction in emissions in 2020, we assume that the 2021 emissions will rebound to their 2019 levels, within the same level estimated by the International Energy Agency (IEA PR 2021). This ensures a smooth transition between historical emissions and the three projections—a Reference case, a 2.0 °C scenario (see Teske et al. 2019) and the OECM 1.5 °C—as well as capturing the impact of COVID and the subsequent recovery efforts. The impact of harmonization on each of the OECM scenarios is illustrated in Fig. 11.3.

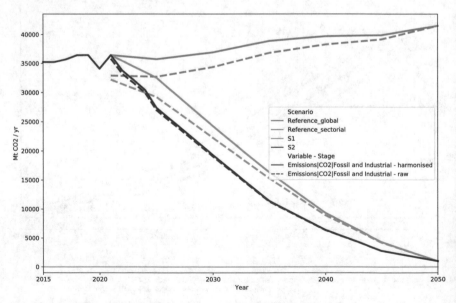

Fig. 11.3 Harmonization of fossil CO_2 emissions with historical emissions from the Global Carbon Project 2020 (GCP 2021b)

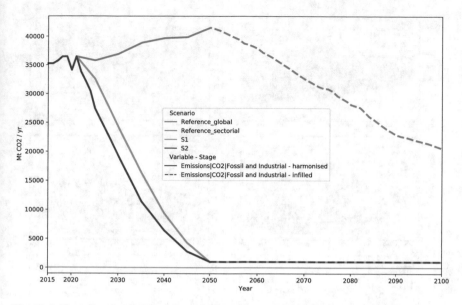

Fig. 11.4 Extending fossil CO_2 emissions from 2050 to 2100

11.3.3   Extending Emissions to 2100

A simple approach is taken to extending emissions to 2100. This process is also called 'infilling'. For the mitigation scenarios OECM 2.0 °C and OECM 1.5 °C, fossil CO_2 emissions are simply held constant from 2050 to 2100. For the reference scenarios, fossil CO_2 emissions are extended forward by assuming that the emissions follow the evolution of other pathways at a similar level of emissions in 2050. This process has been undertaken with the Silicone Software (Lamboll et al. 2020). The other pathways are taken from the SR1.5 database, i.e. the scenarios that underpinned the IPCC's Special Report on 1.5 °C (Huppmann 2018).

The SR1.5 scenarios are, at the time of writing, the most comprehensive set of strong mitigation scenarios available in the literature. Consequently, they provide the best basis for statistical inferences on how emissions will evolve over time (e.g. as we have done here by inferring the post-2050 emissions based on the emissions in 2050) and how the evolution of one set of emissions (e.g. fossil CO_2) is linked to changes in other sets of emissions (e.g. CH_4) (Fig. 11.4).

11.3.4   Infilling Emissions Other Than Fossil CO_2

11.3.4.1   Emissions in the SR1.5 Database

The OECM 1.5 °C fossil CO_2 time series is infilled with non-fossil-fuel CO_2 emissions from the SR1.5 database, whose targets are similar to the OECM 1.5 °C emissions trajectory (Fig. 11.5). This method examines the relationship between fossil

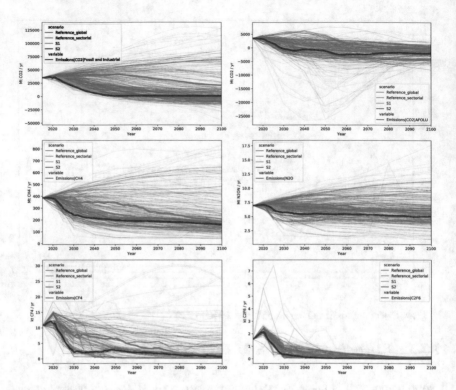

Fig. 11.5 Infilled emissions time series compared with the SR1.5 scenario database

CO_2 based on the OECM 1.5 °C pathways and other emissions of the SR1.5 data-base. This process requires the re-harmonization of the SR1.5 database to match the historical emissions inputs used by MAGICC v7.5.3 in the probabilistic AR6 set-up in 2015. This ensures that all-time series start from a consistent point, so there are no spurious jumps in the complete emissions time series, which are then passed to the climate model MAGICC (see Sect. 11.3.2 and Gidden et al. 2018).

In Fig. 11.9, the four scenarios analysed (thick lines) are shown in the context of the international integrated assessment model (IAM) scenarios, shown with blue thin lines, which represent 411 scenarios taken from the IPCC Special Report on the 1.5 °C warming scenario database. We show the OECM-modelled fossil and indus-trial CO_2 emissions (top left), the inferred CO_2 land-use (AFOLU)-related emis-sions (panel top right), inferred total CH_4 emissions (panel middle left), inferred total N_2O emissions (panel middle right), inferred total CF_4 emissions (panel bottom left), and inferred total C_2F_6 emissions (panel bottom right).

11.3.4.2   Emissions Not in the SR1.5 Database

In addition to the SR1.5 database emissions, as described in the previous section, emissions from the SSP scenarios—see definition in Sect. 11.3.1—are introduced into the analysis, as shown in Fig. 11.6.

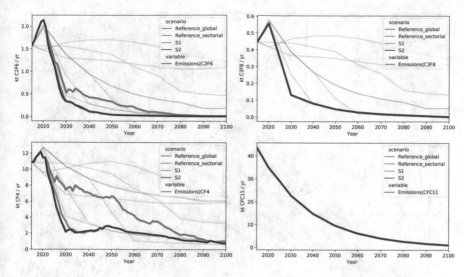

Fig. 11.6 Infilled emissions compared with the SSP scenarios

The SSP emissions pathways chosen for the analysis are those that are closest to the OECM 1.5 °C pathway. The SSP scenarios were selected using the root mean square (RMS) methodology, which measures closeness based on the difference in emissions for gases that have similar applications and uses.

A scenario for the extremely potent GHG octafluoropropane (C_3F_8) emissions, for example, was chosen based on its similarity to C_2F_6 emissions, which is a simplified way of inferring the appropriate emissions.

Hexafluoroethane (C_2F_6) is, like C_3F_8, a substance used in the semiconductor industry. However, this pragmatic technique is appropriate because the climate impact of these species is minor, representing <10% of the total GHG emissions.

In Fig. 11.10, the four scenarios analysed are shown in thick lines in the context of the SSP scenarios, which are marked in thin blue lines, and represent specific SSP scenarios (O'Neill et al. 2016). As examples, C_2F_6 emissions are shown (panel top left) as well as C_3F_8 emissions (panel top right), which follow from the C_2F_6 emissions, together with CF_4 emissions (panel bottom left) and CFC_{11} emissions (panel bottom right), which follow from the CF_4 emissions.

Carbon tetrafluoride (CF_4) and trichlorofluoromethane (CFC_{11}) are both substances used in refrigeration.

11.3.5   Temperature Projections and Exceedance Probabilities

Here, we provide the global mean probabilistic temperature projections, including their medians and 5%–95% ranges, for the OECM scenarios analysed (Fig. 11.7). These probabilistic ranges are sourced from the underlying 600 ensemble members, which are calibrated against the IPCC AR6 WG1 findings.

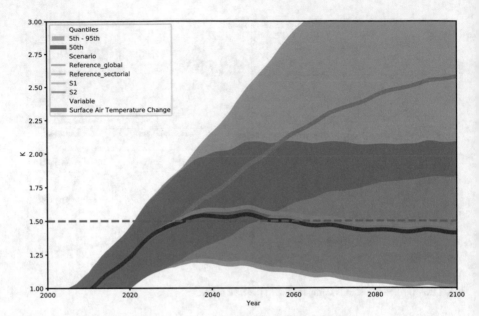

Fig. 11.7 Probabilistic global mean surface air temperature (GSAT) projections relative to 1850–1900

Similar to the SSP1-1.9 scenario in IPCC AR6 WG1, both the OECM 2.0 °C and OECM 1.5 °C pathways slightly overshoot the 1.5 °C pathway in their medians during the middle of the century, before dropping back to below 1.5 °C warming towards the end of the century. These probabilistic temperature projections can also be converted into exceedance probabilities (Fig. 11.8), i.e. the likelihood of exceeding a given temperature threshold at each point in time.

Both the OECM 2.0 °C and OECM 1.5 °C pathways are characterized as 1.5 °C low-overshoot pathways, i.e. pathways that end up below 1.5 °C (with a greater than 50% chance) at the end of century but slightly exceed a 50% chance of 1.5 °C over the course of the century. Both pathways are consistent with what is referred to in the SR1.5 report as '1.5 °C-compatible pathways'. However, the likelihood that the OECM 1.5 °C scenario will stay below 1.5 °C throughout the century, despite strong mitigation actions, does not exceed 67%. Figure 11.7 shows the probabilistic global mean surface air temperature (GSAT) projections relative to 1850–1900 for the scenarios analysed.

11.4   One Earth Summary Graph

The OECM 1.5 °C mitigation scenario limits the global average temperature rise to 1.5 °C using a carbon budget of 400 GtCO$_2$ in cumulative emissions, commencing in January 2020, as defined in the IPCC's Sixth Assessment Report, Working Group 1 (AR6 2021).

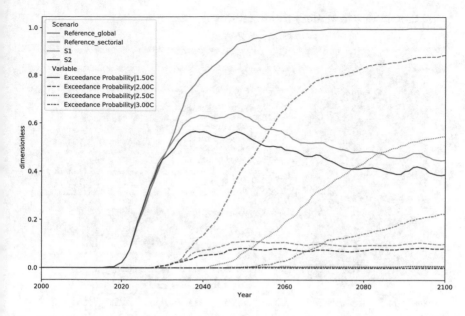

Fig. 11.8 Exceedance probabilities for the analysed scenarios relative to 1.5 °C, 2 °C, 2.5 °C, and 3 °C warming until 2100

The OECM calls for net-zero emissions by 2040, achieved by:

1. A rapid fossil-fuel phase-out for all uses by 2050 and a transition to 100% renew-able energy.
2. Negative emission through nature-based solutions:

    (a) Approximately 400 GtCO$_2$ of additional carbon to be removed by reforesta-tion and land restoration by 2100.
    (b) Natural land carbon sinks to absorb CO$_2$ but which will decline in the second half of the century.
    (c) Natural ocean carbon sinks, which will continue to absorb CO$_2$ throughout the century.

The IPCC AR6 presents the 400 GtCO$_2$ carbon budget as providing a 'good' (67%) chance of limiting warming to 1.5 °C, but it does not incorporate the anthro-pogenic emissions that occurred between the pre-industrial era (1750–1800) and the early industrial era (1850–1900).

If historical emissions between 1750 and 1900 are included, a 400 GtCO$_2$ carbon budget provides a 'fair' (50%) chance of an increase of 1.5 °C. In this case, a 'good' chance to achieve 1.5 °C warming would require an even steeper decline in emis-sions—net zero by 2040, instead of 2050—with the possibility of achieving 1.4 °C by 2100.

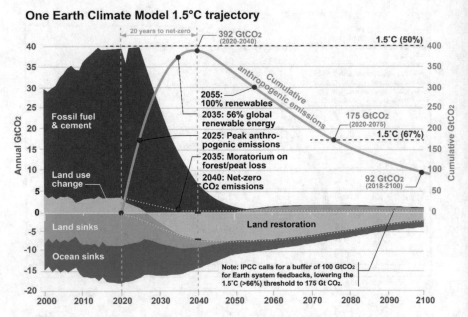

Fig. 11.9 Probability of remaining under 1.5 °C. (Source: Creative Commons: Karl Burkart, One Earth)

Figure 11.9 shows the reduction of energy-related CO_2 emissions (black), the removal of carbon by reforestation and land restoration (yellow), the natural land carbon sinks (green), and ocean carbon sinks (blue).

References

GCP. (2021a). *The global carbon project*. Online database.

GCP. (2021b). *The global carbon project*. Website, viewed December 2021. https://www.global-carbonproject.org/

Gidden, M. J., Fujimori, S., van den Berg, M., Klein, D., Smith, S. J., van Vuuren, D. P., & Riahi, K. (2018). A methodology and implementation of automated emissions harmonization for use in Integrated Assessment Models. *Environmental Modelling & Software, 105*, 187–200., ISSN 1364-8152. https://doi.org/10.1016/j.envsoft.2018.04.002

Gidden, M. J., Riahi, K., Smith, S. J., Fujimori, S., Luderer, G., Kriegler, E., van Vuuren, D. P., van den Berg, M., Feng, L., Klein, D., Calvin, K., Doelman, J. C., Frank, S., Fricko, O., Harmsen, M., Hasegawa, T., Havlik, P., Hilaire, J., Hoesly, R., … Takahashi, K. (2019). Global emissions pathways under different socioeconomic scenarios for use in CMIP6: a dataset of harmonized emissions trajectories through the end of the century. *Geoscientific Model Development, 12*, 1443–1475. https://doi.org/10.5194/gmd-12-1443-2019

Hoesly, R. M., Smith, S. J., Feng, L., Klimont, Z., Janssens-Maenhout, G., Pitkanen, T., Seibert, J. J., Vu, L., Andres, R. J., Bolt, R. M., Bond, T. C., Dawidowski, L., Kholod, N., Kurokawa, J.-I., Li, M., Liu, L., Lu, Z., Moura, M. C. P., O'Rourke, P. R., & Zhang, Q. (2018). Historical (1750–2014) anthropogenic emissions of reactive gases and aerosols from the Community

Emissions Data System (CEDS). *Geoscientific Model Development, 11*, 369–408. https://doi.org/10.5194/gmd-11-369-2018

Huppmann, D., Rogelj, J., Kriegler, E., et al. (2018). A new scenario resource for integrated 1.5 °C research. Nature. *Climate Change, 8*, 1027–1030. https://doi.org/10.1038/s41558-018-0317-4

IEA PR. (2021). *Press release; global carbon dioxide emissions are set for their second-biggest increase in history, 20 April 2021*. https://www.iea.org/news/global-carbon-dioxide-emissions-are-set-for-their-second-biggest-increase-in-history

IPCC. (2001). *TAR climate change 2001: The scientific basis, Radiative Forcing of Climate Change, Ramaswamy et al..* https://www.ipcc.ch/report/ar3/wg1/chapter-6-radiative-forcing-of-climate-change/.

IPCC. (2006). *2006 IPCC guidelines for national greenhouse gas inventories*. https://www.ipcc-nggip.iges.or.jp/public/2006gl/pdf/0_Overview/V0_1_Overview.pdf

IPCC, AR4, Forster, P., Ramaswamy, V., Artaxo, P., Berntsen, T., Betts, R., Fahey, D. W., Haywood, J., Lean, J., Lowe, D. C., Myhre, G., Nganga, J., Prinn, R., Raga, G., Schulz, M., & Van Dorland, R. (2007). Changes in atmospheric constituents and in radiative forcing. In S. Solomon, D. Qin, M. Manning, Z. Chen, M. Marquis, K. B. Averyt, M. Tignor, & H. I. Miller (Eds.), *Climate change 2007: The physical science basis. Contribution of Working Group I to the Fourth Assessment Report of the Intergovernmental Panel on Climate Change*. Cambridge University Press.

IPCC. (2019). *2019 refinement to the 2006 IPCC guidelines for national greenhouse gas inventories*. https://www.ipcc.ch/site/assets/uploads/2019/12/19R_V0_01_Overview.pdf

IPCC AR 6. (2021). *Climate change 2021: The physical science basis*. Contribution of Working Group I to the Sixth Assessment Report of the Intergovernmental Panel on Climate Change [Masson-Delmotte, V., P. Zhai, A. Pirani, S.L. Connors, C. Péan, S. Berger, N. Caud, Y. Chen, L. Goldfarb, M.I. Gomis, M. Huang, K. Leitzell, E. Lonnoy, J.B.R. Matthews, T.K. Maycock, T. Waterfield, O. Yelekçi, R. Yu, and B. Zhou (eds.)]. Cambridge University Press. In Press (see Cross-Chapter Box 7.1), https://www.ipcc.ch/report/ar6/wg1/downloads/report/IPCC_AR6_WGI_Chapter_07.pdf).

Lamboll, R. D., Nicholls, Z. R., Kikstra, J. S., Meinshausen, M., & Rogelj, J. (2020). Silicone v1.0.0: An open-source Python package for inferring missing emissions data for climate change research. *Geoscientific Model Development, 13*(11), 5259–5275.

Meinshausen, M. (2019). Implications of the developed scenarios for climate change. In S. Teske (Ed.), *Achieving the Paris climate agreement goals*. Springer. https://doi.org/10.1007/978-3-030-05843-2_12

Meinshausen, M., & Dooley, K. (2019). Mitigation scenarios for non-energy GHG. In S. Teske (Ed.), *Achieving the Paris climate agreement goals*. Springer. https://doi.org/10.1007/978-3-030-05843-2_4

Nicholls, Z., Meinshausen, M., Lewis, J., Corradi, M. R., Dorheim, K., Gasser, T., et al. (2021). Reduced complexity Model Intercomparison Project Phase 2: Synthesizing Earth system knowledge for probabilistic climate projections. *Earth's Future, 9*, e2020EF001900. https://doi.org/10.1029/2020EF001900

O'Neill, B. C., Tebaldi, C., van Vuuren, D. P., Eyring, V., Friedlingstein, P., Hurtt, G., Knutti, R., Kriegler, E., Lamarque, J.-F., Lowe, J., Meehl, G. A., Moss, R., Riahi, K., & Sanderson, B. M. (2016). The scenario model intercomparison project (ScenarioMIP) for CMIP6. *Geoscientific Model Development, 9*, 3461–3482. https://doi.org/10.5194/gmd-9-3461-2016

Rogelj, J., Popp, A., Calvin, K. V., et al. (2018). Scenarios towards limiting global mean temperature increase below 1.5 °C. *Nature Climate Change, 8*, 325–332. https://doi.org/10.1038/s41558-018-0091-3

Teske, S., et al. (2019). Energy scenario results. In S. Teske (Ed.), *Achieving the Paris climate agreement goals*. Springer. https://doi.org/10.1007/978-3-030-05843-2_8

UNEP. (2021) *United Nations Environment Programme and Climate and Clean Air Coalition (2021). Global methane assessment: Benefits and costs of mitigating methane emissions*. Nairobi: United Nations Environment Programme. ISBN: 978-92-807-3854-4 Job No: DTI/2352/PA, https://www.unep.org/resources/report/global-methane-assessment-benefits-and-costs-mitigating-methane-emissions

UNEP MP. (2021). *UN Environment Programme; Implementing Agency of the Multilateral Fund for the Implementation of the Montreal Protocol/Ozon Action as part of UN Environment Programme's Law Division and serves 147 developing countries through the Compliance Assistance Programme*, website https://www.unep.org/ozonaction/who-we-are/about-montreal-protocol

Vallero. (2019). *Air pollution calculations* (pp. 175–206., ISBN 9780128149348). Vallero, D. A. Elsevier. 10.1016/B978-0-12-814934-8.00008-9.

Part VII
Results

Chapter 12
OECM 1.5 °C Pathway for the Global Energy Supply

Sven Teske and Thomas Pregger

Abstract This chapter summarizes all the calculated energy demands for the industry, service, transport, and building sectors. The supply side results for the OECM 1.5 °C scenario are documented. Electricity generation and the power generation required globally are provided by technology, together with the corresponding renewable and fossil energy shares. A detailed overview of the heat demand by sector, the heat temperature levels required for industrial process heat, and the OECM 1.5 °C heat supply trajectories by technology are presented, in both total generation and installed capacities. The calculated global final and primary energy demands, carbon intensities by source, and energy-related CO_2 emissions by sector are given. Finally, the chapter provides the global carbon budgets by sector.

Keywords Global electricity generation · Final electricity demand · Power plant capacities · Heat generation capacities · Final energy demands of energy-intensive industries · Global carbon budget

12.1 Introduction

The final energy demands for the industries, services, transport, and buildings sectors, including residential buildings, were determined based on the assumed global population and economic development until 2050 (for details see Chap. 2), within the context of increased energy efficiencies across all sectors. All supply scenarios were developed on the basis of a global carbon budget of 400 $GtCO_2$ between 2020 and 2050, in order to qualify as an IPCC Shared Socioeconomic Pathway 1 (SSP1) no- or low-overshoot scenario (IPCC 2021).

The supply side of this 1.5 °C energy scenario pathway builds upon modelling undertaken in an interdisciplinary project led by the University of Technology

S. Teske (✉)
University of Technology Sydney – Institute for Sustainable Futures (UTS-ISF),
Sydney, NSW, Australia
e-mail: sven.teske@uts.edu.au

T. Pregger
German Aerospace Center (DLR), Institute for Engineering Thermodynamics (TT),
Department of Energy Systems Analysis, Stuttgart, Germany

293

Sydney (UTS). The project modelled sectorial and regional decarbonization pathways to achieve the Paris climate goals—to maintain global warming well below 2 °C and to 'pursue efforts' to limit it to 1.5 °C. That project produced the OneEarth Climate Model (OECM), a detailed bottom-up examination of the potential to decarbonize the energy sector. The results of this ongoing research were published in 2019 (Teske et al. 2019), 2020 (Teske et al. 2020), and 2021 (Teske et al. 2021). For this analysis, the 1.5 °C supply scenario has been updated to match the detailed bottom-up analysis documented in Chaps. 5, 6, 7, and 8.

12.2 OECM 1.5 °C Pathway for the Global Electricity Supply

The global electricity demand has grown continuously over the past decades. Global electricity generation more than doubled over the past 30 years, from 12,030 TWh (IEA WEO 1994) in 1991 to 26,942 TWh in 2019 (IEA WEO 2020). The COVID-19 pandemic led to a small reduction of about 2%, or 500 TWh (IEA WEO 2020), equal to Germany's annual electricity demand in 2020. The decline in demand was due to lockdowns and the consequent reductions in industrial manufacturing and services. However, the electricity demand increased again to pre-COVID levels in 2021. Increasing market shares of electric vehicles also increased the electricity demand in the transport sector globally. The OECM 1.5 °C pathways will accelerate this trend and the electrification of the transport sector and the provision of space and process heat to replace fossil fuels will continue to increase the global electricity demand.

12.2.1 Global Final Electricity Demand

Figure 12.1 shows the development of the final electricity demand by sector between 2019 and 2050. The significant increase in the demand is due to the electrification of heat, for both space and process heating, and to a lesser extent for hydrogen and synthetic fuels. The overall global final demand in 2050 will be 2.5 times higher than in the base year, 2019. In 2050, the production of fuels alone will consume the same amount of electricity as the total global electricity demand in 1991. Therefore, the demand shares will change completely, and 47% of all electricity (Fig. 12.2) will be for heating and fuels that are mainly used in the industry and service sectors. Electricity for space heating—predominantly from heat pumps—will also be required for residential buildings.

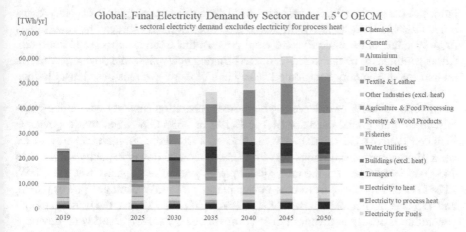

Fig. 12.1 Electricity demand by sector under the OECM 1.5 °C pathway in 2019–2050

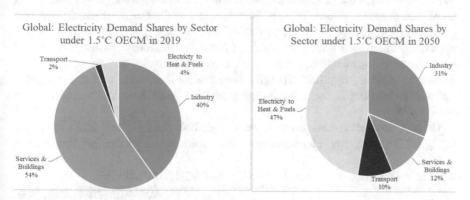

Fig. 12.2 Electricity demand shares by sector under the OECM 1.5 °C pathway in 2019 and 2050

12.2.2   Global Electricity Supply

Just as the electricity demand has changed over the past three decades, the global electricity supply has changed significantly in the same period. In 1994, 63% of electricity was generated from fossil fuels, 19% from hydropower, and 17% from nuclear power (UN 1996). Since 2010, the share of renewable electricity has increased every year. By the end of 2019, renewables contributed 27.3%, and by 2020, the share was expected to have increased to 29%. For the second consecutive year, electricity production from fossil fuels was estimated to have declined, driven mainly by a 2% reduction in coal-based power generation (REN21 GSR 2020).

The global installed capacity, by power plant technology and as a whole, has also changed rapidly. In 2010, just under 50% of all new annual additions to power-generating capacities were renewables, and 10 years later, this share had risen to 83%. Since 2012, net additions of renewable power generation capacity have outpaced the net installation of both fossil fuel and nuclear power capacity combined (REN21 GSR 2020). With the cost competitiveness achieved by renewables (mainly solar photovoltaic PV and wind power), this trend is expected to continue. China continues to be the world leader in bringing new renewable power generation on line, and the country contributed nearly half of all renewables-based installations in 2020.

In 2020, 256 GW of new renewable power generation capacity was added globally, leading to a total capacity of 1668 GW, or 2838 GW when hydro power is included (REN21 GSR 2020). By the end of 2020, the combined capacity of all solar photovoltaic installations was 760 GW, and wind power capacity summed to 743 GW. By comparison, the total global power generation capacity was 7484 GW, 2124 GW of which was from coal power plants, 1788 GW from gas power plants, and 415 GW from nuclear power plants (IEA WEO 2020). Thus, the trend in global generation is clearly in favour of cost-competitive new solar PV and wind power.

Table 12.1 shows the development of the projected global electricity generation shares. Under the OECM 1.5 °C pathway, coal- and lignite-based power plants will be phased out first, followed by gas power plants as the last fossil-fuelled power-generation technology to be taken out of service after 2040. Renewable power plants, especially solar photovoltaic and onshore and offshore wind, are projected to have the largest growth rates, leading to a combined share of 70% of electricity generation globally by 2050. To fully decarbonize the power sector, the overall renewable electricity share will increase from 25% in 2019 to 74% in 2030 and to 100% by 2050.

Global power plant capacities will quadruple between 2019 and 2050, as shown in Fig. 12.3. Capacity will increase more than actual power generation, because the

Table 12.1 Global electricity supply shares under the OECM 1.5 °C pathway

		2019	2025	2030	2035	2040	2050
Coal	[%]	31%	17%	5%	1%	0%	0%
Lignite	[%]	7%	1%	1%	1%	0%	0%
Gas	[%]	24%	20%	15%	8%	4%	0%
Oil	[%]	3%	2%	1%	0%	0%	0%
Nuclear	[%]	10%	7%	4%	2%	0%	0%
Hydrogen (produced with renewable electricity)	[%]	0%	0%	0%	2%	2%	5%
Hydro power	[%]	16%	14%	13%	10%	9%	9%
Wind	[%]	5%	14%	22%	28%	32%	36%
Solar photovoltaic	[%]	2%	18%	30%	37%	36%	34%
Biomass	[%]	1%	3%	2%	2%	1%	1%
Geothermal	[%]	0%	1%	2%	2%	3%	3%
Solar thermal power plants	[%]	0%	1%	4%	8%	10%	10%
Ocean energy	[%]	0%	0%	0%	1%	1%	1%
Renewables share	[%]	25%	52%	74%	89%	95%	100%
Electricity supply: specific CO_2 emissions per kWh	[gCO_2/ kWh]	509	290	136	53	24	0

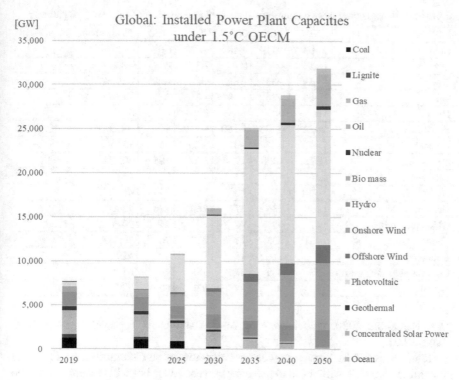

Fig. 12.3 Global installed power plant capacities under the OECM 1.5 °C pathway in 2019–2050

capacity factors for solar photovoltaic and wind power are lower than those for fuel-based power generation. By 2030, solar photovoltaic and wind will make up 70% of the generation capacity, compared with 15% in 2019, and will clearly dominate by 2050, with 78% of the total global generation capacity.

To implement the generation capacity required for the OECM 1.5 °C pathway, the global annual market for solar photovoltaic must increase from 139 GW per year (market in 2020) to an average of 800–1000 GW *additional* capacity per year from 2025 onwards to 2040. Thereafter, the overall additional annual capacity will decrease to under 100 GW to reach the required capacity calculated for 2050. However, solar photovoltaic is likely to remain at an annual market level of around 1000 GW—predominantly to provide the replacement capacity for plants that reach the end of their lifetimes after 25–30 years.

The annual market for onshore wind must increase from 87 GW in 2020 to 134 GW in 2025 and 373 GW in 2035. The total onshore wind capacity will continue to rise by 250 GW per year until 2050—including repowering. The annual onshore wind market is therefore likely to stabilize at around 300 GW per year. The size of the annual offshore wind market must increase from 6 GW in 2020 to 47 GW in 2025 to implement the 1.5 °C pathway and to grow further to around 100 GW per

Table 12.2 Global power plant capacities—annual changes under the OECM 1.5 °C pathway

[GW/yr]	2025	2030	2035	2040	2050	Total change: 2019–2050
Coal	−63	−119	−48	−6	0	−1285
Lignite	−55	−4	0	−3	0	−356
Gas	−59	−38	−62	−92	−52	−1934
Oil	−68	−26	−49	−8	0	−836
Diesel	0	0	0	0	0	0
Fossil fuel	−245	−188	−160	−110	−52	−429
Nuclear	−15	−18	−18	−20	−5	154
Renewables	931	1238	1994	876	485	30,731
Biomass	24	−5	5	0	3	154
Hydro	−30	31	30	21	27	411
Onshore wind	134	236	373	268	178	7002
Offshore wind	47	44	97	72	66	2024
Photovoltaic	721	803	1176	313	70	16,413
Geothermal	5	11	15	20	22	429
Concentrated solar power	21	109	264	158	76	3598
Ocean	9	9	34	23	42	700

year throughout 2050, with increasing market shares for repowering after 2040 (Table 12.2).

However, fossil-fuel-based power generation must be decommissioned and the global total capacity will not increase over current levels but will remain within the greenhouse gas (GHG) emissions limits. By 2025, global capacities of 63 GW from hard coal plants and 55 GW from brown coal power plants must go offline. All coal power plants in the Organization for Economic Cooperation and Development (OECD) must cease electricity generation by 2030, and the last coal plants must finish operation globally by 2045 to remain within the carbon budget for power generation to limit the global mean temperature increase to +1.5 °C. Specific CO_2 emission per kilowatt-hour will decrease from 509 g of CO_2 in 2019 to 136 g by 2030 and 24 g in 2040 to be entirely CO_2 free by 2050 (see Table 12.1, last row).

12.3 OECM 1.5 °C Pathway for Global Space and Process Heat Supply

Analogous to electricity, the energy demand for space and process heat has been determined for the industry and service sectors and for residential and commercial buildings. The specific value for each sub-sector, such as the steel and aluminium industries, has been documented in Chaps. 5, 6, and 7. In this section, we focus on the cumulative heat demand and the supply structure required for the two main sectors, *service and buildings* and *industry*.

Services and buildings usually do not require temperatures over 100 °C. Therefore, the supply technologies are different from those of the industry sector, which requires temperature up to 1000 °C and above. The overall final heat demand will increase globally under the OECM 1.5 °C pathway, but the demand shares will change significantly. With energy efficiency measures for buildings (see Chap. 7), the overall space heating demand will decrease globally, even with increased floor space. However, the industrial process heat demand is projected to increase, because energy efficiency measures will not compensate for the increasing production due to the expected increase in global GDP to 2050. In 2019, the industry sector consumed 43% of the global heat demand and the service and buildings sector the remaining 57%. By 2050, these shares will be exchanged and the industry sector will consume close to 60% of the global heat demand (Fig. 12.4).

Table 12.3 shows the supply structure for the *services and buildings* and *industry* sectors. District heat is projected to remain the smallest part of the global heat supply, followed by cogeneration. Direct heating systems installed on-site will continue to supply the majority of the heat demand. The most important technologies required to implement the OEM 1.5 °C pathway for buildings will be heat pumps and solar thermal heating for buildings, while on-site generation for industry will allow the transition from fossil-fuel-based heating plants to electrical systems, such as electric resistance ovens, electric arc furnaces, and, to a lesser extent, bioenergy or synthetic-fuel-based heating plants.

Cogeneration plants for buildings and the service sector will decline as the on-site heating demand decreases with increased efficiency. For industry, cogeneration will remain an alternative and slightly increase overall generation. However, cogeneration requires fuel, and after the phase-out of fossil fuels, only biofuels, hydrogen, or synthetic fuels will be an option for CO_2-free operation. The limited sustainable potential for bioenergy-based fuels and the relatively high costs of synthetic fuels will allow only minor growth of cogeneration plants or heating plants for the industry sector.

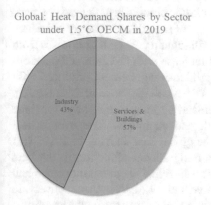

Fig. 12.4 Electricity demand shares by sector under the OECM 1.5 °C pathway in 2019 and 2050

Table 12.3 Global heat demand by sectors under the OECM 1.5 °C pathway

		2019	2025	2030	2035	2040	2050
Service and buildings							
District heat	[PJ/yr]	6044	6413	6771	6631	6351	5717
Direct heat	[PJ/yr]	78,834	88,834	93,794	91,845	87,978	79,191
Cogeneration	[PJ/yr]	9801	4693	4324	4813	5287	5016
Total service and buildings	[PJ/yr]	94,679	99,940	104,889	103,288	99,616	89,924
Industry							
District heat	[PJ/yr]	4996	5321	5883	6548	7192	8529
Direct heat	[PJ/yr]	66,424	72,904	80,602	89,714	98,540	116,868
Cogeneration	[PJ/yr]	1172	1191	1317	1466	1610	1910
Total industry	[PJ/yr]	72,592	79,415	87,801	97,728	107,342	127,307
Final heat demand							
District heat	[PJ/yr]	5691	11,734	12,654	13,178	13,543	14,246
Direct heat	[PJ/yr]	150,607	161,737	174,395	181,559	186,518	196,060
Cogeneration	[PJ/yr]	10,974	5885	5641	6279	6897	6926
Total final heat demand	[PJ/yr]	167,272	179,356	192,690	201,016	206,958	217,231

To develop the 1.5 °C pathways for process heat based on a renewable energy supply, it is necessary to separate the temperature levels for the required process heat, because not all renewable energy technologies can produce high-temperature heat. Whereas the heat generation for low-temperature heat can be achieved with renewable-electricity-supplied heat pumps or solar collectors, temperatures over 500 °C are assumed to be generated predominantly by combustion processes based on bioenergy up until 2030. After 2030, the share of electric process heat from electric resistance heat and electric arc furnaces is projected to increase to replace fossil fuels. Hydrogen and synthetic fuels will also play increasing roles in supplying high-temperature process heat.

The OECM model differentiates four temperature levels: low (<100 °C), medium low (100–500 °C), medium high (500–1000 °C), and high (>1000 °C).

Figure 12.5 shows the development of the industry process heat demand by temperature level. Whereas the values will increase over time despite energy efficiency measures, the shares of the temperature levels will remain constant. This arises from the assumption that all industry products will increase with the assumed development of the global GDP. No replacement of products, e.g. cement produced with alternative materials, is assumed because this was beyond the scope of this research.

Table 12.4 shows the total process heat demand by temperature level for three major industries combined: aluminium, steel, and chemicals. The overall heat demand of these sectors represented 20% of the global heat demand in 2019. This share is projected to increase to 37% due to a significant reduction in the heat demand in the building sector (see Chap. 7). The steel and chemical industries had similar process heat demands in 2019, at 13 EJ/year and 12 EJ/year, respectively. In contrast, the process heat demand of the aluminium industry was a quarter of this, at 3 EJ/year. Most of the process heat required by the aluminium industry is

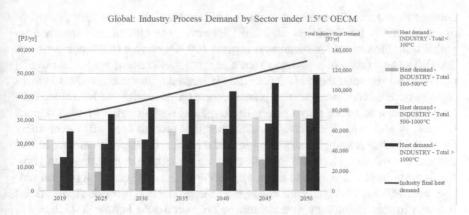

Fig. 12.5 Global: industry process demand by sector under the OECM 1.5 °C pathway in 2019–2050

Table 12.4 Total global process demand, by temperature level, in the aluminium, steel, and chemical industries

	Units	2019	2025	2030	2035	2040	2050
Total (process) heat demand Aluminium, steel, and chemical industries	[PJ/yr]	32,725	36,676	39,253	42,530	45,611	52,154
Share of global (process) heat demand	[%]	20%	22%	24%	27%	30%	37%
Share of global (process) heat demand by temperature level							
Heat demand <100 °C		14%	28%	27%	26%	25%	23%
Heat demand 100–500 °C		26%	48%	47%	46%	45%	44%
Heat demand 500–1000 °C		62%	66%	65%	65%	64%	64%
Heat demand >1000 °C		53%	44%	42%	40%	39%	37%

high-temperature heat (72%), whereas the iron ore and steel industry require only 57% high-temperature heat. The majority of the process heat required by the chemical industry is in the medium–high level (48%), between 500 °C and 1000 °C (Table 12.4).

12.3.1 Global Heat Supply

The process heat supply in 2019 relied heavily on fossil fuels (83%), mainly coal (33%) and gas (36%). Renewables played a minor role and the majority of renewable process heat was from biomass. To increase the renewable energy shares—especially for high-temperature heat—is more challenging than for the electricity sector. The fuel switch from coal and gas to biomass requires fewer technical changes than a transition towards geothermal energy, all forms of heat pumps, or direct electricity use (Keith et al. 2019). However, the OECM assumes that the global limit for sustainable biomass is around 100 EJ per year (Seidenberger et al. 2008). The generation of high-temperature heat requires concentrated solar thermal plants. However, solar thermal process heat is limited to low temperatures in most regions, because concentrated solar plants require direct sunlight with no cloud coverage, and can therefore only operate in the global sunbelt range in most regions (Farjana et al. 2018). Therefore, it is assumed that process heat will increasingly derive from electricity-based technologies: heat pumps, for low-temperature levels, and direct resistance electricity and electric arc furnaces, for medium- and high-temperature levels. However, to adapt appliances to generate electricity-based process heat will require significant changes in the production process. A significant increase in this technology is assumed to be unavailable before 2025 but will increase rapidly between 2026 and 2030. Hydrogen and synthetic fuels produced with renewable electricity will increase after 2030, especially for processes that cannot be electrified.

A global phase-out of coal for heat production is a priority objective to reduce specific CO_2 emissions. To replace fuel-based heat production, electrification, especially for low- (<100 °C) and medium-level (100–500 °C) process heat, is extremely important in achieving decarbonization.

Table 12.5 shows the assumed trajectory for the generation of industry process heat between 2019 and 2050. In 2019, gas and coal dominated global heat production. Renewables only contributed 9%—mainly biomass—and electricity had a minor share of 1%. District heat—mainly from gas-fired heating plants—contributed the remaining 7% of the process heat supply, whereas hydrogen and synthetic fuels contributed no measurable proportion. The global OECM 1.5 °C pathway phases out coal and oil for process heat generation between 2035 and 2040, and gas is phased-out as the last fossil fuel by 2050. The most important process heat supply technologies are electric heat systems, such as heat pumps, direct electric resistance heating, and arc furnace ovens for process heat; the share will increase to 22% by 2030 and 60% by 2050. Bioenergy will remain an important source of heat, accounting for 25% in 2050—2.5 times more than in 2019. Synthetic fuels and hydrogen are projected to grow to 8% of the total industry heat supply by 2050.

Table 12.5 Heat supply under the OECM 1.5 °C pathway

Industry process heat supply, including industry combined heat and power (CHP)	Units	2019 (%)	2025 (%)	2030 (%)	2035 (%)	2040 (%)	2050 (%)
Coal	[%]	33	18	11	6	0	0
Oil	[%]	14	5	3	1	0	0
Gas	[%]	36	38	25	22	17	0
Renewable heat (bioenergy, geothermal, and solar thermal)	[%]	9	24	32	27	21	25
Electricity for heat	[%]	1	8	22	36	49	60
Heat (district)	[%]	7	6	6	7	7	7
Hydrogen and synthetic fuels	[%]	0	0	1	2	6	8

Table 12.6 Calculated global capacities for renewable and electric heat generation under the OECM 1.5 °C pathway

Calculated capacities for heat generation	Units	2019	2025	2030	2035	2040	2050
Bioenergy	[GW$_{thermal}$]	5221	6586	8924	6262	5476	3817
Geothermal	[GW$_{thermal}$]	0	461	743	1059	1469	2426
Solar thermal	[GW$_{thermal}$]	388	2463	4087	5402	6173	8154
Heat pumps	[GW$_{thermal}$]	157	2126	2875	6808	8093	9821
Electricity-based process heat systems	[GW$_{thermal}$]	0	97	427	654	816	1238

Based on the annual heat demand, the generation capacities required by the three renewable heating technologies (solar thermal, geothermal, and bioenergy) have been calculated with average capacity factors.

A *capacity factor* is defined as 'the overall utilization of a power or heat-generation facility or fleet of generators. The capacity factor is the annual generation of a power plant (or fleet of generators) divided by the product of the capacity and the number of hours of operation over a given period. In other words, it measures a power plant's actual generation compared to the maximum amount it could generate in a given period without any interruption. As power or heating plants sometimes operate at less than full output, the annual capacity factor is a measure of both how many hours in the year the power plant operated and at what percentage of its entire production' (Pedraza 2019).

The same annual capacity factors are assumed for solar thermal as for photovoltaic, at around 1000 hours per year (h/yr), whereas 3000 h/yr is estimated for geothermal energy, and 4500 h/yr. for bioenergy. For electrical systems, 4500 h/yr. is assumed. The capacities shown in Table 12.6 are indicative; the actual installed capacity required under industrial conditions is dependent on a variety of factors, one of which is the production volume of a specific manufacturing plant.

12.4 OECM 1.5 °C Final and Primary Energy Balances

The European statistics bureau, EUROSTAT, defines 'final energy consumption' as 'total energy consumed by end users, such as households, industry and agriculture. It is the energy which reaches the final consumer's door and excludes that which is used by the energy sector itself' (EUROSTAT 2021). Therefore, final energy is the energy actually used from the analysed industry, service sector, or building, or for transport.

Figure 12.6 shows the global final energy trajectory for *Industry* as a whole and for five analysed sectors, as well as for *transport* and *service and buildings*. The overall *Industry* demand will increase significantly—as the only sector— from 120 EJ in 2019 to 177 EJ in 2050 (over 40%), whereas the *transport* energy demand will increase by more than 50%, due mainly to electrification and the introduction of strict efficiency standards for all vehicles. The demand of the *service and buildings*

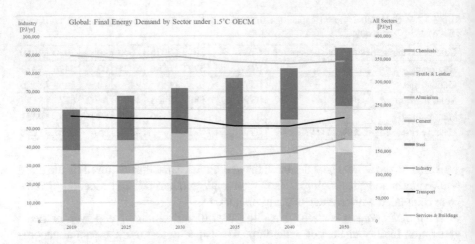

Fig. 12.6 Global final energy demand by sector under the OECM 1.5 °C pathway in 2019–2050

Table 12.7 Total global final energy demand of the aluminium, steel, and chemical industries

	Unit	2019	2025	2030	2035	2040	2050
Total final energy demand of the Aluminium, steel, and chemical industries	[PJ/ year]	41,434	48,312	51,922	56,820	61,660	72,172
Share of global final energy demand for *industry*	[%]	34%	41%	40%	41%	42%	41%
Share of global final energy demand	[%]	11%	14%	15%	17%	19%	22%

sector will decrease by just over 10%, leading to a global total final energy demand in 2050 that is 7% lower than in 2019. A combination of ambitious efficiency measures and the replacement of a significant amount of fuels for transport and heating with electrification will reduce the global energy demand despite a growing population and constant economic growth. The energy demands of energy-insensitive industries—chemicals, cement, steel, and aluminium—will increase continuously throughout the entire modelling period to 2050, but specific energy demands per production unit will decrease, decoupling economic growth from energy demands.

12.4.1 Final Energy Demands of Energy-Intensive Industries: Aluminium, Steel, and Chemicals

A closer look at the energy-intensive industries shows that the aluminium, steel, and chemical industries combined accounted for 34% of the global industrial final energy demand and 11% of the total final energy demand in 2019 (Table 12.7). The combined energy share of these three sectors will increase to 41% by 2050 under the

OECM scenario, in response to the assumed higher efficiencies in other industry sectors, such as construction and mining. The overall energy demand of the three sectors will increase from 41 EJ/year to 72 EJ/year in this period, driven mainly by the projected increase in the global GDP and therefore their production volumes.

A comparison of the consumption shares of *industry*, *transport*, and *service and buildings* shows that their shares in the OECM 1.5 °C pathway will shift very much in favour of industry. The technical energy efficiency potential in the buildings sector (Chap. 7) and the transport sector (Chap. 8) will be significant, whereas the energy demand of the service sector (Chap. 6) is projected to increase further—mainly with the growing population and therefore the growing demand for products produced by this sector.

Figure 12.7 shows that the energy demand for transport will decrease by more than half (to 16%) and that this share will be taken up by the industry sector. The demand of the *service and buildings* sector will remain at the same level, because the reduced energy demand for buildings—mainly achieved by climatization—will be compensated by the increase in the energy demand of service industries, mainly for food production.

In the next section, we present the generation components for the three main sectors *industry*, *transport*, and *service and buildings*. The latter group is called *other sectors* by the International Energy Agency (IEA). Table 12.8 shows the total final energy demand for each of the three sectors and their supply by technology. The *transport* energy demand is almost exclusively supplied by oil, whereas natural gas and electricity provide only minor contributions, and coal is not used at all for transport. *Industry* uses the majority of coal and almost half the global demand for gas.

The data show the transition towards renewable energy and an increased electricity demand between 2025 and 2050. The renewable energy share in the *other*

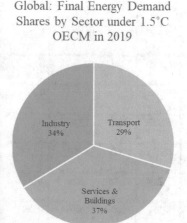

Global: Final Energy Demand Shares by Sector under 1.5°C OECM in 2019

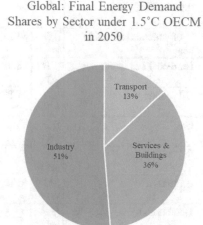

Global: Final Energy Demand Shares by Sector under 1.5°C OECM in 2050

Fig. 12.7 Global final energy demand shares by sector under the OECM 1.5 °C pathway in 2019 and 2050

Table 12.8 Global final energy demand and supply under the OECM 1.5 °C pathway

		2019	2025	2030	2035	2040	2050
Total (including non-energy use)	[PJ/yr]	400,278	400,693	403,245	379,724	375,313	373,805
Total energy use	[PJ/yr]	**360,974**	**362,934**	**364,237**	**341,125**	**336,184**	**333,337**
Transport	[PJ/yr]	**105,978**	**102,279**	**88,392**	**64,912**	**57,213**	**45,566**
Oil products	[PJ/yr]	97,777	94,337	74,080	28,304	10,039	0
Natural gas	[PJ/yr]	2149	851	710	629	284	0
Biofuels	[PJ/yr]	3810	4499	7939	11,787	17,135	15,299
Synfuels	[PJ/yr]	6	0	296	1310	4023	5517
Electricity	[PJ/yr]	1437	2901	4199	16,977	18,179	17,050
of which is renewable electricity	[PJ/yr]	353	1510	3116	15,038	17,251	17,050
Hydrogen	[PJ/yr]	0	541	1877	6534	7837	7700
Renewables energy share of transport	[%]	4%	6%	15%	53%	81%	100%
Industry	[PJ/yr]	**120,884**	**119,361**	**132,074**	**140,330**	**147,271**	**177,417**
Eectricity	[PJ/yr]	34,511	38,826	56,655	76,699	96,013	128,790
of which is renewable electricity	[PJ/yr]	8470	20,212	42,043	67,937	91,109	128,790
Public district heat	[PJ/yr]	6075	5321	5883	6548	7192	8529
of which is renewable heat	[PJ/yr]	478	439	1,179	2,488	4,926	8529
Hard coal and lignite	[PJ/yr]	28,989	15,786	10,479	5786	0	0
Oil products	[PJ/yr]	11,890	4800	3051	928	0	0
Gas	[PJ/yr]	31,263	33,350	23,957	21,671	17,421	0
Solar	[PJ/yr]	17	3129	6054	7460	8459	12,540
Biomass	[PJ/yr]	8139	16,898	23,228	15,559	7826	10,003
Geothermal	[PJ/yr]	0	1173	1989	3562	4546	7524

(continued)

Table 12.8 (continued)

		2019	2025	2030	2035	2040	2050
Hydrogen	[PJ/yr]	0	78	778	2118	5815	10,032
Renewable energy share for industry	[%]	14%	35%	57%	71%	83%	100%
Service and buildings ('other sectors')	[PJ/yr]	**134,111**	**141,294**	**143,771**	**135,883**	**131,700**	**110,354**
Electricity	[PJ/yr]	46,315	51,709	53,072	66,661	69,171	64,159
of which is renewable electricity	[PJ/yr]	11,367	26,919	39,383	59,046	65,638	64,159
Public district heat	[PJ/yr]	6044	6413	6771	6631	6351	5717
of which is renewable heat	[PJ/yr]	476	529	1357	2520	4351	5717
Hard coal and lignite	[PJ/yr]	6299	4482	0	0	0	0
Oil products	[PJ/yr]	16,139	9808	3472	1140	329	0
Gas	[PJ/yr]	30,183	27,592	25,636	17,116	10,930	0
Solar	[PJ/yr]	1229	4762	7040	9848	11,320	13,585
Biomass	[PJ/yr]	27,903	34,622	44,763	30,549	27,938	15,515
Geothermal	[PJ/yr]	0	1905	3017	3939	5660	9340
Hydrogen	[PJ/yr]	0	0	0	0	0	2038
Renewable energy share of 'other sectors'	[%]	31%	49%	66%	78%	87%	100%
Total renewable energy share	[PJ/yr]	62,248	117,216	184,061	239,693	283,834	333,337
Renewable energy share	[%]	17%	32%	51%	70%	84%	100%

sectors group will increase fastest, whereas the renewable energy supply for *transport* will grow slowly.

12.4.2 Global Primary Energy Demand: OECM 1.5 °C Pathway

The global primary energy demand under the OECM 1.5 °C pathway is shown in Table 12.9. Primary energy includes all losses and defines the total energy content of a specific energy source. In 2019, coal and oil made the largest contributions to

Table 12.9 Global primary energy demand and supply under the OECM 1.5 °C pathway

		2019	2025	2030	2035	2040	2050
Total (including non-energy use)	[PJ/yr]	564,549	536,105	513,593	488,696	473,300	465,433
Fossil (excluding non-energy use)	**[PJ/yr]**	418,757	330,140	235,489	136,409	72,398	0
Hard coal	[PJ/yr]	138,615	80,288	33,924	13,234	3000	0
Lignite	[PJ/yr]	20,955	5724	3278	3069	704	0
Natural gas	[PJ/yr]	121,586	117,698	104,028	84,014	55,245	0
Crude oil	[PJ/yr]	137,600	126,431	94,259	36,093	13,450	0
Nuclear	[PJ/yr]	30,156	24,148	17,219	10,350	3119	0
Renewables	**[PJ/yr]**	76,332	144,057	221,876	303,338	358,654	424,966
Hydro	[PJ/yr]	15,534	15,601	17,640	19,665	21,214	23,345
Wind	[PJ/yr]	4694	14,626	26,764	44,573	58,591	81,707
Solar	[PJ/yr]	3433	30,123	68,644	134,896	166,698	192,489
Biomass	[PJ/yr]	52,300	79,302	100,722	90,526	92,287	94,159
Geothermal	[PJ/yr]	366	4113	7499	11,931	17,338	28,596
Ocean energy	[PJ/yr]	4	293	608	1748	2525	4669
Total renewable energy share, including electricity and synfuel imports	[PJ/yr]	76,329	144,057	221,876	303,338	358,654	424,966
Renewable energy share	[%]	15%	30%	49%	69%	83%	100%
Non-energy use	[PJ/yr]	39,304	37,760	39,008	38,599	39,129	40,468
Coal	[PJ/yr]	2205	2429	2445	2483	2453	0
Gas	[PJ/yr]	8302	7433	7757	7753	7938	8371
Oil	[PJ/yr]	28,798	27,898	28,806	28,363	28,739	32,097

the global energy supply, followed by natural gas, whereas renewable energies contributed only 15%. The table also provides the projected trajectories for supplies for non-energy uses, e.g. oil for the petrochemical industry. The OECM does not phase-out fossil fuels for non-energy use, because their direct replacement with biomass is not always possible. A detailed analysis of the feedstock supply for non-energy uses was beyond the scope of this research.

12.5   Global CO_2 Emissions and Carbon Budget

In the last step, we calculated the energy-related carbon emissions. The OECM 1.5 °C net-zero pathway is based on efficient energy use and a renewable energy supply only, leading to full energy decarbonization by 2050. No negative emission technologies are used and the OECM results in zero energy-related carbon emissions. However, process emissions are compensated by nature-based solutions, such as increased forest coverage. The details are documented by Meinshausen and Dooley (2019) and in Chaps. 11 and 14.

The global carbon budget identifies the total amount of energy-related CO_2 emissions available to limit global warming to a maximum of 1.5 °C with no or only a low overshoot. The Intergovernmental Panel on Climate Change (IPCC) is the United Nations body that assesses the science related to climate change. In August 2021, the IPCC published a report that identified the global carbon budget required to achieve a global temperature increase of 1.5 °C with 67% likelihood as 400 $GtCO_2$ between 2020 and 2050 (IPCC 2021).

12.5.1   Global CO_2 Emissions by Supply Source

The CO_2 emissions per petajoule (PJ) of energy depend upon the quality and energy content of the energy source, e.g. coal. The German Environment Agency (UBA) has reported the specific CO_2 emissions for a variety of fossil fuels in order to calculate Germany's annual carbon emissions. In terms of coal, the UBA reports that 'most varieties of hard coal have a carbon content (with respect to the original substance) between 60 and 75%. The average content, which can vary from year to year, ranges between 65 and 66%. Hard coal within the lower range, up to a carbon content of about 56%, and a net calorific value of no more than 22 MJ/kg, is referred to as low-grade coal. Hard coal within the upper range is of coking-coal quality. The highest carbon content, reaching values over 30%, is found in anthracite coal' (UBA 2016). The OECM uses global average emission factors for hard coal, brown coal, oil, and gas, as shown in Table 12.10.

Table 12.10 Emission factors: primary energy relative to energy-supply-related CO_2 emissions

Primary energy source	Unit	Emission factor
Hard coal	[ktCO_2/PJ]	93.0
Lignite/brown coal	[ktCO_2/PJ]	111.0
Crude oil	[ktCO_2/PJ]	75.0
Natural gas	[ktCO_2/PJ]	56.0
Natural gas transport	[ktCO_2/PJ]	2.8
Refinery fuel oil	[ktCO_2/PJ]	4.0
Refinery gasoline/diesel/ kerosene	[ktCO_2/PJ]	4.0
Coal transformation	[ktCO_2/PJ]	4.0

Based on the development of the primary energy supply from fossil fuels, as defined in Table 12.9, the annual energy-related CO_2 emissions are calculated as the average global emission factors. Table 12.11 shows the calculated CO_2 emissions for fossil power generation and cogeneration and for specific sectors. The sectorial breakdown provided follows the IEA sectorial breakdown and therefore varies from the values provided for end-use sectors in Table 12.12. The specific CO_2 intensities for power generation are made available in grams of CO_2 per kilowatt-hour across total electricity generation—and therefore include carbon-free electricity

Table 12.11 Global energy-supply-related CO_2 emissions under the OECM 1.5 °C pathway

Global supply-related CO_2 emissions		2019	2025	2030	2035	2040	2050
Condensation power plants	[$MtCO_2$]	10,147	6997	3636	1696	840	0
Hard coal (and non-renewable waste)		5333	3758	1124	116	0	0
Lignite		1864	263	164	150	71	0
Gas		2294	2059	1759	1330	747	0
Oil + diesel		657	918	588	100	22	0
Combined heat and power (CHP) plants	[$MtCO_2$]	3610	1696	1405	1178	678	0
Hard coal (and non-renewable waste)		2472	792	523	337	187	0
Lignite		258	137	130	147	0	0
Gas		830	731	717	669	471	0
Oil		50	37	35	24	20	0
CO_2 emissions power and CHP plants	[$MtCO_2$]	13,758	8693	5040	2874	1518	0
Hard coal (and non-renewable waste)		7805	4549	1647	454	187	0
Lignite		2121	399	294	297	71	0
Gas		3125	2790	2476	1999	1218	0
Oil + diesel		707	955	623	124	42	0
CO_2 intensity (g/kWh)	[$MtCO_2$]						
CO2 intensity fossil elec. generation		780	711	624	536	505	0
CO2 intensity total elec. generation		509	291	135	52	24	0
CO_2 emissions by sector	[$MtCO_2$]	35,225	25,045	17,395	9924	5031	0
Industry		6685	4328	3133	2378	1113	0
Other sectors[a]		7620	5239	3951	2823	1714	0
Transport		7490	7095	5565	2127	754	0
Power generation		13,430	8383	4746	2596	1450	0
Other conversion[2]		706	706	606	433	167	62
Population	[million]	7713	8184	8548	8888	9199	9735
CO_2 emissions per capita (t/capita)	[tCO_2/ capita]	4.6	3.1	2.0	1.1	0.5	0.0

[a]Includes CHP auto producers; [2]district heating, refineries, coal transformation, gas transport

generation, such as from renewables—and for fossil-fuel-generated power only. The reduction in CO_2 intensity for fossil-fuel-based power generation between 2019 and 2040 indicates that the share of natural gas will increase and that power plants will become more efficient and therefore generate more units of electricity per unit of fuel.

Table 12.11 shows the energy-related CO_2 from the supply side and therefore defines the carbon budgets for coal, lignite, oil, and gas. The OECM also determines the carbon budget for end-use sectors.

12.5.2 Global CO₂ Budget

The remaining carbon budget for each of the following sectors has been defined based on the bottom-up demand analysis of the 12 main industry and service sectors, as documented in Chaps. 5, 6, 7, and 8. Each of those industry and service sectors must complete the transition to fully decarbonized operation within the carbon budget provided. It is very important that the carbon budget shows the cumulative emissions up to 2050, and not the annual emissions. A rapid reduction in annual emissions is therefore vital.

The shares of the cumulative carbon budget required to achieve the 1.5 °C net-zero target are shown in Fig. 12.8. Table 12.12 shows the remaining cumulative CO_2 emissions in gigatonnes. The total energy-related CO_2 for the aluminium industry between 2020 and 2050 is calculated to be 6.1 Gt, 1.5% of the total budget. For the

Table 12.12 Global carbon budget by end-use sector under the OECM 1.5 °C pathway ($GtCO_2$)

	2020–2030	2020–2050
Cement	6.5	9.2
Steel	13.8	19.2
Chemicals	16.6	24.7
Textile and leather	2.9	4.3
Aluminium	4.8	6.1
Buildings	69.1	87.8
Fisheries	0.3	0.6
Agriculture and food	10.1	13.6
Forestry and wood	3.7	5.2
Water utilities	0.7	1.0
Aviation	15.5	20.3
Shipping	28.1	48.1
Road transport	64.5	82.1
Remaining fuels	28.4	49.5
Fossil fuel production	4.5	5.2
Remaining electricity	9.3	14.7
Utilities (operation)	0.6	0.6
Other	6.8	8.8
	286.2	401.1

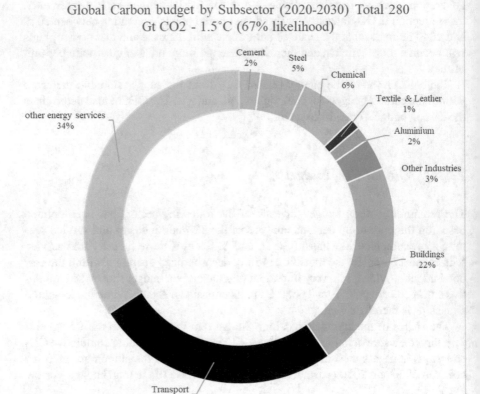

Fig. 12.8 Global carbon budget by sub-sector under 1.5 °C OECM pathway in 2020–2050

steel industry, the remaining budget is 19.1 Gt of CO_2 (4.8%), whereas the chemical industry has the highest carbon budget of 24.8 $GtCO_2$ or 6.2% of the total carbon budget. All other remaining industries can emit 27.1 $GtCO_2$ (6.8%), and all other energy-related activities, such as for buildings, transport, and residential uses, have a combined remaining emissions allowance of 323 $GtCO_2$, or 80.7% of the budget.

References

EUROSTAT. (2021). *EUROSTAT – Statistics explained, online database viewed December 2021.* https://ec.europa.eu/eurostat/statistics-explained/index.php?title=Glossary:Final_energy_consumption

Farjana, S. H., Huda, N., Mahmud, M. A. P., Saidur, R. (2018) Solar process heat in industrial systems—A global review. *Renewable and Sustainable Energy Reviews, 82,* 2270–2286. https://doi.org/10.1016/J.RSER.2017.08.065

IEA WEO. (1994). *International Energy Agency—World Energy Outlook 1994,* IEA, 9 rue de la Fédération, 75739 Paris Cedex 15, France.

IEA WEO. (2020). *International Energy Agency—World Energy Outlook 2020,* IEA, 9 rue de la Fédération, 75739 Paris Cedex 15, France.

IPCC. (2021). Climate change 2021: The physical science basis. In V. Masson-Delmotte, P. Zhai, A. Pirani, S. L. Connors, C. Péan, S. Berger, N. Caud, Y. Chen, L. Goldfarb, M. I. Gomis, M. Huang, K. Leitzell, E. Lonnoy, J. B. R. Matthews, T. K. Maycock, T. Waterfield, O. Yelekçi, R. Yu, & B. Zhou (Eds.), *Contribution of Working Group I to the sixth assessment report of the intergovernmental panel on climate change*. Cambridge University Press.

Keith, L., Dani, A., Roman, B., Stephen, E., Michael, L., Ahmad, M., Jay, R., Hugh, S., Cameron, S., Kali, U., Muriel, W. (2019). *Renewable energy options for industrial process heat*. Australian Reneable Energy Agency (ARENA). https://arena.gov.au/assets/2019/11/renew-able-energyoptions-for-industrial-process-heat.pdf

Meinshausen, M., Dooley, K. (2019). Mitigation scenarios for non-energy GHG. In: Teske, S. (ed.), Achieving the Paris climate agreement goals global and regional 100% renewable energy scenarios with non-energy GHG pathways for +1.5°C and +2°C. Springer Open. https://elib.dlr.de/126810/1/Teskeetal.-Achieving theParisClimateAgreementGoals-Globalandregiona l100percentrenewableenergyscenarios-2019.pdf

Pedraza. (2019). *Conventional energy in North America*, ISBN 978-0-12-814889-1, https://doi.org/10.1016/B978-0-12-814889-1.00004-8

REN21 GSR. (2020). *Renewables 2021, Global status report*, REN21 Secretariat, c/o UN Environment Programme; 1 rue Miollis, Building VII, 75015 Paris, France, ISBN 978-3-948393-03-8

Seidenberger, T., Thrän, D., Offermann, R., Seyfert, U., Buchhorn, M., Zeddies, J. (2008). *Global biomass potentials*. Report prepared for Greenpeace International, German Biomass Research Center, Leipzig. https://www.proquest.com/docview/816398809

Teske, S., Pregger, T., Simon, S., Naegler, T., Pagenkopf, J., van den Adel, B., Meinshausen, M., Dooley, K., Briggs, C., Dominish, E., Giurco, D., Florin, N., Morris, T., Nagrath, K., Deniz, Ö., Schmid, S., Mey, F., Watari, T., & McLellan, B. (2019). Achieving the Paris climate agreement goals global and regional 100% renewable energy scenarios with non-energy GHG pathways for +1.5°C and +2°C. In S. Teske (Ed.), *Achieving the Paris climate agreement goals: Global and regional 100% renewable energy scenarios with non-energy GHG pathways for +1.5C and +2C*. Springer International Publishing. https://doi.org/10.1007/978-3-030-05843-2

Teske, S., Niklas, S., Atherton, A., Kelly, S., & Herring, J. (2020). *Sectoral pathways to net zero emissions*. https://www.uts.edu.au/sites/default/files/2020-12/OECMSectorPathwaysReportFINAL.pdf

Teske, S., Pregger, T., Simon, S., Naegler, T., Pagenkopf, J., Deniz, Ö., van den Adel, B., Dooley, K., & Meinshausen, M. (2021). It is still possible to achieve the paris climate agreement: Regional, sectoral, and land-use pathways. *Energies, 14*(8), 1–25. https://doi.org/10.3390/en14082103

UBA. (2016). *CO2 emission factors for fossil fuels, climate change 28/2016, Kristina Juhrich, Emissions Situation (Section I 2.6), German Environment Agency (UBA) June 20*.

UN. (1996). *United Nations Press release, 9th October 1996, New United Nations Handbook Indicates Global Energy Production and Consumption Pattern in Years 1991–1994* https://www.un.org/press/en/1996/19961009.dev2120.html

Chapter 13
Scopes 1, *2*, and *3* Industry Emissions and Future Pathways

Sven Teske, Kriti Nagrath, Sarah Niklas, Simran Talwar, Alison Atherton, Jaysson Guerrero Orbe, Jihane Assaf, and Damien Giurco

Abstract The *Scope 1*, *2*, and *3* emissions analysed in the OECM are defined and are presented for the 12 sectors analysed: (1) energy, (2) power and gas utilities, (3) transport, (4) steel industry, (5) cement industry, (6) farming, (7) agriculture and forestry, (8) chemical industry, (9) aluminium industry, (10) construction and buildings, (11) water utilities, and (12) textiles and leather industry. The interconnections between all energy-related CO_2 emissions are summarized with a Sankey graph.

Keywords Scope 1, 2, and 3 emissions · Industry · Service · Transport · Buildings steel · Cement · Aluminium · Chemicals agriculture · Forests water utilities · Textile and leather

13.1   Introduction

The OECM methodology has been presented in previous chapters, based on which energy consumption and supply concepts for sectorial pathways were developed. All 12 sectors analysed have been described, the assumptions presented, and the derivations of the energy pathways explained in detail. The resulting energy-related CO_2 levels for the sectors are described in Chap. 12. The present chapter focuses on the results of the calculated *Scope 1*, *2*, and *3* emissions for all the sectors analysed.

The industry-specific emission budgets are further subdivided into so-called *Scope 1*, *2*, and *3* emissions, which define the responsibility for those emissions. So far, this system has only been applied to companies, and not yet to entire industry

S. Teske (✉) · K. Nagrath · S. Niklas · S. Talwar · A. Atherton
J. G. Orbe · J. Assaf · D. Giurco
University of Technology Sydney – Institute for Sustainable Futures (UTS-ISF),
Sydney, NSW, Australia
e-mail: sven.teske@uts.edu.au

© The Author(s) 2022
S. Teske (ed.), *Achieving the Paris Climate Agreement Goals*,
https://doi.org/10.1007/978-3-030-99177-7_13

sectors or regions. For a better overview, the OECM definitions of *Scopes 1*, *2*, and *3*, which are explained in detail in Chap. 4, are shown again in Box 13.1.

Box 13.1: OneEarth Climate Model: Definitions of Scope 1, 2, and 3 Emissions

Scope 1 – All direct emissions from the activities of an organization or under their control. Including on-site fuel combustion, such as gas boilers, fleet vehicles, and air-conditioning leaks. For this analysis only, the economic activities covered under the sector-specific GICS classification that are counted under the sector are included. All the energy demands reported by the International Energy Agency (IEA) Advanced World Energy Balances (IEA, 2020, 2021) for a specific sector are included.

Scope 2 – Indirect emissions from electricity purchased and used by an organization. Emissions are created during the production of this energy, which is eventually used by the organization. For reasons of data availability, the calculation of these emission focuses on the electricity demand and 'own consumption', e.g. reported for power generation.

Scope 3 – Greenhouse gas (GHG) emissions caused by the analysed industry, limited to sector-specific activities and/or products, as classified in the GICS. The OECM only includes sector-specific emissions. Traveling, commuting, and all other transport-related emissions are reported under *transport*. The lease of buildings is reported under *buildings*. All other finance activities, such as 'capital goods', are excluded because no data are available for the GICS industry sectors, and their inclusion would lead to double counting. The OECM analysis is limited to energy-related CO_2 and energy-related methane (CH_4) emissions. All other GHGs are calculated outside the OECM model by Meinshausen and Dooley (2019).

The results and key parameters for the primary and secondary energy sectors are presented first, followed by those for the *industries* and *services*, *buildings*, and *transport* sectors.

13.2 Scope 1, 2, and 3: Energy and Utilities

The *energy* sector includes the primary production of energy from oil, gas, hard coal, and lignite, and all renewable energies. This includes the exploration for all types of fossil fuels; the operation of oil and gas drilling facilities, mining equipment, and fossil fuel transport to refineries; and further processing facilities, as defined under GICS Sector *10 Energy*. To remain within the defined carbon budget, no new oil, gas, or coal-mining projects can be opened up, an assumption that is

Table 13.1 GICS Sector 10 *energy*

10 Energy	
1010 10	Energy equipment and services
	1010 1010 Oil and gas drilling
	1010 1020 Oil and gas equipment and services
1010 20	Oil, gas, and consumable fuels
	1010 2010 Integrated oil and gas
	1010 2020 Oil and gas exploration and production
	1010 2030 Oil and gas refining and marketing
	1010 2040 Oil and gas storage and transportation
	1010 2050 Coal and consumable fuels

consistent with the recommendations of the IEA NetZero by 2050 report (IEA-NZ 2050) (Table 13.1).

As documented in section 10, the OECM 1.5 °C trajectory requires a phase-out of brown coal (lignite) and hard coal by 2030 in all Organization for Economic Cooperation and Development (OECD) countries and in all other regions thereafter by 2050, at the very latest. The phase-out of brown coal has priority over that of hard coal, because its specific CO_2 emissions are higher. For the oil and gas sector, it is assumed that existing mines will wind-down, with an average decline in production of minus 2% per year for coal, minus 4% per year for onshore oil fields and 6% per year offshore oil fields, and minus 4% per year on- and offshore gas fields, which represent the average industry standards on the global scale (see Chap. 10). However, the production decline rates will differ significantly by region and geological formation. It is assumed that natural gas will be phased out by 2050 and partly replaced by alternative fuels, such as hydrogen and/or synthetic fuels, from 2025 onwards. The *energy* sector is also assumed to transition to utility-scale renewable energy projects and therefore to maintain its core business of energy production. Utility-scale renewables are defined as power plants that produce bulk power that is sold to utilities or end-use customers in the *industry* or *service* sector, such as offshore and onshore wind farms, solar farms, and geothermal and biomass power plants (including combined heat and power) with over 1 megawatt installed capacity.

A significant part of the renewable energy production by this sector under the 1.5 °C pathway will be from offshore wind, both to supply utilities with electricity and to produce hydrogen and other synthetic fuels. Figure 13.1 shows the global amounts of annual energy production in petajoules (PJ). The renewable energy production level will reach parity with those of oil and gas by 2030 and will continue to grow throughout the next two decades. The remaining oil and gas production shown for 2050 is for non-energy use.

Energy—Scope 1 emissions are defined as the direct emissions related to the extraction, mining, and burning of fossils fuels. This analysis covers both the

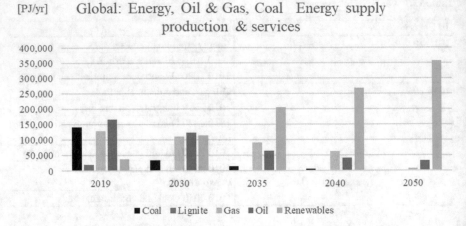

Fig. 13.1 Global primary energy sector—energy production under the OECM 1.5 °C pathway

energy-related CO_2 emissions and non-energy GHGs, such as methane (CH_4) emissions from mining and fossil fuel production.

Energy—Scope 2 emissions are indirect emissions from the electricity used for the operation of mining equipment, oil and gas rigs, refineries, and other equipment required in the primary energy sector. Their calculation is based on statistical information ('own consumption') from the IEA Advanced Energy Balances. The OECM assumes the global average carbon intensity of electricity generation for each calculated year according to the OECM power scenario, which will reach 100% renewables by 2050 (for details, see Chap. 12).

Energy—Scope 3 emissions are embedded CO_2 emissions, which occur when the fossil fuel produced by the primary energy industry is burnt by end users.

Table 13.2 shows the *scope 1, 2,* and *3* emissions for coal, oil, and gas and the development of the fuel intensity of the global economy. In 2019, as a global average, 1.25 PJ of coal was used for each billion US$ of gross domestic product ($GDP). This coal intensity is assumed to halve by 2025 and to drop by 85% by 2030. The global economy will grow independently of coal use under the OECM 1.5 °C pathway (Table 13.3).

The *utilities* sector covers energy transport and the operation and maintenance of power- and heat-generating equipment and is responsible for the energy transport infrastructure, such as power grids and pipelines to the end user. In this analysis, the *utilities* sector is a secondary energy service provider, whose core function is the generation and distribution of electricity and the distribution of natural gas, as well as hydrogen and synthetic fuels, beyond 2030. It operates and maintains power and cogeneration plants, power grids (all voltage levels), and pipelines and provides energy services, such as balancing, demand-side management, and storage. 'Utilities' are energy service companies linking the primary energy supply with consumers.

Electricity and gaseous fuel supplies are the core commodities of gas and power utilities. With the increased electrification of the transport and heating sectors, the

Table 13.2 Global *energy* sector—*scopes 1, 2*, and *3* for coal, oil, and gas

Sub-sector	Units	2019	2025	2030	2035	2040	2050
			Projections				
Coal (hard and brown)							
Coal *Scope 1*:	[MtCO$_2$eq/yr]	2434	1419	583	270	93	0
Compared with 2019	[%]		−42%	−76%	−89%	−96%	−100%
Coal Scope 2:	[MtCO$_2$eq/yr]	260.4	127.3	56.5	24.4	7.9	0.0
Compared with 2019	[%]		−51%	−78%	−91%	−97%	−100%
Coal *Scope 3*:	[MtCO$_2$eq/yr]	15,328.7	7864.3	3272.1	1327.6	290.7	0.0
Compared with 2019	[%]						
Total non-energy GHG:	[MtCO$_2$eq/yr]	111	67	37	27	12	0
Compared with 2019	[%]		−40%	−67%	−76%	−89%	−100%
Coal intensity of economy	[PJ/bn$GDP]	1.25	0.52	0.19	0.07	0.02	0.00
Compared with 2019	[%]		−59%	−85%	−94%	−98%	−100%
Oil							
Oil *Scope 1*:	[MtCO$_2$eq/yr]	990.3	918.5	732.3	383.6	251.1	191.0
Compared with 2019	[%]		−94%	−95%	−97%	−98%	−99%
Oil *Scope 2*:	[MtCO$_2$eq/yr]	165.6	153.5	122.4	64.1	42.0	31.9
Compared with 2019	[%]		−41%	−53%	−75%	−84%	−88%
Oil *Scope 3*:	[MtCO$_2$eq/yr]	10,416.1	9580.3	7162.7	2766.7	1051.1	0.0
Compared with 2019	[%]		−8%	−31%	−73%	−90%	−100%
Total non-energy GHG:	[MtCO$_2$eq/yr]	96	98	94	60	43	0
Compared with 2019	[%]		156%	147%	58%	11%	−100%
Gas							
Gas *Scope 1*:	[MtCO$_2$eq/yr]	1103.2	1062.8	949.0	778.5	535.2	71.1
Compared with 2019	[%]		−93%	−94%	−95%	−96%	−100%
Gas *Scope 2*:	[MtCO$_2$eq/yr]	60.7	58.4	52.2	42.8	29.4	3.9
Compared with 2019	[%]		−78%	−80%	−84%	−89%	−98%
Gas *Scope 3*:	[MtCO$_2$eq/yr]	8,082.4	7515.9	6608.3	5428.2	3792.6	478.2
Compared with 2019	[%]		−7%	−18%	−33%	−53%	−94%
Total non-energy GHG:	[MtCO$_2$eq/yr]	85	91	104	140	175	0
Compared with 2019	[%]		139%	172%	267%	357%	−100%

Table 13.3 Global *energy* sector—*scopes 1, 2*, and *3*

Total energy, gas, oil, and coal sector						
Energy sector—*scope 1*:	[MtCO$_2$eq/yr]	4527	3400	2265	880	262
	[%]	0%	−25%	−50%	−81%	−94%
Energy sector—*scope 2*:	[MtCO$_2$eq/yr]	487	339	231	79	36
	[%]	0%	−30%	−53%	−84%	−93%
Energy sector—*cope 3*:	[MtCO$_2$eq/yr]	33,827	24,960	17,043	5134	478
	[%]	0%	−26%	−50%	−85%	−99%

electricity demand—and therefore the potential market value of power utilities—will increase significantly. Renewable electricity will overtake global coal- and gas-fuelled power generation combined by 2025. By 2045, the market volume of hydrogen and synthetic fuels will be as high as that of natural gas for gas utilities, making them important new products.

Utilities—Scope 1 emissions are defined as the direct emissions from fuels related to the generation and transmission of electricity and the distribution of fossil fuels and/or renewable gas.

Utilities—Scope 2 emissions are indirect emissions from the electricity used for the production of a sector's core product. This includes the electricity consumption of power plants, losses by power grids, and the operation of pumps for gas pipelines, etc. Their calculations are based on statistical information listed under 'self-consumption' of the IEA Advanced Energy Balances plus the global average power grids losses, which are assumed to be 7.5%.

Utilities—Scope 3 emissions are embedded CO_2 emissions that occur with the use of electricity or gaseous fuels by end users. Table 13.4 shows all *scope 1, 2*, and *3* emissions for the *utilities* sector by sub-sector and in total.

Table 13.4 Global *utilities* sector—*scopes 1, 2,* and *3*

Sub-sector	Units	2019	2025	2030	2035	2040	2050
			Projections				
Power							
Power—*scope 1*:	[MtCO₂eq/yr]	1292	741	596	522	479	469
Compared with 2019	[%]		−43%	−54%	−60%	−63%	−64%
Power—*scope 2*:	[MtCO₂eq/yr]	112	86	54	24	11	0
Compared with 2019	[%]		−23%	−52%	−78%	−91%	−100%
Power—*scope 3*:	[MtCO₂eq/yr]	14,722	9124	5337	3110	1919	469
Compared with 2019	[%]		−38%	−64%	−79%	−87%	−97%
Gas							
Gas—*scope 1*:	[MtCO₂eq/yr]	1243	917	694	522	341	43
Compared with 2019	[%]		−26%	−44%	−58%	−73%	−97%
Gas—*scope 2*:	[MtCO₂eq/yr]	175	140	125	103	71	9
Compared with 2019	[%]		−20%	−29%	−41%	−60%	−95%
Gas—*scope 3*:	[MtCO₂eq/yr]	7,183	7009	6278	5151	3414	24
Compared with 2019	[%]		−2%	−13%	−28%	−52%	−100%
Total CH₄ emissions	[MtCH₄/yr]	14.9	14.4	12.9	10.5	7.3	1.0
Compared with 2019	[%]		−4%	−14%	−29%	−51%	−94%
Utilities							
Utilities—*scope 1*:	[MtCO₂eq/yr]	2535	1659	1289	1044	819	512
Compared with 2019	[%]		−35%	−49%	−59%	−68%	−80%
Utilities—*scope 2*:	[MtCO₂eq/yr]	287	226	179	127	81	10
Compared with 2019	[%]		−21%	−38%	−56%	−72%	−97%
Utilities—*scope 3*:	[MtCO₂eq/yr]	21,905	16,134	11,615	8,261	5333	493
Compared with 2019	[%]		−26%	−47%	−62%	−76%	−98%

13.3   *Scopes 1, 2,* **and** *3: Industry*

All results for the *scope* 1, 2, and *3* emissions for the five main energy-intensive industry sectors are based on the energy demand assessment documented in Chap. 5.

13.3.1   Scopes 1, 2, *and 3: Chemical Industry*

The *chemical industry* is the most complex industry of all the sectors analysed, and the data available on its energy demand are less detailed than for, for example, the steel industry. Furthermore, the production of chemical commodities (see Sect. 5.1) is energy intensive, and they are used not only across the chemical industry but also in other sectors. Therefore, the calculation results shown in Table 13.5 are subject to uncertainties resulting from the paucity of detailed data. The global energy demand data for, for example, the pharmaceuticals industry are not available, and the calculations are based upon sector-specific energy intensities and the market shares of the pharmaceuticals industry in 2019 (see Sect. 5.1.3).

Chemicals—Scope 1 emissions are defined as the direct emissions related to the production of raw materials for the chemical industry from natural gas, ethane, oil-refining by-products (such a propylene), and salt, which are used to manufacture bulk chemicals, such as sulfuric acid, ammonia, chlorine, industrial gases, and basic polymers, such as polyethylene and polypropylene.

Chemicals—Scope 2 emissions are indirect emissions from the electricity used for the production and processing of chemical products and the manufacture of goods that fall under *chemicals*, as classified under GICS 1510 10.

Chemicals—Scope 3 emissions are all non-energy-related GHG emissions and aerosols that fall under the Montreal Protocol (UNEP MP, 2021). Montreal Protocol gases are mainly propellants, foams, or liquids and gases used for cooling and refrigeration that are produced by the chemical industry. More details about these gases are given in Chap. 11.

Scope 1 and 2 emissions will reach zero by 2050, whereas *Scope 3* emissions will only be reduced by 73% compared with 2019 due to the nature of those substances.

13.3.2   Scope 1, 2 *and 3:* **Cement Industry**

The energy intensity of the cement production processes is well-documented, and data for the energy demands and process emissions are available. This analysis includes all steps in cement production, from quarrying the raw materials to its storage in cement silos. However, the further processing of cement for construction, for example, is not included but is included in the *buildings and construction* sector (Sect. 13.4).

Table 13.5 Global *scope 1*, 2, and *3* emissions of the *chemical industry*

Chemical industries	Units	2019	2025	2030	2035	2040	2050
Scope 1:	[MtCO$_2$eq/ yr]	1257	994	707	554	323	0
Compared with 2019	[%]		−21%	−44%	−56%	−74%	−100%
Scope 2:	[MtCO$_2$eq/ yr]	761	499	263	114	57	0
Compared with 2019	[%]		−34%	−66%	−85%	−93%	−100%
Scope 3:	[MtCO$_2$eq/ yr]	2520	1852	1220	991	775	682
Compared with 2019	[%]		−27%	−52%	−61%	−69%	−73%
Chemical industries—sub-sectors	**Units**	**2019**	**2025**	**2030**	**2035**	**2040**	**2050**
Pharmaceutical industry— *scope 1*	[MtCO$_2$eq/ yr]	230	181	129	101	59	0
Pharmaceutical industry— *scope 2*	[MtCO$_2$eq/ yr]	202	133	70	31	15	0
Pharmaceutical industry— *scope 3*	[MtCO$_2$eq/ yr]	0	0	0	0	0	0
Agricultural chemicals—*scope 1*	[MtCO$_2$eq/ yr]	270	213	151	119	69	0
Agricultural chemicals—*scope 2*	[MtCO$_2$eq/ yr]	111	73	38	17	8	0
Agricultural chemicals—*scope 3*	[MtCO$_2$eq/ yr]	0	0	0	0	0	0
Inorganic chemicals and consumer products—*scope 1*	[MtCO$_2$eq/ yr]	229	181	128	101	59	0
Inorganic chemicals and consumer products—*scope 2*	[MtCO$_2$eq/ yr]	205	134	71	31	15	0
Inorganic chemicals and consumer products—*scope 3*	[MtCO$_2$eq/ yr]	0	0	0	0	0	0
Manufactured fibres and synthetic rubber—*scope 1*	[MtCO$_2$eq/ yr]	301	237	169	132	77	0
Manufactured fibres and synthetic Rubber—*scope 2*	[MtCO$_2$eq/ yr]	39	25	13	6	3	0
Manufactured fibres and synthetic rubber—*scope 3*	[MtCO$_2$eq/ yr]	0	0	0	0	0	0
Bulk petrochemicals and intermediates, plastic resins— *scope 1*	[MtCO$_2$eq/ yr]	229	181	129	102	59	0
Bulk petrochemicals and intermediates, plastic resins— *scope 2*	[MtCO$_2$eq/ yr]	205	133	70	30	15	0
Bulk petrochemicals and intermediates, plastic resins— *scope 3*	[MtCO$_2$eq/ yr]	0	0	0	0	0	0

Table 13.6 Global *scope 1, 2, and 3* emissions for the *cement industry*

Total materials/cement	Units	2019	2025	2030	2035	2040	2050
Scope 1:	[MtCO$_2$eq/yr]	1731	1609	1388	1217	1044	734
Compared with 2019	[%]		−7%	−20%	−30%	−40%	−58%
Scope 2:	[MtCO$_2$eq/yr]	248	116	54	21	10	0
Compared with 2019	[%]		−53%	−78%	−92%	−96%	−100%
Scope 3:	[MtCO$_2$eq/yr]	9685	6085	3039	1416	746	0
Compared with 2019	[%]		−37%	−69%	−85%	−92%	−100%

Cement—Scope 1 emissions are defined as the direct energy-related CO$_2$ emissions related to all steps of cement production, from mining to the final raw product that is used in further processes and applications. The fuels for mining vehicles are included, as well as the process heat for clinker production in kilns, etc. Emissions from the calcination process—the decomposition of limestone into quick lime and carbon dioxide (Kumar et al., 2007)—are also included.

Cement—Scope 2 emissions are the indirect emissions from the electricity used across all steps of the value chain of the cement industry.

Cement—Scope 3 emissions of the cement industry are *scope 2* emissions of the *buildings* sector, according to the World Business Council for Sustainable Development's Cement Sector Reporting Guidance (WBCSD, 2016).

By 2050, there will be no energy-related CO$_2$ emissions from the cement industry under the OECM 1.5 °C pathway (Sect. 5.2). Process emissions from calcination are assumed to decline from 0.4 tCO$_2$ per tonne of clinker to 0.24 tCO$_2$—an assumption based on the IEA Technology Roadmap (IEA, 2018). Table 13.6 and Fig. 13.2 show the calculated results for the *scope 1, 2, and 3* emission of the global cement industry.

13.3.3   Scopes 1, 2, *and* 3: Aluminium Industry

As for the cement industry, all aluminium production processes are well-documented. The processes and their energy demand for each step of aluminium production, from bauxite mining to aluminium sheets or aluminium blocks, which are then delivered to other industries for further processing, are available in the literature. The recycling of aluminium for the production of secondary aluminium is also included. All assumptions for the projected development of the aluminium industry—including bauxite mining—are documented in Sect. 5.3.

Aluminium—Scope 1 emissions are defined as the direct energy-related CO$_2$ emissions related to the use of fuels for bauxite mining, alumina processing, and all steps of the production of primary and secondary aluminium. The process emissions from anode or paste (IAI, 2006) consumption, which lead to CO$_2$ emissions that are not energy related, are included.

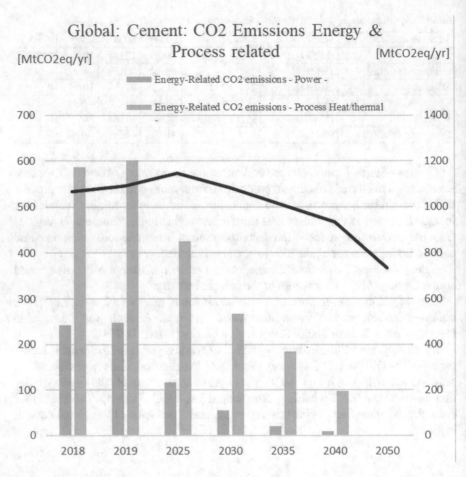

Fig. 13.2 Global *cement* sector—energy- and process-related CO_2 under the OECM 1.5 °C pathway

Aluminium—Scope 2 emissions are the indirect emissions from the electricity used across all the steps of the value chain of the aluminium industry.

Aluminium—Scope 3 emissions are solely those emissions caused by tetrafluoromethane, a strong GHG that is produced in certain aluminium production processes. A recent study published in Nature highlights the increased emissions of this gas, which probably derive from aluminium production facilities in Asia (Nature 8/2021). We decided to include tetrafluoromethane emissions in this OECM analysis to highlight the importance of this finding.

By 2050, all energy-related CO_2 emissions of the aluminium industry will be zero and the industry will be fully decarbonized. However, process-related GHG emissions are not expected to be completely phased out (Table 13.7).

Table 13.7 Global *scope 1, 2*, and *3* emissions for the *aluminium industry*

Aluminium industry	Units	2019	2025	2030	2035	2040	2050
Scope 1:	[MtCO$_2$eq/yr]	401	337	308	297	282	270
Compared with 2019	[%]		−16%	−23%	−26%	−30%	−33%
Scope 2:	[MtCO$_2$eq/yr]	522	305	144	57	26	0
Compared with 2019	[%]		−42%	−72%	−89%	−95%	−100%
Scope 3:	[MtCO$_2$eq/yr]	72	47	15	13	14	18
Compared with 2019	[%]		−35%	−79%	−82%	−81%	−75%

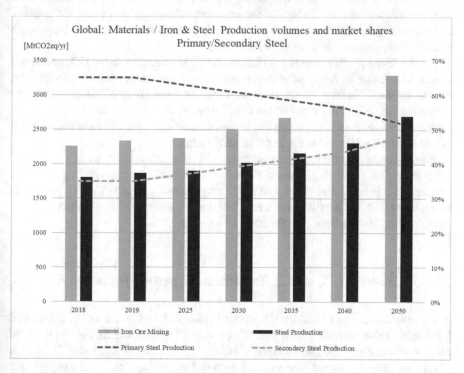

Fig. 13.3 Global *steel* sector—iron-ore mining and steel production under the OECM 1.5 °C pathway

13.3.4   Scope 1, 2, *and* 3: Steel Industry

Global and regional steel industry emissions are among the most discussed of all industry emissions. Various industry- and science-based working groups have developed relevant scenarios over the past decade. However, most of them are consistent with the Iron and Steel Technology Roadmap of the IEA (IEA, 2020). The OECM 1.5 °C pathway for the *steel industry* is based to a large extent on the IEA assumptions for the energy demand side but has added a more ambitious decarbonization scenario for the energy *supply* side. Figure 13.3 shows the development of iron-ore mining and primary and secondary steel production assumed for the global market between 2019 and 2050. The increase in secondary steel—recycled steel, mainly from scrap—will increase from 35% in 2019 to 48% in 2050, leading to a reduction in the iron and mining demands and the process emissions that are only related to primary steel production. Therefore, a high recycling rate will directly affect process emissions, which are not related to the actual energy supply but to the steel-making process itself. Further information about the assumptions for the *steel industry* is documented in Chap. 5 (Sect. 5.4).

The OECM analysis includes energy-related CO_2 emissions that occur from iron-ore mining across all steps of the steel manufacturing processes for primary and secondary steel but exclude manufacturing processes that use steel for product manufacture, such as the automotive industry.

Steel—Scope 1 emissions are defined as the direct energy-related CO_2 emissions related to the use of fuels for iron-ore mining and the production of primary and secondary steel. Process emissions from anode or paste (IAI, 2006) consumption, which lead to CO_2 emissions that are not energy related, are included.

Steel—Scope 2 emissions are the indirect emissions from the electricity used across all steps of the value chain of the steel industry.

Steel—Scope 3 emissions are only process-related emissions, as defined in the 2006 IPCC Guidelines for National Greenhouse Gas Inventories (IPCC, 2006). It is assumed that process emissions will decline significantly from 0.92 tCO_2 per tonne currently to 0.08 tCO_2 by 2050 as a result of the transition to electric-furnace-based steel production (see Sect. 5.4.3) (Table 13.8).

13.3.5   Scopes 1, 2, *and* 3: Textile and Leather Industry

The *textile and leather industry* is part of the IEA *industry* sector but is not part of the GICS (15) *materials* group (see Chap. 4). The *textile and leather industry* is closely associated with the *chemicals industry*, from which synthetic fibres and plastic for accessories are sourced, and with the *agriculture* sector, for cotton and other natural fibres. The production of leather depends on animal farms, especially those that produce meat. The assumptions made for the calculation of the energy-related CO_2 emissions of this industry are documented in Sect. 5.5.

Table 13.8 Global *scope 1*, 2, and 3 emissions for the *steel* industry

Total materials/steel	Units	2019	2025	2030	2035	2040	2050
Scope 1:	[MtCO$_2$eq/yr]	1073	762	489	353	187	0
Compared with 2019	[%]		−29%	−54%	−67%	−83%	−100%
Scope 2:	[MtCO$_2$eq/yr]	645	459	222	95	48	0
Compared with 2019	[%]		−29%	−66%	−85%	−93%	−100%
Scope 3:	[MtCO$_2$eq/yr]	1980	1757	1219	804	542	216
Compared with 2019	[%]		−11%	−38%	−59%	−73%	−89%

Table 13.9 Global *scope 1, 2,* and *3* emissions for the *textile and leather industry*

Textile and leather industry	Units	2019	2025	2030	2035	2040	2050
Scope 1:	[$MtCO_2eq/yr$]	178	151	109	87	51	0
Compared with 2019	[%]		−15%	−39%	−51%	−71%	−100%
Scope 2:	[$MtCO_2eq/yr$]	181	127	68	30	15	0
Compared with 2019	[%]		−30%	−62%	−83%	−92%	−100%
Scope 3:	[$MtCO_2eq/yr$]	38	30	24	22	22	20
Compared with 2019	[%]		−23%	−37%	−43%	−44%	−48%

Textile and Leather—Scope 1 emissions are defined as the direct energy-related CO_2 emissions associated with all the steps of textile and leather production that require process heat or fuels. It covers leather production, but not the production of fibres, which is part of the *chemicals* sector. The calculation of these emissions includes the value chain until delivery to retail.

Textile and Leather—Scope 2 emissions are the indirect emissions from the electricity used for the production of textile and leather products, excluding fibres manufacture and retail.

Textile and Leather—Scope 3 emissions include 25% of all CH_4 emissions from the agricultural sector to reflect the non-energy-related GHG emissions associated with the production of natural fibres and animal skins (Table 13.9).

13.4 *Scope 1, 2,* and *3*: *Services*

All the results for the *scope 1, 2,* and *3* emissions of the four *service* sectors analysed are based on the energy demand assessment documented in Chap. 6. Non-energy-related GHG emissions form the majority of the *service* sector emissions, whereas energy-related CO_2 is a relatively small component compared with that in other sectors, such as *industry* and *transport*. These non-energy-related GHG emissions—referred to as agriculture, forestry, and other land-uses (AFOLU) in climate science—are among the main emitters of non-energy CO_2, CH_4, and nitrous oxide (N_2O). The *service* sectors analysed, *agriculture and food, forestry and wood, fisheries,* and *water utilities,* are described and the assumptions are documented in Chaps. 6, 11, and 14. Therefore, in this section, we focus solely on the presentation of their calculated *scope 1, 2,* and *3* emissions.

13.4.1 Scope 1, 2, *and* 3: Agriculture and Food *Sector*

The *agriculture and food* sector includes all economic activities from 'the field to the supermarket'. With regard to the energy used, this sector is a combination of the service sector *agriculture* and the industry sub-sector *food and tobacco.* Therefore, it includes crop and animal farming and the processing of all commodities into food, beverages, and tobacco products.

Agriculture and Food—Scope 1 emissions are related to fuel used in agricultural vehicles, such as tractors, machinery for harvesting and other equipment used on farms, as well as heat for food and tobacco processing and packaging.

Agriculture and Food—Scope 2 emissions include those for electricity purchased from utilities for either farming or any step in food processing or packaging. On-site electricity generation (e.g. on farms via solar photovoltaic, wind power, or bioenergy from residuals) will reduce *scope 2* emissions, but sub-sector-specific on-site generation is not assumed in this analysis.

Agriculture and Food—Scope 3 emissions include AFOLU emissions, N_2O, and ammonia emissions from fertilizers and CH_4 emissions (see Chaps. 11 and 14).

All energy-related CO_2 emissions of the *agriculture and food* sector will be reduced by half by 2030 and phased out entirely by 2050. However, it is assumed that AFOLU emissions from agriculture cannot be reduced to zero, because the demand for food for the growing global population will increase (Table 13.10).

13.4.2 Scopes 1, 2, *and* 3: Forestry and Wood *Sector*

Like the *agriculture and food* sector, the *forestry and wood* sector contains to subsectors: *forestry*, which is part of the IEA's *other sectors*, and the IEA industry subsector *wood and wood products*, which includes the pulp and paper industry. Details of the energy demand of this sector are provided in Sect. 6.2.

Forestry and Wood—Scope 1 emissions include those from heavy machinery for wood harvesting, all-terrain vehicles, power tools, chainsaws, etc.

Forestry and Wood—Scope 2 emissions are the indirect emissions from electricity. Like the agricultural sector, the forestry sector has significant potential for on-site power and heat generation, e.g. from forestry residuals, which can lower its *scope 2* emissions, but this is not assumed under the OECM 1.5 °C pathway.

Table 13.10 Global *scope 1*, *2*, and 3 emissions for the *agriculture and food* sector (including tobacco)

Agriculture, food, and tobacco	Units	2019	2025	2030	2035	2040	2050
Agriculture, food processing—*scope 1*:	[MtCO₂eq/yr]	355	272	184	134	95	0
Compared with 2019	[%]			−24%	−48%	−62%	−73%	−100%
Agriculture, food processing—*scope 2*:	[MtCO₂eq/yr]	975	632	324	138	67	0
Compared with 2019	[%]			−35%	−67%	−86%	−93%	−100%
Agriculture, food processing—*scope 3*:	[MtCO₂eq/yr]	6,837	5413	4515	4243	4205	3994
Compared with 2019	[%]			−21%	−34%	−38%	−38%	−42%

Forestry and Wood—Scope 3 emissions are forestry-related AFOLU emissions. The transition towards sustainably managed forests, the cessation of deforestation, and the commencement of reforestation are integral parts of the OECM 1.5 °C pathway as 'carbon sinks'. Therefore, *scope 3* emissions will become negative by 2030 (see Chaps. 11 and 14) (Table 13.11).

13.4.3 Scopes 1, 2, *and* 3: Fisheries *Sector*

The majority of all energy-related *scope 1* and 2 emissions in this industry are from fishing vessels and other equipment directly related to wild catches and aquaculture fish farms. Whereas the energy demand for fishing vessels is documented in the literature (see Sect. 6.3), no statistical data on the global energy demand for aquaculture and fish farming are available. Instead, only accumulated data on the GHG emissions for the global aquaculture sector have been published and have been used to calculate the *scope 3* emissions (MacLeod et al., 2020). Therefore, the energy demand of the fishing industry in 2019 and its projection until 2050 are estimates with uncertainties.

Fisheries—Scope 1 emissions are defined as the direct energy-related CO_2 emissions related to the use of fuels for fishing vessels and directly related to the infrastructure, such as refrigerators and freezers for fish on board fishing vessels.

Fisheries—Scope 2 emissions are the indirect emissions from the electricity used for cooling devices as part of the cooling chain for fish, from 'dock to supermarket'.

Fisheries—Scope 3 emissions are emissions from aquaculture as defined by MacLeod et al. (2020) as 'emissions arising from fishmeal production, feed blending, transport … and non-feed emissions from the nitrification and denitrification of nitrogenous compounds in the aquatic system ('aquatic N_2O')'. Also included are the estimated energy-use emissions, mainly for pumping water.

Table 13.12 shows the results for all the calculated emissions in this industry. It is assumed that about one-quarter of aquaculture *scope 3* emissions are directly related to energy use and will therefore be reduced to zero with the use of 100% renewable energy.

Table 13.11 *Global scope 1, 2, and 3 emissions for the forestry and wood sector*

Forestry and wood	Unit	2019	2025	2030	2035	2040	2050
Forestry, wood products—*scope 1*:	[MtCO₂eq/yr]	196	155	105	76	54	0
Compared with 2019	[%]			−21%	−47%	−61%	−73%	−100%
Forestry, wood products—*scope 2*:	[MtCO₂eq/yr]	344	184	97	42	21	0
Compared with 2019	[%]		−46%	−72%	−88%	−94%	−100%
Forestry, wood products—scope 3:	[MtCO₂eq/yr]	2648	1164	−619	−1241	−835	−1359
Compared with 2019	[%]		−56%	−123%	−147%	−132%	−151%

Table 13.12 Global *scope 1, 2,* and *3* emissions for the *fisheries* sector

Fishery	Units	2019	2025	2030	2035	2040	2050
Fishing industry—*scope 1*:	[MtCO$_2$eq/yr]	29	28	25	21	16	0
Compared with 2019	[%]	0%	−4%	−15%	−29%	−47%	−100%
Fishing industry—*scope 2*:	[MtCO$_2$eq/yr]	4	3	1	1	0	0
Compared with 2019	[%]	0%	−32%	−63%	−82%	−88%	−100%
Fishing industry*scope 3*:	[Mt CO$_2$eq/yr]	250	239	227	215	202	178
Compared with 2019	[%]	0%	−4%	−9%	−14%	−19%	−29%

Table 13.13 Global *scope 1, 2,* and *3* emissions for *water utilities*

Water utilities	Unit	2019	2025	2030	2035	2040	2050
Water utilities—*scope 1*:	[MtCO$_2$eq/yr]	77	53	33	22	15	0
Compared with 2019	[%]	0%	−32%	−57%	−71%	−81%	−100%
Water utilities—*scope 2*:	[MtCO$_2$eq/yr]	27	14	7	3	1	0
Compared with 2019	[%]	0%	−47%	−74%	−90%	−95%	−100%
Water utilities—*scope 3*:	[MtCO$_2$eq/yr]	830	881	925	971	1020	1125
Compared with 2019	[%]	0%	6%	11%	17%	23%	35%

13.4.4 Scopes 1, 2, *and* 3: Water Utilities

Only 13% of the GHG emission from *water utilities* are related to energy use. The bulk of GHG emissions are related to CH_4 and N_2O emission from sewers or the treatment of biological wastewater and the resulting sludge. Chapter 6 documents all the assumptions and input data used to calculate the *scope 1, 2,* and *3* emissions for water utilities.

Water Utilities—Scope 1 emissions are defined as the direct energy-related CO_2 emissions associated with the supply of the low- and medium-temperature process heat used in all steps of wastewater treatment.

Water Utilities—Scope 2 emissions are the indirect emissions from the electricity used across all steps of wastewater treatment processes.

Water Utilities—Scope 3 emissions are the CH_4 and N_2O emissions from sewers or biological wastewater treatment. They are calculated with average global emission factors of 0.17 kg CO_2 equivalents per cubic metre ($kgCO_2eq/m^3$) for CH_4 emissions and 0.033 $kgCO_2eq/m^3$ for N_2O emissions.

Water utilities have significant potential to use the CH_4 from sewage and wastewater treatment for on-site power and heat generation. The identified *scope 2* emissions for water utilities do not include the implementation of this technology. The *scope 3* emissions shown in Table 13.13 are entirely related to CH_4 and N_2O emissions and are projected to increase with the growing global population. The use of on-site CH_4 emissions with a global warming potential (GWP) of 25 (see Chap. 11) for electricity and heat generation would result in CO_2 (GWP = 1), instead of CH_4 emissions, and would therefore significantly reduce the *scope 3* emissions.

Therefore, we strongly recommend the utilization of on-site CH_4 emissions for energy generation.

13.5   *Scopes 1, 2*, and *3*: *Buildings*

The *buildings* sector is further broken down into residential and commercial buildings and is based on calculations that include construction. The energy demand for construction is taken from the IEA World Energy Balances, and the demand includes the *construction of buildings* (ISIC Rev. 4, Div. 41), *civil engineering* (ISIC Rev. 4, Div. 42), and *specialized construction activities* (Div. 43), as documented in Chap. 4. It is assumed that 60% of the energy used for construction is for buildings. The energy demands calculated for residential and commercial buildings are based on a separate research project under the leadership of the Central European University (Chatterjee et al., 2021) and are documented in Chaps. 3 and 7.

Buildings—Scope 1 emissions are defined as direct energy-related CO_2 emissions associated with the construction of those buildings.

Buildings—Scope 2 emissions are indirect emissions from the residential and commercial use of electricity and energy for space heating. The commercial electricity demand is the remaining electricity that is not allocated elsewhere in the *service*, *industry*, *transport*, or *residential* sectors, to avoid double counting.

Buildings—Scope 3 emissions are the *scope 1* emissions of the *cement industry* to capture the embedded building emissions from construction materials.

There are no *scope 3* emissions calculated for construction to avoid double counting with the remaining *buildings* sector. Table 13.14 shows the global *scope 1, 2*, and *3* emissions for all sub-sectors and for the overall *buildings* sector.

13.6   Scope *1, 2*, and *3*: *Transport*

The *transport* sector includes all travel modes (aviation, shipping, and road transport), and passenger and freight transport have been calculated separately on the basis of current and projected passenger-kilometres (pkm) and tonne-kilometres (tkm), as documented in Chap. 8. The *transport* sector includes the manufacture of vehicles and other transport equipment, as defined in GICS group 2030 (see Chap. 4) and documented in Sect. 8.9.

Transport—Scope 1 emissions are defined as the direct energy-related CO_2 emissions associated with the manufacture of road and rail vehicles, planes, and ships.

Transport—Scope 2 emissions are the indirect emissions from electricity used for all from the electric drives in vehicles and the electricity required for hydrogen or synthetic fuel production. The emission factors for this electricity—as in all other *scope 2* emission calculations—are based on the OECM 1.5 °C pathway for power

Table 13.14 Global *scope 1*, *2*, and *3* emissions for *buildings*

Residential and commercial Buildings and construction	Units	2019	2025	2030	2035	2040	2050
Buildings—*scope 1*:	[MtCO$_2$eq/yr]	128	81	54	38	23	0
Compared with 2019	[%]	0%	−37%	−58%	−71%	−82%	−100%
Buildings—*scope 2*:	[MtCO$_2$eq/yr]	9685	6085	3039	1416	746	0
Compared with 2019	[%]	0%	−37%	−69%	−85%	−92%	−100%
Buildings—*scope 3*:	[MtCO$_2$eq/yr]	1,690	1959	1609	1388	1217	884
Compared with 2019	[%]	0%	16%	−5%	−18%	−28%	−48%
Buildings—sub-sectors							
Residential buildings—*scope 1*:	[MtCO$_2$eq/yr]	83	52	34	23	14	0
Compared with 2019	[%]	0%	−38%	−60%	−72%	−83%	−100%
Residential buildings—*scope 2*:	[MtCO$_2$eq/yr]	4578	2830	1343	605	320	0
Compared with 2019	[%]	0%	−38%	−71%	−87%	−93%	−100%
Residential buildings—*scope 3*:	[MtCO$_2$eq/yr]	1120	1023	860	739	623	428
Compared with 2019	[%]	0%	−9%	−23%	−34%	−44%	−62%
Commercial buildings—*scope 1*:	[MtCO$_2$eq/yr]	45	30	21	15	9	0
Compared with 2019	[%]	0%	−35%	−54%	−67%	−80%	−100%
Commercial buildings—*scope 2*:	[MtCO$_2$eq/yr]	5107	3255	1696	810	425	0
Compared to 2019	[%]	0%	−36%	−67%	−84%	−92%	−100%
Commercial buildings—*scope 3*:	[MtCO$_2$eq/yr]	611	586	528	478	421	305
Compared with 2019	[%]	0%	−4%	−13%	−22%	−31%	−50%
Construction of buildings—*scope 1*:	[MtCO$_2$eq/yr]	128	81	54	38	23	0
Compared with 2019	[%]	0%	−37%	−58%	−71%	−82%	−100%
Construction buildings—*scope 2*:	[MtCO$_2$eq/yr]	63	36	20	9	5	0
Compared with 2019	[%]	0%	−43%	−69%	−86%	−93%	−100%

generation, with an emission factor of 0.5 kg CO_2 per kilowatt-hour in 2019, which will decline to zero by 2050.

Transport—*Scope 3* emissions are all the emissions caused by the utilization of all vehicles, planes, and ships for passenger and freight transport by end users. These emissions are not further allocated to other sectors in which vehicles are used to avoid double counting. Data are unavailable on how freight kilometres are distributed to, for example, the cement or steel industry.

Table 13.15 provides the global *scope 1*, *2*, and *3* emissions for the *transport* sector. Specific emissions from, for example, airports or single airline offices, as

Table 13.15 Global *scope* 1, 2, and 3 emissions for the transport sector

Total transport sector	Units	2019	2025	2030	2035	2040	2050
Aviation: transport —*scope 1*:	[MtCO$_2$eq/yr]	16	10	6	4	2	0
Compared with 2019	[%]	0	−39%	−62%	−76%	−86%	−100%
Aviation: transport—*scope 2*:	[MtCO$_2$eq/yr]	0.4	14.7	25.9	62.0	35.2	0.0
Compared with 2019	[%]	0	0%	0%	0%	0%	0%
Aviation: transport—*scope 3*:	[MtCO$_2$eq/yr]	936.1	1439.0	1193.3	459.4	96.0	0.0
Compared with 2019	[%]	0	54%	27%	−51%	−90%	−100%
Navigation: transport—*scope 1*:	[MtCO$_2$eq/yr]	23	14	9	6	3	0
Compared with 2019	[%]	0	−38%	−61%	−75%	−86%	−100%
Navigation: transport—*scope 2*:	[MtCO$_2$eq/yr]	1.7	20.4	144.9	56.4	68.9	0.0
Compared with 2019	[%]	0%	0%	0%	0%	0%	0%
Navigation: transport—*scope 3*:	[MtCO$_2$eq/yr]	793.8	3645.2	2601.8	2624.9	510.6	0.0
Compared with 2019	[%]	0%	359%	228%	231%	−36%	−100%
Road: transport—*scope 1*:	[MtCO$_2$eq/yr]	183	111	70	44	25	0
Compared with 2019	[%]	0	−39%	−62%	−76%	−86%	−100%
Road: transport—*scope 2*:	[MtCO$_2$eq/yr]	34.5	157.8	119.7	287.8	138.1	0.0
Compared with 2019	[%]	0%	0%	0%	0%	0%	0%
Road: transport—*scope 3*:	[MtCO$_2$eq/yr]	7223.4	6010.9	4769.2	1410.2	643.8	0.0
Compared with 2019	[%]	0%	−17%	−34%	−80%	−91%	−100%
	Units	2019	2025	2030	2035	2040	2050
Transport—*scope 1*:	[MtCO$_2$eq/yr]	223	136	85	53	30	0
Compared with 2019	[%]	0	−38%	−61%	−75%	−86%	−100%
Transport—*scope 2*:	[MtCO$_2$eq/yr]	36.6	192.8	290.4	406.2	242.2	0.0
Compared with 2019	[%]	0	445%	721%	1049%	585%	−100%
Transport—*scope 3*:	[MtCO$_2$eq/yr]	8953.2	11095.2	8564.3	4494.5	1250.5	0.0
Compared with 2019	[%]	0	27%	−2%	−49%	−86%	−100%

defined under GICS 2030 5010, cannot be assessed on a global scale because of lack of data. Furthermore, these emissions are allocated under 'commercial buildings'. *Scope 3* emissions are the 'classic' emissions when consumers drive a car or use a plane. The OECM deliberately includes electricity emissions from, for example, electric cars under *scope 2* emissions, because car manufacturers today include the charging infrastructure in their value chain and are therefore responsible for it.

13.7   *Scopes 1, 2*, and *3*: Global Summary

A global assessment of *scopes 1, 2*, and *3* for the whole *industry* sector is a new research area, and changes had to be made to the method for determining those emissions, which was originally developed by the World Resource Institute (WRI), as documented in Chap. 4.

The OECM methodology differs from the original concept primarily insofar as the interactions between industries and/or other services are kept separate. A primary class is defined for the primary energy industry, a secondary class for the supply utilities, and an end-use class for all the economic activities that consume energy from the primary- or secondary-class companies, to avoid double counting. All the emissions by defined industry categories (e.g. with GICS) are also separated, streamlining the accounting and reporting systems. The volume of data required is reduced, and reporting is considerably simplified with the OECM methodology.

Figure 13.4 shows the global energy-related *scope 1, 2*, and *3* CO_2 emissions in 2019 as a Sanky flow chart. The primary energy emissions are on the left and the end-use-related emissions are on the right. The carbon budgets remain constant, from production to end use, apart from losses and statistical differences. A simplified description is that all *scope 1* emissions are on the left, with the primary energy industry as the main emitter, and all *scope 3* emissions are on the right, with the consumers of all forms of energy and for all purposes as the main emitters. In the secondary energy industry, utilities are the link between the demand of end users

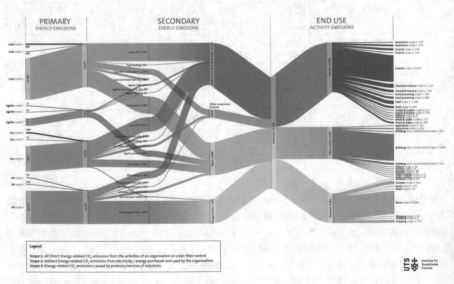

Fig. 13.4   Global *scope 1, 2*, and *3* energy-related CO_2 emissions in 2019

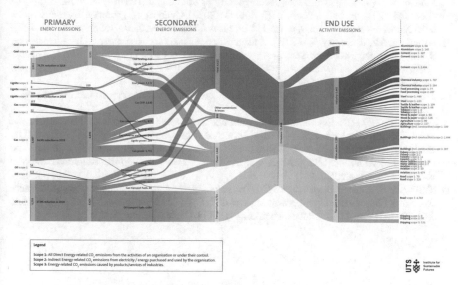

Fig. 13.5 Global *scope 1, 2,* and *3* energy-related CO_2 emissions in 2030 under the OECM 1.5 °C pathway

and the supply by the primary energy industry. The figure also shows the complex interconnections between demand and supply.

Figure 13.5 shows the energy-related CO_2 emissions and the interconnections between various sectors and consumers in 2030 under the global 1.5 °C pathway.

References

Chatterjee, S., Kiss, B., & Ürge-Vorsatz, D. (2021). *How far can building energy efficiency bring us towards climate neutrality?* Central European University.

IAI. (2006). *International Aluminium Institute, the aluminium sector greenhouse gas protocol, greenhouse gas emissions monitoring and reporting by the aluminium industry,* October 2006, https://ghgprotocol.org/sites/default/files/aluminium_1.pdf

IEA. (2018). *Technology roadmap—Low-carbon transition in the cement industry.* Paris. Available at: https://www.iea.org/reports/technology-roadmap-low-carbon-transition-in-the-cement-industry

IEA. (2020). *Iron and steel technology roadmap—Towards more sustainable steelmaking.* https://iea.blob.core.windows.net/assets/eb0c8ec1-3665-4959-97d0-187ceca189a8/Iron_and_Steel_Technology_Roadmap.pdf

IEA. (2021). *Net Zero by 2050.* IEA. https://www.ica.org/reports/net-zero-by-2050

IPCC. (2006). *2006 IPCC guidelines for national greenhouse gas inventories* https://www.ipcc-nggip.iges.or.jp/public/2006gl/pdf/3_Volume3/V3_4_Ch4_Metal_Industry.pdf

Kumar, G. S., Ramakrishnan, A., & Hung, Y. T. (2007). Lime calcination. In L. K. Wang, Y. T. Hung, & N. K. Shammas (Eds.), *Advanced physicochemical treatment technologies. Handbook of environmental engineering* (Vol. 5). Humana Press. https://doi.org/10.1007/978-1-59745-173-4_14

MacLeod, M. J., Hasan, M. R., Robb, D. H. F., et al. (2020). Quantifying greenhouse gas emissions from global aquaculture. *Sci Rep, 10*, 11679. https://doi.org/10.1038/s41598-020-68231-8

Meinshausen, M., & Dooley, K. (2019). Mitigation Scenarios for Non-energy GHG. In: Teske, S. (Ed.), *Achieving the Paris Climate Agreement Goals Global and Regional 100% Renewable Energy Scenarios with Non-energy GHG Pathways for +1.5°C and +2°C*. SpringerOpen; https://link.springer.com/chapter/10.1007/978-3-030-05843-2_4

Nature (8/2021). Research Highlight, 18 August 2021, What's the mystery source of two potent greenhouse gases? The trail leads to Asia.

UNEP MP. (2021). UN Environment Programme; Implementing Agency of the Multilateral Fund for the Implementation of the Montreal Protocol/Ozon Action as part of UN Environment Programme's Law Division and serves 147 developing countries through the Compliance Assistance Programme, website https://www.unep.org/ozonaction/who-we-are/about-montreal-protocol

WBCSD. (2016). *Cement sector scope 3 GHG accounting and reporting guidance*. World Business Council for Sustainable Development Maison de la Paix, Chemin Eugène-Rigot 2,Case postale 246, 1211 Geneve 21, http://docs.wbcsd.org/2016/11/Cement_Sector_Scope3.pdf

Chapter 14
Nature-Based Carbon Sinks: Carbon Conservation and Protection Zones

Kriti Nagrath, Kate Dooley, and Sven Teske

Abstract Basic information on ecosystem-based approaches to climate mitigation is provided, and their inclusion in international climate and nature conservation treaties is discussed. Key concepts around net-zero emissions and carbon removal are examined, as are the roles they play in the One Earth Climate Model, which develops a 1.5 °C-compatible scenario by combining ecosystem restoration with deep decarbonization pathways. The carbon removal potentials of the five ecosystem restoration pathways—forests and agricultural lands, forest restoration, reforestation, reduced harvest, agroforestry, and silvopasture—are provided. Land-use management options, including the creation of 'carbon conservation zones' (CCZ), are discussed.

Keywords Ecosystem-based approaches · Ecosystem restoration pathways · Forest restoration · Reforestation · Reduced harvest · Agroforestry · Silvopasture · 'Carbon conservation zones' (CCZ)

14.1 Ecosystem Approaches to Climate Action

This section looks at the variety of ecosystem approaches available for implementation as climate solutions. It also follows the global developments of ecosystem and nature outcomes from the recent climate summit, the 26th Conference of the Parties (COP26) to the United Nations Framework Convention on Climate Change (UNFCCC).

K. Nagrath · S. Teske (✉)
University of Technology Sydney – Institute for Sustainable Futures (UTS-ISF), Sydney, NSW, Australia
e-mail: sven.teske@uts.edu.au

K. Dooley
University of Melbourne, Melbourne, VIC, Australia

S. Teske (ed.), *Achieving the Paris Climate Agreement Goals*,
https://doi.org/10.1007/978-3-030-99177-7_14

337

14.1.1   Understanding Ecosystem Approaches

Climate change and climate action can no longer be discussed without reference to their environmental impacts, in particular the crises of biodiversity loss and ecosystem decline. Ecosystem approaches to climate management that restore degraded ecosystems and focus on nature will play important roles in climate solutions, for both mitigation and adaptation. These approaches aim to maintain and increase the resilience of people and the ecosystems upon which they rely and to reduce their vulnerability (Lo, 2016). Healthy, well-managed ecosystems have climate change mitigation potential, through the sequestration and storage of carbon in healthy forests, wetlands, and coastal ecosystems (IPBES, 2019).

Approaches to protecting and restoring nature can take a variety of forms. These include initiatives such as the sustainable management, conservation, and restoration of ecosystems. The ecosystem approach is a strategy for the integrated management of land, water, and living resources, which promotes their conservation and sustainable use in an equitable way, as defined by the Convention on Biological Diversity (CBD). The convention also defines 'ecosystem-based adaptation' (EbA) as the use of biodiversity and ecosystem services as part of an overall adaptation strategy to help people adapt to the adverse effects of climate change.

Ecosystem services are the benefits people obtain from ecosystems, which have been classified by the Millennium Ecosystem Assessment as supporting services, such as seed dispersal and soil formation; regulating services, such as carbon sequestration, climate regulation, water regulation and filtration, and pest control; provisioning services, such as the supply of food, fibre, timber and water; and cultural services, such as recreational experiences, education, and spiritual enrichment.

The International Union for Conservation of Nature (IUCN) defines nature-based solutions (NbS) as 'actions to protect, sustainably manage, and restore natural or modified ecosystems, that address societal challenges effectively and adaptively, simultaneously providing human well-being and biodiversity benefits'. These solutions include ecosystem restoration strategies, such as ecological restoration, ecological engineering, and forest landscape restoration; issue-specific ecosystem-related strategies, such as ecosystem-based adaptation or mitigation and disaster risk reduction; infrastructure-related strategies; ecosystem-based management strategies; and area-based ecosystem protection strategies.

'Landscape restoration' refers to the improvement of degraded land on a large scale, to rebuild ecological integrity and enhances people's lives. It involves restoring degraded forests and agricultural lands by reducing the intensity of use or improving productivity with mixed-use approaches, such as agroforestry and climate-smart agriculture (Winterbottom, 2014).

'Ecological engineering' is defined as the design of sustainable ecosystems that integrate human society with its natural environment for the benefit of both. It includes ecosystem rehabilitation (actions that repair the structures and functions of indigenous ecosystem), nature engineering, and habitat reconstruction and reclamation (stabilization, amelioration, increases in utilitarian or economic value). However, indigenous ecosystems are rarely used as models (Mitsch, 2012).

Therefore, there is a diversity of approaches that can be adopted to protect and restore the natural world.

14.1.2 Ecosystem and Nature Outcomes at COP26

The 2021 Glasgow Climate Pact recognizes the critical role of protecting, conserving, and restoring nature, while ensuring social and environmental safeguards, through the following text:

> Emphasizes the importance of protecting, conserving and restoring nature and ecosystems to achieve the Paris Agreement temperature goal, including through forests and other terrestrial and marine ecosystems acting as sinks and reservoirs of greenhouse gases and by protecting biodiversity, while ensuring social and environmental safeguards. (*Decision-/CP.26 Glasgow Climate Pact*, 2021).

Food, land, and nature were popular topics at COP26 and featured in a series of pledges, speeches, initiatives, and coalitions (Chandrasekhar & Viglione, 2021). These included deforestation pledges, new climate pledges, the methane pledge, and other agricultural innovation and policy announcements. The key pledges included:

- 141 countries containing >90% of global forests signed the Glasgow Leaders' Declaration on Forests and Land Use and committed to working collectively to halt and reverse forest loss and land degradation by 2030. Their efforts will include agricultural policies and programmes to incentivize sustainable agriculture, promote food security, and benefit the environment (2021). This declaration has mobilized over US$22 billion of public and private finance (gov.uk, 2021).
- 28 governments, representing 75% of global trade in key commodities that can threaten forests, signed a new Forest, Agriculture and Commodity Trade (FACT) Statement, which will reduce pressures on forests and deliver sustainable trade (gov.uk, 2021).
- The Global Methane Pledge was signed by 110 countries, responsible for nearly half the global methane emissions, with the aim of reducing their methane emissions by 30% from 2020 levels by 2030, using emissions mitigation strategies.
- The Agriculture Innovation Mission for Climate saw US$4 billion in new public-sector investment pledged for agricultural innovation, including climate-resilient crops and regenerative solutions to improve soil health.
- Canada announced CDN$1 billion in international support for nature-based solutions, a fifth of its climate finance budget.

Although these declarations are signs that we are moving in the right direction, there are concerns regarding the uncertainty around key definitions and the transition from promise to action, which must be resolved soon.

A World Wild Fund for Nature (WWF) study found that 92% of countries' new climate action plans now include measures to tackle nature loss. One hundred and five of the 114 enhanced Nationally Determined Contributions (NDCs) submitted by 12 October included nature in their climate mitigation or adaptation plans. Of the 96 NDCs that cited using nature for climate mitigation, 69 quantified these as numerical targets, mostly in the forest sector (Bakhtary et al., 2021).

14.1.3   Concepts of Consequence

This section discusses key concepts around net-zero emissions and carbon removal that we must understand to model the pathways in Chap. 11.

'Net zero' refers to the balance between the amount of greenhouse gases (GHGs) produced by humans and the amount removed from the atmosphere. This means that for any remaining emissions produced, an equivalent amount must also be removed through processes such as planting new forests, which reduce the GHGs accumulating in the atmosphere, to reach net-zero emissions.

As discussed in previous chapters, it is imperative for the various industry sectors to reduce their energy emissions (which primarily arise from fossil fuels) to zero. Given the temporal differences between the fossil and terrestrial carbon cycles, any essential residual emissions arising from non-energy sources and processes must be removed via geological storage, to go beyond net-zero emissions and eventually achieve net-negative emissions to reduce atmospheric GHG concentrations.

Carbon dioxide (CO_2) removal (CDR) is the process of removing CO_2 from the atmosphere and locking it away in a carbon 'sink' for a long period of time. A carbon sink is a natural or human-made reservoir that accumulates and stores carbon and thus lowers the concentration of CO_2 in the atmosphere. Forests and oceans are natural carbon sinks and absorb more CO_2 from the atmosphere than they emit. CDR requires that we enhance the ability of these natural sinks to remove and store carbon, or store this carbon geologically.

There are both natural and technological strategies to remove carbon from the atmosphere and store it in a sink. Natural strategies include reforestation and the ecosystem restoration approaches discussed above, where carbon is removed from the atmosphere by photosynthesis and stored in vegetation and soil. Although natural solutions, such as restorative agriculture and reforestation, can help remove carbon, they must be thoroughly monitored and balanced against competing demands on land use. Technological strategies, such as direct air capture and enhanced mineralization, that capture carbon underground or under the ocean or in products such as concrete, are also being explored but are yet to be commercialized on a large scale.

The OECM model focuses on natural strategies for carbon removal. The different land management pathways for achieving this are discussed in the next section.

14.2   Ecosystem Restoration Pathways

The OECM model presents a 1.5 °C-compatible scenario combined with ecosystem-based approaches. The ecosystem restoration pathways outlined in this section have been published as Littleton et al. (2021) and have been built on previous work by Meinshausen and Dooley (2019) .

14.2.1   Pathways

The five pathways involve forests and agricultural lands: forest restoration, reforestation, reduced harvest, agroforestry, and silvopasture. The first three pathways focus on the forestry sector, and the latter two are most relevant for the agriculture and food sector. In all three forest pathways, the intervention is natural regeneration with no active planting of trees (Littleton et al., 2021).

Forest restoration sets aside natural (secondary) forest areas that are partly deforested or degraded for conservation purposes. This pathway is applied to all biomes. Reforestation includes the reforestation of mixed native species maintained for conservation purposes. It is limited to biomes that would naturally support forests, after the identification of previously forested land in close proximity within 70–105 km for tropical forests and within 11–18 km of temperate forests. Reforestation in boreal biomes is excluded because the albedo effect accompanies changes from deforested to forested land types, specifically at high latitudes, which can potentially increase warming. The reforestation pathway is the only land-management intervention in this scenario that requires a change in land use. Reduced harvest describes a reduction in harvest intensity by 25% in commercial forests in boreal and temperate biomes. In tropical and subtropical biomes, commercial timber extraction is halted completely, given the lack of evidence that any form of reduced-impact logging leads to increased carbon stocks. These management interventions only apply to natural managed forests and not to plantations. Areas of shifting cultivation are excluded from consideration for reduced harvesting, to avoid impacting communities dependent on subsistence agriculture (Littleton et al., 2021).

The pathways involving the regeneration of agricultural areas—agroforestry and silvopasture—allow for existing land uses to continue. Temperate, subtropical, and tropical cropland and grazing areas with mean annual precipitation ranging from 400 to 1000 mm per year were targeted for these two pathways. Agroforestry can be implemented in many different ways, but here it is assumed to be the integration of additional trees into agricultural landscapes, which will result in significant sequestration across large areas of temperate and tropical croplands. Silvopasture, defined here as a reduction in grazing intensity on managed pastures, results in almost twice the level of carbon sequestration over a similar land area.

14.2.2   Methodology

Spatial distribution for the five pathways was identified using WRI's global map of forest condition and the ESA-CCI land cover maps for the forest and agriculture pathways, respectively. The areas identified for ecosystem restoration were simulated in a community land surface model, the Joint UK Land Environment Simulator (JULES) to get the carbon sequestration potential. JULES also incorporates the dynamic global vegetation model TRIFFID to simulate vegetation and carbon cycle processes. JULES simulations were run using meteorological forcing output from HadGEM2-ES, covering the period 1880–2014 (historical) and 2015–2100 (SSP1–2.6) on a 3-hour timestep at the N96e grid size.

For temperature projections, MAGICC, a reduced-complexity probabilistic climate emulator, was used, which reflects updated climate science knowledge. The scenarios are consistent with limiting warming to 1.5 °C by the end of the century, although at best, with a roughly 50–50 chance of staying below this limit.

A no-removal baseline scenario is modelled under a shared socioeconomic sustainability future (SSP1) and represents CO_2 emissions from forestry and land use (including land-use changes) in the absence of the ecosystem restoration measures considered here. To minimize the risk of double-counting sequestration, all carbon sequestration reported in this baseline scenario are set to zero from 2050 onwards.

14.2.3   Results

The results of Littleton et al. (2021) showed the median gross cumulative potential of additional CO_2 removal with the five ecosystem restoration pathways to be 93 Gt of carbon (C) until 2100, as shown in Table 14.1. The peak annual sequestration rate for all ecosystem restoration pathways (forest restoration, reforestation, reduced harvest, agroforestry, and silvopasture) is 3.1 GtC per year in 2041, as forests reach maturity. Then on, the flux declines, with an average annual sequestration rate of 1.1 GtC per year from 2050 to 2100. This removal will be offset by ongoing net land-use emissions but still has a significant contribution to temperature reduction. Combined with a 100% renewable energy scenario by 2050 in the OECM, this additional carbon uptake reduced 2100 temperature by a further 0.12 °C when compared to a no-removal scenario (Littleton et al., 2021).

The most successful restoration pathway identified in terms of carbon sequestration is *reduced harvest* in the tropics, with carbon gains of 80–100 tC per hectare by 2100 in Southeast Asia and the Amazon basin. Reduced harvest means that less timber is harvested. The pathway assumes that harvest intensity in temperate and

Table 14.1 Summary statistics for the cumulative uptake of CO_2 in all pathways

Pathway	Cumulative uptake (GtCO$_2$ in 2020–2100) (global average)	Land area (million ha) (land-use change in bold)
Forest restoration	21	541
Reforestation	29	**344**
Reduced harvest	33	1047
Agroforestry	5.2	849
Silvopasture	1.6	478
All pathways	93	3259
(of which land-use change)		344

boreal biomes is decreased, and commercial harvesting is completely stopped in tropical forests. Reduction in harvest can be achieved through either reduced harvest intensity in natural forests or doubling the length of rotation of managed forests. This will have a significant impact on timber supply and on the economics of forestry industries. Strategies to continue to meet the timber demand include shifting away from wood products, increasing efficiency, and recycling of wood-based products, to avoid the expansion of plantation forests.

The next highest gains are seen in reforestation in China, Latin America, and Southeast Asia in the decade leading up to 2050. Reforestation target areas are adjacent to existing intact forests and are consistent with targets in national policies and international commitments like the Bonn Challenge. The analysis acknowledges that natural succession to native vegetation is more cost-effective and has a greater success rate than planting new forests. As the carbon sequestration potential of full regrowth of deforested land to forested land is higher than in recovering carbon stocks in degraded forests, this is the only pathway that requires a land-use change of 344 Mha converted from deforested areas to reforested land. All other pathways maintain the existing land uses.

As seen in Table 14.2, the largest concentrations of carbon storage occur where humid tropical and warm temperate forests are allowed to regenerate. Higher rates of sequestration will be seen in Asia, Latin America, and Africa, where tropical biomes have higher net primary productivity than elsewhere, but also because greater land areas are forested in the tropics.

The pathways were designed to ensure that they do not negatively impact agriculture production; it does not completely eliminate the competition for land. Agroforestry should enhance agricultural productivity and has wide geographic applicability. Silvopasture could enhance it or could require reduced stocking rates. Silvopasture results in lower uptake, due to higher initial soil carbon content in temperate pasture lands compared to croplands. Both pathways result in rapid but temporary increase in carbon stocks (Littleton et al., 2021).

Table 14.2 Gross regional carbon sequestration rates in ten world regions as categorized for the RCP database (Littleton et al., 2021)

World region	Carbon uptake in 2050 (GtCO$_2$)	Cumulative uptake by 2100 (GtCO$_2$)
Africa	0.35	5.8
China+	0.57	18.8
India +	0.13	4.0
Latin America	0.71	18.7
Middle East	0.00	0.0
Northern America	0.55	12.1
Pacific OECD	0.13	2.8
Reforming economies	0.67	12.7
Rest of Asia	0.57	14.3
Western Europe	0.24	3.9
World total	3.9	93.0

Importantly, all ecosystem restoration pathways presented here reach the full extent in terms of area by 2040 and then held constant over the rest of the study period. This is coupled with the assumption that net deforestation will be halted by 2030. Without firm action to stop deforestation, gains made through the proposed ecosystem pathways will be offset by forest loss.

It is also important to realize that these ecosystem pathways do not and should not be used to offset fossil fuel emissions. Carbon uptake from land-based mitigation is slow and offers long-term temperature reduction. However, this approach needs to be implemented in conjunction with net-zero targets for other sectors not as a substitute. While removing more carbon from the atmosphere than is emitted into it would begin to reverse some aspects of climate change, some changes would still continue in their current direction for decades to millennia. The reversal of global surface warming lags the decrease in the atmospheric CO_2 concentration by a few years (IPCC, 2021).

14.3 Managing Land Use

This section discusses the impact of the ecosystem restoration pathways on existing land use and the land-use changes required for agriculture to meet the future food demand.

14.3.1 Mapping Land Use for Agriculture

One of the biggest challenges in managing land use is the agricultural expansion required to feed 9 billion people in 2050. Based on the 2012 Food and Agriculture Organization (FAO) projections, the overall demand for agricultural products is expected to grow at 1.1% per year from 2005/2007 to 2050, which will result in a 60% increase globally by 2050 to meet the increased demand. Meeting this demand will require additional land for agriculture, but there is no consensus in the literature on how much more land will be required. The FAO projections indicate that about 70 million ha of additional land will be required for agricultural use in 2050 (Alexandratos & Bruinsma 2012). Several studies have discussed doubling production to meet the 2050 demand, particularly given the shift towards protein-rich diets and the consequent need for land to grow animal feed (Ray et al. 2013). Scenarios that do not link production with health and nutrition involve the expansion of agricultural lands into forests (Maggio et al. 2018). However, Hunter et al. (2017) disagree with the call to double agriculture production, largely because of recent production gains and because it is claimed that an increase of approximately 25%–70% above the current production levels should be sufficient to meet the 2050 demand. Conijn et al. (2018) noted that the planetary boundary for agricultural land was already exceeded in 2010, and a 2050 scenario without efficiency gains to meet the increased demand for food would require an increase of >3.5 Gha in agricultural land (grassland and cropland areas would increase by 78% and 67%, respectively). The FAO's latest alternative pathways to 2050 estimate that arable land must increase by 86 million ha from 2012 in the sustainability scenario and by 165 million hectares in the business-as-usual scenario.

Therefore, projections of the increased land required for agriculture range from 70 million ha to 3.5 billion ha. The FAO (2018) has identified a global reserve of at least 400 million ha of suitable and unprotected land that could be brought under rain-fed cultivation. However, when losses to urbanization and degradation are considered, less than half of this reserve will be available. Data from the FAO–International Institute for Applied Systems Analysis (IIASA) Global Agro-ecological Zones (GAEZ v4) suggest that around 360 million ha of additional and unprotected areas and areas that are highly suitable for rain-fed crop production will be available by 2050. The majority of this land is situated in low- and medium-income countries.

All these scenarios involve increasing agricultural land at the expense of forests, and the resulting deforestation will have drastic consequences for the emission intensity of the sector. However, if a small expansion is coupled with the other strategies discussed in Chap. 6, there may be enough land to feed the 9 billion people estimated to exist in 2050 (FAO Forecast).

14.3.2   Mapping Land Use for Forestry

Unlike agricultural land, forested land has been declining over time, and in 2020, 4 billion ha were recorded as under forest. An estimated 420 million ha of forest was lost through deforestation between 1990 and 2020, although the rate slowed over that period and the net reduction in the global forest area was about 178 million ha (FAO 2020a). Agriculture has driven an estimated 80% of the deforestation worldwide (FAO 2017). The global expansion of agricultural land has stabilized over the last 20 years at around 4.9 billion ha (FAO 2017).

The rate of net forest loss has been decreasing substantially as deforestation declines in some countries, whereas an increase in forest area has been seen in other countries, with both afforestation and the natural expansion of forests. However, there has been a reduction in the rate of forest expansion in the last decade (FAO 2020a).

Regional inequalities are not reflected in this global overview. In tropical and subtropical regions, annual forest losses still amounted to 7 million ha in 2000–2010, whereas the agricultural area expanded by 6 million ha per year in the same period (FAO 2018). The largest reductions were observed in Brazil (down 53.2 million ha) and Indonesia (down 27.5 million ha). However, small increases were seen in Europe and the United States. The largest increase was in China, where the forest area was 51.2 million ha larger in 2015 than in 1990 (EUROSTAT 2020).

14.3.3   Implications for Decarbonization

As seen in previous chapters, the *services* and *industry* sectors can decarbonize their energy emissions (i.e. *Scope 1* and *Scope 2* emissions) by incorporating energy efficiency and transitioning to a 100% renewable energy source. The electrification of industry process heat, although harder to achieve, is another key step in the decarbonization pathway, and there is increasing innovation and technological development to support this. The largest challenge in reaching net-zero emissions remains the management of non-energy process emissions. The OECM model estimates 2.2 GtC will be released in unavoidable emissions annually in 2050 from the nine industrial sectors modelled in this study.

Ecosystem approaches can potentially remove CO_2 from the atmosphere at the gigatonne scale, with potentially significant co-benefits, as discussed above (Meinshausen & Dooley, 2019). To achieve 93 GtC sequestration until 2100, land use must shift towards forest on over 350 million ha of land (Littleton et al., 2021).

The annual peak uptake calculated by Littleton et al. (2021) for all five ecosystem pathways is 3.1 Gt/year in 2041 and 1.1 Gt C per year from 2050 to 2100. While in the short term this appears to provide an opportunity to offset non-energy-related industrial process emissions (e.g. from cement and steel production) that are difficult to avoid with currently available technologies by using ecosystem approaches, in the long term these emissions must be reduced to zero or removed and stored geologically to prevent further warming.

Decarbonization pathways are being developed at the global level. At this level, there is little conflict between the competing uses of cropland, pastureland, and forests for carbon removal. Adopting ecosystem approaches, such as agroforestry or silvopasture, where trees are integrated into cropland or grazing lands, will help to increase the carbon stock while meeting the increasing demand for forestry and agricultural products. It should be noted that a lot of deforestation and the capacity and demand for increased agricultural and livestock products will occur in tropical and subtropical regions, often in developing countries. At the local level, there must be a nuanced approach to addressing the balance between environmental, economic, and well-being outcomes.

14.4   Creating Carbon Conservation Zones (CCZ)

The role of nature and ecosystem services as climate solutions is gaining increasing attention. As well as their climate mitigation and carbon sequestration potential, ecosystem approaches have co-benefits that contribute to sustainable development goals in terms of livelihoods, productivity, biodiversity conservation, health, and ecosystem services. However, it is important to note that even with ambitious land-use restoration, carbon removal can still only compensate for a small part of current emissions. The vast majority of emissive activities must cease if we are to achieve an approximately 1.5 °C target, and all the available removal strategies are required to achieve net-negative emission pathways and reduce the atmospheric concentrations of CO_2.

Feasible approaches to CDR using land-based mitigation options cannot be implemented in a vacuum but must address broader social and environmental objectives. Carbon conservation zones, which implement different ecosystem approaches, must address these broader objectives:

- Respecting indigenous rights and knowledge of land

Indigenous peoples and their connection to land play an important role in protecting and conserving nature and advancing climate solutions. This connection and their stewardship in protecting nature is particularly important in forested areas around the world. Several studies have found that the best forest protection is provided by people with collective legal titles to their land, i.e. by indigenous people (Fa et al., 2020; FAO and FILAC, 2021), and have recognized the contributions of indigenous knowledge to ecosystem-based climate solutions. For the first time, COP26 formally acknowledged the roles and contributions of indigenous people's culture and knowledge in climate action and nominated indigenous peoples to engage directly with governments as knowledge holders and experts (2021a; UN Climate Change News, 2021).

Assisted natural regeneration strategies based on indigenous knowledge are promising ways to restore degraded lands (Schmidt et al., 2021). Formal recognition of indigenous people's rights over their forested lands can slow deforestation (Ricketts et al. 2010; Ceddia et al. 2015). These efforts must be supported by policies and actions that recognize collective territorial rights, provide compensation for environmental services, and allow community forest management, the revitalization of ancestral knowledge, and the strengthening of grassroots organizations and mechanisms for territorial governance (FAO and FILAC, 2021).

- Understanding financial implications

A study investigating the benefits of investing in ecosystem restoration found that tropical forests offered one of the highest value for restoration investment (after coastal and inland wetlands) (De Groot et al., 2013). Case studies across the world have also established that natural regeneration is significantly cheaper than tree planting, while simultaneously providing much higher carbon sequestration, but need to be incentivized by long-term funding mechanisms (Di Sacco et al., 2021). Much of the restoration opportunity identified in this study lies in tropical forested developing countries, and financing incentives and support will be critical to ongoing success.

Reducing Emissions from Deforestation and Forest Degradation (REDD) is an effort to provide incentives through payment for results, allowing developing countries to reduce emissions from forested lands. REDD+ goes beyond addressing deforestation and forest degradation and fosters conservation, the sustainable management of forests, and the enhancement of forest carbon stocks. Initiatives like the Reforestation Accelerator are working with impact investment funds and innovative incubation ideas to provide seed funding to unlock ecosystem-based solutions (The

Nature Conservancy, 2022). Such mechanisms can address the lack of financial support that is a major barrier to implementing ecosystem approaches.

- Protecting and conserving biodiversity

Reversing land degradation and limiting climate change depend upon retaining forests with high ecological integrity. A wide diversity of values and services tends to be found at higher levels in the more-intact forests of a given type. Biomass carbon stocks are a good example (Keith et al. 2009; Mackey et al. 2020), and forests and other ecosystems with no history of significant disturbance collectively absorb around 30% of anthropogenic carbon emissions annually (Friedlingstein et al. 2020).

Ambitious policies that prioritize the retention of forest integrity, especially in the most-intact areas, are now urgently required, in parallel with the current efforts to halt deforestation and restore the integrity of forests globally (Grantham et al., 2020). Higher levels of biodiversity generally support greater levels of ecosystem service production (e.g. carbon sequestration) than lower biodiversity levels, and ecosystem properties, such as resilience, are important considerations when managing human-modified ecosystems (Ferreira et al., 2012). It is necessary to build on the synergies between climate action and activities directed towards conserving biodiversity.

- Influencing supply chains and investment portfolios

Over the last decade, there has been a swell of industry-led commitments to zero-deforestation supply chains, but they are not yet implemented and many companies are yet to act (NYDF Assessment Partners 2020). The Carbon Disclosure Project's (CDP) Investor Report flagged that industry targets for net-zero deforestation are unlikely to be met unless commodity producers in the supply chain manage of their deforestation risk. This highlights the issue that certification is not enough and that companies require initiatives, such as education and financing, to promote sustainable agriculture and demonstrate strong policy commitments to end deforestation (Sin et al., 2020).

Forests and forest products are important parts of a number of supply chains for food, consumer goods, transport, etc., and companies and investors can play an important role in protecting and conserving nature through corporate commitments and by influencing their downstream supply chains.

References

Alexandratos, N., Bruinsma, J. (2012). World Agriculture towards 2030/2050: the 2012 revision. In WORLD AGRICULTURE (No. 12; 03). http://www.fao.org/3/ap106e/ap106e.pdf

Bakhtary, H., Haupt, F., & Elbrecht, J. (2021). *NDCs – a force for nature?* (4th ed.).

Chandrasekhar, A., Viglione, G. (2021). *COP26: Key outcomes for food, forests, land use and nature in Glasgow*. Carbon Br. https://www.carbonbrief.org/cop26-key-outcomes-for-food-forests-land-use-and-nature-in-glasgow. Accessed 12 Dec 2021.

Ceddia, M. G., Gunter, U., Corriveau-Bourque, A. (2015). Land tenure and agricultural expansion in Latin America: The role of Indigenous Peoples' and local communities' forest rights,

Global Environmental Change, 35, 316–322, ISSN 0959-3780. https://doi.org/10.1016/j.
gloenvcha.2015.09.010

Conijn, J. G., Bindraban, P. S., Schröder, J. J., Jongschaap, R. E. E. (2018). Can our global food system meet food demand within planetary boundaries? *Agriculture, Ecosystems & Environment,
251*, 244–256. https://doi.org/10.1016/J.AGEE.2017.06.001

De Groot, R. S., Blignaut, J., Van Der Ploeg, S., et al. (2013). Benefits of investing in ecosystem
restoration. *Conservation Biology, 27*, 1286–1293. https://doi.org/10.1111/COBI.12158

Decision-/CP.26 Glasgow Climate Pact. (2021). Glasgow.

Di Sacco, A., Hardwick, K. A., Blakesley, D., et al. (2021). Ten golden rules for reforestation to
optimize carbon sequestration, biodiversity recovery and livelihood benefits. *Global Change
Biology, 27*, 1328–1348. https://doi.org/10.1111/GCB.15498

Eurostat. (2020, September). Agri-environmental indicator – energy use. Eurostat. https://
ec.europa.eu/eurostat/statistics-explained/index.php/Agri-environmental_indicator_-_energy_
use#Data_sources

Fa, J. E., Watson, J. E. M., Leiper, I., et al. (2020). Importance of indigenous peoples' lands for the
conservation of intact forest landscapes. *Frontiers in Ecology and the Environment, 18*.

FAO. (2017). The future of food and agriculture: Trends and Challenges. http://www.fao.org/3/
i6583e/i6583e.pdf

FAO. (2018). The future of food and agriculture Alternative pathways to 2050. http://www.fao.
org/3/I8429EN/i8429en.pdf

FAO. (2020a). Global Forest Resources Assessment 2020 Main Report. https://www.fao.org/3/
ca9825en/ca9825en.pdf

FAO and FILAC. (2021). *Forest governance by indigenous and tribal peoples an opportunity for
climate action in Latin America and the Caribbean.* Santiago.

Ferreira, J., Gardner, T., Guariguata, M., et al. (2012). Chapter 2 Forest biodiversity, carbon and
other ecosystem services: Relationships and impacts of deforestation and forest degradation. In J. A. Parrotta, C. Wildburger, & S. Mansourian (Eds.), *Understanding relationships
between biodiversity, carbon, forests and people: The key to achieving REDD+ objectives. A
global assessment report prepared by the Global Forest Expert Panel on Biodiversity, Forest
Management, and REDD+*.

Friedlingstein, P., O'Sullivan, M., Jones, M. W., et al. (2020). Global Carbon Budget 2020. *Earth
System Science Data 12*, 3269–3340. https://doi.org/10.5194/essd-12-3269-2020

Glasgow Leaders' Declaration on Forests and Land Use. (2021). UNFCCC. https://ukcop26.org/
glasgow-leaders-declaration-on-forests-and-land-use/. Accessed 12 Dec 2021.

Gov.uk. (2021). *World leaders summit on 'Action on forests and land use' – GOV.UK.* Gov.
UK. https://www.gov.uk/government/publications/cop26-world-leaders-summit-on-action-on-
forests-and-land-use-2-november-2021/world-leaders-summit-on-action-on-forests-and-land-
use. Accessed 12 Dec 2021.

Grantham, H. S., Duncan, A., Evans, T. D., et al. (2020). Anthropogenic modification of forests
means only 40% of remaining forests have high ecosystem integrity. *Natural Communications,
111*(11), 1–10. https://doi.org/10.1038/s41467-020-19493-3

Hunter, M. C., Smith, R. G., Schipanski, M. E., Atwood, L. W., Mortensen, D. A. (2017).
Agriculture in 2050: recalibrating targets for sustainable intensification. *BioScience, 67*(4),
386–391. https://doi.org/10.1093/BIOSCI/BIX010

IPBES. (2019). *Summary for policymakers of the global assessment report on biodiversity and
ecosystem services of the intergovernmental science-policy platform on biodiversity and ecosystem services.* IPBES secretariat.

IPCC. (2021). *Climate change 2021: The physical science basis. Contribution of Working Group
I to the sixth assessment report of the intergovernmental panel on climate change.* Cambridge
University Press.

Keith, H., Mackey B. G., Lindenmaye, D. B. (2009). Re-valuation of forest biomass carbon stocks
and lessons from the world's most carbon-dense forestsr. The Fenner School of Environment
and Society, Australian National University, Canberra, ACT 0200, Australia.. https://www.
pnas.org/cgi/doi/10.1073/pnas.0901970106

Littleton, E. W., Dooley, K., Webb, G., et al (2021). *Dynamic modelling shows substantial contribution of ecosystem restoration to climate change mitigation.*

Lo, V. (2016). *Synthesis report on experiences with ecosystem-based approaches to climate change adaptation and disaster risk reduction.*

Mackey, B., Kormos, C. F., Keith, H., et al. (2020). Understanding the importance of primary tropical forest protection as a mitigation strategy. *Mitigation and Adaptation Strategies for Global Change, 25,* 763–787. https://doi.org/10.1007/s11027-019-09891-4

Maggio, A., Scapolo, F., van Criekinge, T., Serraj, R. (2018). Global drivers and megatrends in agri-food systems. In R. Serraj & P. Pingali (Eds.), *Agriculture & Food Systems To 2050: Global Trends, Challenges and Opportunities* (pp. 47–83). World Scientific Publishing Co. https://doi.org/10.1142/9789813278356_0002

Meinshausen, M., & Dooley, K. (2019). Mitigation scenarios for non-energy GHG. In S. Teske (Ed.), *Achieving the Paris climate agreement goals: Global and regional 100% renewable energy scenarios with non-energy GHG pathways for +1.5C and +2C* (pp. 79–91). Springer.

Mitsch, W. J. (2012). What is ecological engineering? *Ecological Engineering, 45,* 5–12. https://doi.org/10.1016/J.ECOLENG.2012.04.013

Ray, D. K., Mueller, N. D., West, P. C., Foley, J. A. (2013). Yield trends are insufficient to double global crop production by 2050. *PLoS ONE, 8*(6), 66428. https://doi.org/10.1371/JOURNAL.PONE.0066428

Ricketts, T. H., Soares-Filho, B., da Fonseca G. A. B., Nepstad, D., Pfaff, A., Petsonk, A., et al. (2010). Indigenous lands, protected areas, and slowing climate change. *PLoS Biology, 8*(3), e1000331. https://doi.org/10.1371/journal.pbio.1000331

Schmidt, M. V. C., Ikpeng, Y. U., Kayabi, T., et al. (2021). Indigenous knowledge and forest succession management in the Brazilian Amazon: Contributions to reforestation of degraded areas. *Frontiers in Forests and Global Change, 4,* 31. https://doi.org/10.3389/FFGC.2021.605925/BIBTEX

Sin, L., Lam, F., Crocker, T., et al. (2020). *Zeroing-in on deforestation.*

The Nature Conservancy. (2022). *The reforestation accelerator.* Nat. Conserv. https://www.nature.org/en-us/what-we-do/our-insights/perspectives/reforestation-accelerator-driving-natural-climate-solutions/. Accessed 26 Jan 2022.

UN Climate Change News. (2021). *COP26 strengthens role of indigenous experts and stewardship of nature | UNFCCC.* UNFCCC. https://unfccc.int/news/cop26-strengthens-role-of-indigenous-experts-and-stewardship-of-nature. Accessed 17 Dec 2021.

Winterbottom, R. (2014). *Restoration: It's about more than just the trees.*

Part VIII
Synthesis

Chapter 15
Discussion, Conclusions, and Policy Recommendations

Sven Teske, Thomas Pregger, Sarah Niklas, Kriti Nagrath, Simran Talwar, Souran Chatterjee, Benedek Kiss, and Diana Ürge-Vorsatz

Abstract This section summarizes the main findings of all parts of the research, with priority given to the most important findings to avoid the repetition of previous chapters. The key findings for the industry, services, buildings, and transport sectors, including the 12 sub-sectors analyzed, are provided and discussed. Policy recommendations for each sector and recommendations for the actions for governments, industries, the real economy, and financial institutions are offered.

Keywords Conclusion · Policy recommendations · Industry · Services · Financial institutions · Government policies

15.1 Background: Discussion of the Results with Academia, Industry, Government Agencies, and Financial Institutions

In this section, we focus on the outcomes and conclusions of qualitative research rather than on the quantitative results documented in previous chapters. The most important technical measures are highlighted for each sector, followed by policy recommendations. This section reflects extensive discussions and workshops with

S. Teske (✉) · S. Niklas · K. Nagrath · S. Talwar
University of Technology Sydney – Institute for Sustainable Futures (UTS-ISF),
Sydney, NSW, Australia
e-mail: sven.teske@uts.edu.au

T. Pregger
German Aerospace Center (DLR), Institute for of Networked Energy Systems (VE),
Department of Energy Systems Analysis, Stuttgart, Germany

S. Chatterjee · B. Kiss · D. Ürge-Vorsatz
Central European University, Department of Environmental Sciences and Policy,
Budapest, Hungary

S. Teske (ed.), *Achieving the Paris Climate Agreement Goals*,
https://doi.org/10.1007/978-3-030-99177-7_15

stakeholders from various industries and includes the recommendations of Teske et al. (2019). This chapter documents the key outcomes of two key research projects conducted between 2020 and late 2021:

1. The development of sectorial targets for industry and services with the Net-Zero Asset Owner Alliance (NZAOA), financed by the European Climate Foundation and the United Nations Principles for Responsible Investment
2. The development of the global and regional transport scenario conducted with the German Corporation for International Cooperation GmbH (GIZ) and the Transformative Urban Mobility Initiative (TUMI)

15.2   Conclusion: High-Level Summary

To comply with the Paris Climate Agreement and limit the global mean temperature rise to +1.5 °C, rapid decarbonization of the energy sector with currently available technologies is necessary and is possible.

However, to achieve the transformation to a fully renewable energy supply, all available efficiency potentials must be combined to reduce the total demand. To reach *Net Zero* by 2050, the complete phase out of fossil fuels for all combustion processes is essential.

For the *industry sector*, the transition from fossil-fuel-based process heat to renewable energy or electrical systems is the single most important measure. The further reduction of non-energy-related process emissions—mainly from cement and steel manufacture—by altering or optimizing manufacturing processes is also essential. The remaining process emissions might be compensated by natural carbon sinks, so the *industry* sector must actively support the *service* sector in terms of soil regeneration and reforestation measures.

For the *service sector*, especially agriculture and forestry, reducing GHG emissions must clearly involve reducing the greenhouse gas (GHG) emissions arising from land-use changes. Increasing yield efficiency to avoid the further expansion of *agricultural land* at the expense of forests and other important ecosystems is key. However, feeding the growing world population without increasing the area committed to agriculture will require more than just an increase in technical efficiency. Moreover, there seems to be no alternative to reducing the consumption of meat and dairy products.

The *forestry sector* is the single most important sector for the implementation of nature-based carbon sinks. Deforestation must cease immediately. Reforestation with native trees and plants that are typical of specific regions and climate zones must replace the forest areas that have been lost since 1990.

To reduce the demand of the *transport sector*, a shift from resource-intensive air and road transport to more-efficient and electrified means of transport is required, together with an overall reduction in transport activity, especially in high-income countries. Phasing out the production of combustion engines for passenger cars by

2030 and introducing synthetic fuels for long-distance freight transport are essential elements for the future transportation sector. Even with this ambitious goal, the full decarbonization of the road transport sector will not be achieved before 2050, because cars are used, on average, for 15–20 years. There is also significant potential for efficiency gains in shipping and aviation. However, due to the foreseeable further growth in traffic volume and the lack of alternatives, the large-scale use of synthetic fuels from renewable electricity will also be necessary for these modes of transport. Since not all regions will be able to produce this with domestic resources at reasonable costs, a global trade of these new energy sources must be established.

The decarbonization of the *buildings sector* will require a significant reduction in the energy demand for climatization—heating and cooling—per square metre. The key result of our research is that the global energy demand for buildings can be halved with currently available technologies. The utilization of this efficiency potential will require high renovation rates and changed building codes for new constructions. The widespread use of heat pumps and heat grids is an important element on the supply side. In some areas, however, the supply of renewable gases can substitute today's natural gas consumption with a long-term perspective, especially where there is an industrial gas demand. The conversion of today's gas networks and the local/regional availability of resources for the production of green gases play a decisive role here.

Significant electrification across all sectors before 2030—especially for heating, to process heat, and to replace combustion engines in the *transport* sector—is the decisive and most urgent step. Increased electrification will require sector coupling, demand-side management, and multiple forms of storage (for heat and power), including hydrogen and synthetic fuels. Accelerating the implementation of renewable heat technologies is equally important because half the global energy supply must derive from thermal processes by 2050.

The transition of the global energy sector will only be possible with significant policy changes and reforms in the energy market.

The complete restructuring of the *energy and utilities sectors* is required. The primary *energy* sector—the oil, gas, and coal industry—must wind down all fossil-fuel extraction and mining projects and move towards utility-scale renewable energy projects, such as offshore wind and the production of hydrogen and synthetic fuels.

Power utilities will play a key role in providing the rapidly increasing electricity demand, generated from renewable power. The nexus of the global energy transition will be the power grid. Replacing oil and gas with electricity means that power grids must transport most energy, instead of oil and gas pipelines.

Therefore, the expansion of power grid capacities is one of the most important and also most overlooked measures required. In addition, converting existing gas pipelines and using them for the long-range transport of hydrogen and synthetic methane can significantly reduce the infrastructural demands on the power system and increase efficiency.

According to the scenario, global transmission and distribution grids must transport at least three times more electricity by 2050 than in 2020. The upgrades and expansion of power grids must start immediately because infrastructure projects, such as new power lines, can take 10 years or more to implement. Conversions of existing gas pipelines will be possible first where industrial users need large quantities of hydrogen for decarbonized processes.

Limiting the global mean temperature rise to +1.5 °C cannot be achieved by the decarbonization of the energy sector alone. As stated earlier, it will also require significant changes in land use, including the rapid phase out of deforestation and significant reforestation. These measures are not alternative options to the decarbonization of the energy sector but must be implemented in parallel. If governments fail to act and mitigation is delayed, we face the serious risk of exceeding the carbon budget. In the *One Earth Climate Model* (OECM) 1.5 °C pathway, the land-use sequestration pathways complement very ambitious energy-mitigation pathways and should therefore be regarded as necessary to reduce the CO_2 concentrations that have arisen from the overly high emissions in the past and not as compensatory measures that can be extended indefinitely into the future.

15.3   Industry Sector

Policies to achieve the implementation of new highly efficient technologies and to replace fossil-fuel use in industry must be defined region-wide or even on the global level and will require stringent and regulated implementation. Economic incentives, national initiatives, and voluntary agreements with branches of industry will probably not, by themselves, achieve rapid technological change. Concrete standards and requirements must be defined in great detail, covering as far as possible all technologies and their areas of application. The systematic implementation of already-identified best-available technologies should begin immediately.

Mandatory energy management systems must be introduced to identify efficiency potentials and to monitor efficiency gains. The sustainability features of process chains and material flows must also be considered when designing political measures. Particular attention must be paid to the material efficiency of both production processes and their products, because this can open up major energy efficiency potentials and reduce other environmental effects. Public procurement policies and guidelines will help to establish new markets and to introduce new, more-efficient products and opportunities. The effectiveness of policy interventions must be assessed by independent experts, and the further development of efficiency programs and measures will require ongoing co-ordination by independent executive agencies. The public provision of low-interest loans, investment risk management, and tax exemptions for energy-efficient technologies and processes will significantly support technological changes and incentivize the huge investments required. Knowledge transfer between sectors and countries can be achieved through networks initiated and co-ordinated by governments. Public funding for research and development activities with regard to technological innovation, low-carbon solutions, and their process integration will be vital to push the technological limits further. Innovative approaches to the realization of material cycles and recycling options, the recovery of industrial waste heat, and low-carbon raw materials, and process routes in industry must also be identified and implemented.

15.3.1 Steel Industry

There are two key policy recommendations for the steel industry:

1. The decarbonization of the thermal and electrical energy supplies must be supported until 2030.
2. The expansion of new production processes to decarbonize steel manufacturing must be supported, including for:

   - EAF processes
   - Hydrogen-based steel production

Although policies to support the transition towards a renewable energy supply are identical to those described for the *energy* and *utilities* sectors, support for mainstreaming steel production processes to reduce process emissions must be developed specifically for the regional steel industry.

Research and development grants are required, as well as product certification schemes, to financially encourage changes towards new production lines. Steel-processing industries, such as the automotive and construction sectors, require binding purchase quotas for CO_2-neutral steel. CO_2-intensive steel should gradually be made more expensive with a special 'steel tax', to further promote the production of 'green steel'.

15.3.2 Cement Industry

Just as in the steel industry, the decarbonization of energy production for the cement industry has the highest priority in achieving short-term emission reductions. Reducing process emissions requires increased efficiency along all steps of the production line. However, to date, no processes are available for the production of emissions-free cement. Therefore, nature-based carbon sinks must be established to compensate for the residual process emissions.

The Global Cement and Concrete Association (GCCA 2020) published a 2050 road map that set a 'long-term vision for the industry' that covers the following topics:

- Emissions reductions in cement and concrete production
- Savings delivered by concrete during its lifetime
- Reduced demand by promoting design and different materials (e.g. wood)
- Material and construction efficiencies and improved standards

- Re-use of whole-concrete structures
- Designs for the disassembly and re-use of elements

15.3.3   Chemical Industry

The production of the main feedstocks for the chemical industry, such as ammonia, methanol, ethylene, and propylene, is almost entirely based on oil and gas but also on some biomass and coal. The refinery and production processes are very energy intensive. The production facilities are significantly different in each country and depend upon the company's product range. Therefore, universal policy recommendations are not possible.

However, the decarbonization of the chemical industry must focus on the following key areas:

- Developing alternatives to fossil-based feedstock for the production of high-value chemicals, such as ethylene, propylene, benzene, toluene, and xylene
- Expansion of renewable-energy-based ammonia production
- Transition from coal- and gas-fuelled process heat generation to predominantly electrical systems

The electrification of process heat will significantly increase the electric load for the production side. Therefore, in the transition from fossil- to electricity-based process heat generation, upgrading the power grids must also be considered, and planning must involve the local power-grid operator.

15.4   Land-Use and Non-energy GHGs in the Service Sector

The key recommendations for the *service* sector focus on non-energy GHG emissions and especially the emissions associated with changes in land use (agriculture, forestry, and other land use, AFOLU). Although the transition to a renewable energy supply is a prerequisite for the decarbonization of the *service* sector, deforestation and other forms of land conversion must decline much more rapidly. Moreover, reductions in methane and nitrogen must also be achieved in the agriculture sector. Without nature-based solutions, the 1.5 °C limit will not be possible, even with a rapid decline in fossil-fuel emissions.

Four main natural sequestration pathways are utilized in the OECM, divided into temperate and tropical zones—reforestation, natural forest restoration, sustainable forest management, and cropland afforestation (trees in croplands):

1. *Wild lands* cover approximately 50% of the Earth's terrestrial area and are vital to the world's carbon cycle, sequestering up to one-quarter of anthropogenic carbon emissions and storing approximately 450 gigatonnes of solid carbon (Erb

et al. 2018). Preserving these land and forest intact is key to maintaining our global carbon sinks, making the 1.5 °C limit possible.

2. *Ending deforestation*: Today, land-use changes account for more than 10% of global CO_2 emissions (approximately 4 $GtCO_2$ per year), resulting largely from the clearing of forests for agriculture or other forms of development. Rapidly phasing out the practice of deforestation will greatly increase the chance of achieving the 1.5 °C limit.

3. *Large-scale reforestation*: The most important sequestration measure identified is large-scale reforestation, particularly in the subtropics and tropics. Under the 1.5 °C model, 300 megahectares (Mha) of land area will be reforested in the tropics, and an additional 50 Mha will be reforested in temperate regions.

4. *Natural restoration*: The second most important pathway for carbon removal relies upon natural forest restoration or 'rewilding', increasing the carbon density within approximately 600 Mha of existing forests. Reduced logging and better forestry practices in managed forests will also contribute significantly to reducing total carbon removal.

Planting Trees on Croplands

Tree cropping—a strategy in which trees are planted within croplands—can significantly increase carbon storage on agricultural lands. The OECM estimates that planting trees on 400 Mha of cropland will achieve approximately 30 Gt of carbon removal by 2100.

The four sequestration pathways occur in all countries and regions, although we have excluded reforestation in the boreal forest zone because of the albedo effect.

All four sequestration pathways commence in 2020 but have different phase-in and phase-out rates, which also differ between the boreal/temperate and tropical/subtropical biomes.

1. *Forest restoration*: Boreal/temperate—full potential by 2035, saturation by 2065 (decline to zero around 2100). Tropical/subtropical—full potential by 2030, saturation by 2045 (decline to zero around 2100).

2. *Reforestation*: Boreal/temperate—full potential by 2045, saturation by 2075 (decline to zero around 2150). Tropical/subtropical—full potential by 2040, saturation by 2065 (decline to zero around 2120).

3. *Sustainable use of forests*: Boreal/temperate—full potential by 2040, saturation by 2070 (decline to zero around 2150). Tropical/subtropical—full potential by 2035, saturation by 2055 (decline to zero around 2100).

4. *Agroforestry*: Boreal/temperate—full potential by 2040, saturation by 2060 (decline to zero around 2080). Tropical/subtropical—full potential by 2030, saturation by 2050 (decline to zero around 2080).

15.5   Transport Sector

There are actionable measures in three key areas to decarbonize *transport* in line with the 1.5 °C target: *avoiding* or reducing the need to travel, *shifting* to more-efficient transport modes, and *improving* efficiency through vehicular technology. The implementation of these measure must take place until 2030 in order to reduce emissions sufficiently rapidly.

1. *Phasing out of internal combustion engines by 2030*: To achieve the global decarbonization of transport, it is essential to transition to electric mobility powered by renewable energy. To facilitate the shift to electric mobility, the phasing out of new vehicles (passenger cars, vans, 2–3 wheelers, city buses, etc.) with international combustion engines (ICE) by 2030 is vital. By setting targets, governments can send strong signals to markets and customers to adopt the new technology. Efficiency standards for all vehicle types, with an annual efficiency improvement target of 1%, should also be mandated.
2. *Increasing walking and cycling to optimal levels of international leaders in sustainable mobility*: The large-scale expansion of quality infrastructure for bicycles and walking is required to maintain and extend access to these activities around the globe, while curbing the increase in passenger transport. Compact regional and urban planning principles will support the greater uptake of active mobility. Under the 1.5 °C pathway, up to 50% of trips will be made by foot or cycling, as exemplified by international leaders in sustainable mobility, such as Amsterdam and Copenhagen.
3. *Doubling the public transport capacity by 2030*: Although public transport has seen massive reductions in use during the COVID-19 pandemic, it continues to play a key role as the backbone of urban and inter-urban mobility. To leverage its potential, the capacity of public transport must be doubled, with attention given to service quality and convenience to ensure its acceptance. The integration of shared mobility and 'last mile' transport services will support intermodality between public transport and individual mobility.
4. *Almost full electrification of rail by 2030*: Freight transported by trucks must be shifted to rail transport systems. The share of electric trains must increase, and all diesel locomotives must be phased out by 2050 across all regions. Therefore, the full electrification of rail transport (via overhead- or battery-powered electric trains) must be achieved.
5. *Introduction of hydrogen and synthetic fuels before 2030* as a complementary solution for modes of transport and technologies that cannot be electrified such as shipping and aviation and to some extent long-distance freight transport by road.

15.6 Buildings Sector

The in-depth HEB analysis (Chap. 7) demonstrates the potential to reduce the energy demand in the *buildings* sector with state-of-the-art high-efficiency buildings, implemented worldwide. The findings of the HEB analysis show that with a greater proportion of high-efficiency renovations and construction, it will be possible to reduce the final thermal energy use globally in the building sector by more than half by 2060. For some regions, such as the EU and the Pacific OECD, it will even be possible to achieve net-zero status for the thermal energy demand. However, this pathway towards high-efficiency or net-zero emissions in the *buildings* sector is ambitious in its assumptions and requires strong policy support. On the contrary, if policy support to implement more high-efficiency buildings is not in place, then the total thermal energy demand of the building sector will increase significantly over coming decades. Furthermore, if the use of energy efficiency measures continues at the present rate, 67–80% of the final global thermal energy savings will be locked in by 2060 in the world building infrastructure. This lock-in effect in the *buildings* sector also means that if the present moderate energy performance levels become standard in new and/or retrofitted buildings, it will be almost impossible to further reduce thermal energy consumption in these buildings for many decades to come.

Therefore, to realize the immense potential of the *buildings* sector, strong and ambitious policies are required. The findings of our study are translated into the following policy recommendations:

1. The building energy demand can be harnessed by implementing the advanced retrofitting of existing and historical buildings in developed nations. To promote advanced retrofitting, ambitious building codes and standards must be introduced and effectively enforced. To effectively reinforce these advanced retrofitting strategies, positive incentives, such as subsidies or tax deductions, can be given to both developers and owners. If retrofitting is not performed at an advanced level, then the increased floor area means that the global energy demand will also increase. Furthermore, with the substantial carbon lock-in, the energy demand cannot be reduced substantially in subsequent years. This study shows that strict policies in building energy efficiency measures and their urgent implementation are even more important than an increased retrofitting rate in achieving low-energy building stock.
2. Most of the future thermal energy demand will come from developing nations, such as India. In developing regions, new building stock will play a dominant role in reducing the energy demand, so the construction of new energy-efficient buildings should follow a stringent building code that requires a high level of energy performance in all new construction. Building certification and labelling, technology transfer, training of building specialists, and financial incentives should also be considered to achieve adequate high-efficiency construction.
3. Together with stringent energy-efficient building codes and performance standards, behavioural and lifestyle changes will help to limit floor space growth, especially in urban areas. This will increase the efficiency of energy systems in

buildings. Therefore, more education about low-carbon lifestyles must be provided.

4. Even with ambitious policy assumptions, the building sector will still consume substantial thermal energy globally, which may hinder the transition towards climate neutrality. Therefore, reducing the building energy demand must be accompanied by the promotion of building-integrated solar electric production. The findings of the nearly net-zero scenario show that in developed regions, it will be possible to achieve net-zero status by 2060. Therefore, positive incentives should be given for on-site building-integrated solar energy production.

15.7   Energy and Utilities Sector

The *energy* and *utilities* sectors may constitute separate categories for the financial sector, but for the energy sector, they are two sides of the same coin. The 1.5 °C pathway will lead to a 100% renewable electricity supply, with a significant share of variable power generation. The framework of the traditional electricity market has been developed for central suppliers operating dispatchable and limited dispatchable (base-load) thermal power plants. However, the electricity markets of the future will be dominated by variable generation, with no marginal or fuel costs. The power system will also require the build-up and economic operation of a combination of dispatch generation, storage, and other system services, the operation of which will be conditioned by renewable electricity feed-ins. For both reasons, a significantly different market framework is urgently required, in which the technologies can be operated economically and refinanced. Renewable electricity must be guaranteed priority access to the grid. Access to the exchange capacity available at any given moment should be fully transparent, and the transmission of renewable electricity must always have preference. Furthermore, the design of distribution and transmission networks, particularly for interconnections and transformer stations, should be guided by the objective of facilitating the integration of renewables and achieving a 100% renewable electricity system.

To establish fair and equal market conditions, the ownership of electrical grids should be completely disengaged from the ownership of power-generation and supply companies. To encourage new businesses, relevant grid data must be made available by the operators of transmission and distribution systems. This will require establishing communication standards and data protection guidelines for smart grids. Legislation to support and expand demand-side management is required to create new markets for the integration services for renewable electricity. Public funding for research and development is required to further develop and implement technologies that allow variable power integration, such as the smart-grid technology, virtual power stations, low-cost storage solutions, and responsive demand-side management. Finally, a policy framework that supports the electrification and sector coupling of the *heating* and *transport* sectors is urgently required to ensure a successful and cost-efficient transition process.

15.8 Policy Recommendations

The OECM is an integrated energy assessment tool for the development of science-based targets for all major global industries in a granularity. It includes the key performance indicators (KPIs) required to make informed investment decisions that will credibly align with the global net-zero objective in the short, medium, and long term. The key finding of our work on the OECM 1.5 °C cross-sectorial pathway is that it is still possible to remain with the 1.5 °C limit if governments, industries, and the financial sector act immediately. The technology required to decarbonize the energy supply with renewable energy is available, market ready, and in most cases, already cost competitive. The energy efficiency measures needed to reduce the energy demand have also been understood for years and can be introduced without delay. Finally, the finance industry—for instance, the Net-Zero Asset Owner Alliance—is committed to implementing carbon targets for its investment portfolios. However, policies are required to ensure that all measures are implemented in the rather short time frame required.

Implementing Short-Term Targets for 2025 and 2030
To implement the documented short-term targets for 2025 and 2030, the following actions are required:

Government Policies:

1. Immediate cessation of public and private investment in new oil, coal, and gas projects.
2. Implementation of carbon pricing with a reliable minimum CO_2 price, consistent with the underlying OECM emissions caps.
3. All OECD countries must phase out coal by 2030.
4. The automobile industry must phase out internal combustion engines for passenger cars by 2030.
5. Legally binding efficiency standards must be instituted for all electrical applications, vehicles, and buildings.
6. Renewable energy targets must be based on IPCC-carbon-budget-based 1.5 °C scenarios or detailed country-specific master plans.
7. Mandatory transparent forward-looking and historic disclosure of the most relevant KPIs: energy intensity, share of renewable energy supply, energy demand, carbon emissions, and carbon intensities per production unit.
8. A global phase out of all fossil-fuel subsidies by 2025.
9. Pursuing a nationally and internationally to globally integrated and coordinated policy with the aim of creating investment security and incentives for the necessary transformation processes
10. Conduct a comprehensive scientific analysis of feasible national pathways and formulate corresponding NDCs for 2025/2030 and beyond.
11. Establish global governance of the transformation of energy systems, including monitoring of the necessary political, social, economic, environmental, and legal requirements.

Actions Needed by Industry and Financial Institutions

Industry:

1. Setting and implementing a climate strategy consistent with 1.5 °C no- or low-overshoot sector models.
2. Immediate cessation of investments in new oil, coal, and gas projects.
3. Utilities must rapidly upscale renewable electricity to provide logistical support for reducing *Scope 2* emissions for all industries and services. This is a huge market opportunity for utilities.
4. Development of efficient technologies to implement electric mobility.
5. Mandatory transparent forward-looking and historic disclosure of the most relevant KPIs, such as carbon emissions, energy demand, and carbon intensities per production unit.

Financial Institutions:

1. Setting and implementing decarbonization targets for investment, lending, and underwriting portfolios that are consistent with the 1.5 °C no- or low-overshoot sector models.
2. Cessation of investment in new oil, coal, and gas projects.
3. Ensured coal phase out in OECD countries by 2030 and in all regions between 2030 and 2040.
4. Scaled climate solution investments, especially in emerging economies.
5. Disclosure of:

   • Climate mitigation strategies
   • Short- and mid-term target setting
   • Target achievements
   • Progress of climate solution investments
   • Engagement outcomes

References

Erb, K.-H., Kastner, T., Plutzar, C., et al. (2018). Unexpectedly large impact of forest management and grazing on global vegetation biomass. *Nature, 553*, 73–76. https://doi.org/10.1038/nature25138

GCCA. (2020). Climate ambition: 2050 roadmap. In: *Global cement and concrete association.* https://gccassociation.org/climate-ambition/. Accessed on 20th Dec 2021.

Teske, S., Pregger, T., Simon, S., et al. (2019). *Achieving the Paris climate agreement goals global and regional 100% renewable energy scenarios with non-energy GHG pathways for +1.5°C and +2°C.* Springer International Publishing.

Annex

[TWh/yr]	2019	2025	2030	2035	2040	2045	2050
Power plants	**23,000**	**27,298**	**34,902**	**52,647**	**62,508**	**69,101**	**74,334**
- Hard coal (& non-renewable waste)	6,074	4,334	1,352	146	0	0	0
- Lignite	1,680	271	177	170	84	0	0
- Gas	5,093	4,604	4,055	3,156	1,824	755	0
of which from H2	0	27	169	728	1,122	1,511	1,945
- Oil	757	488	338	61	14	0	0
- Diesel							
- Nuclear	2,764	2,113	1,521	923	281	151	0
- Biomass (& renewable waste)	348	867	760	874	876	943	1,012
- Hydro	4,315	4,334	4,900	5,462	5,893	6,044	6,485
- Wind	1,304	4,257	8,278	15,402	20,953	25,960	29,814
of which wind offshore	*0*	*910*	*1,773*	*3,695*	*5,145*	*6,785*	*8,098*
- PV	568	5,490	11,240	20,141	23,324	24,069	23,730
- Geothermal	86	271	676	1,214	1,964	2,417	3,242
- Solar thermal power plants	12	244	1,436	4,370	6,173	7,252	8,106
- Ocean energy	1	81	169	486	702	907	1,297
Combined heat and power plants	**4,253**	**3,023**	**2,920**	**3,250**	**3,570**	**3,630**	**3,623**
- Hard coal (& non-renewable waste)	2,336	845	551	352	193	0	0
- Lignite	207	136	129	143	0	0	0
- Gas	1,427	1,499	1,433	1,310	872	381	0
of which from H2							
- Oil	55	50	44	27	20	0	0
- Biomass (& renewable waste)	223	488	718	1,242	2,151	2,519	2,764
- Geothermal	5	4	44	58	74	91	122
- Hydrogen	0	1	1	118	260	638	737
CHP by producer							
- Main activity producers	3,592	2,287	2,107	2,346	2,576	2,541	2,444
- Autoproducers	662	735	813	905	994	1,089	1,179
Total generation	**27,027**	**29,910**	**37,229**	**55,082**	**64,555**	**71,028**	**76,368**
- Fossil	17,629	12,227	8,079	5,364	3,008	1,137	0
- Hard coal (& non-renewable waste)	8,410	5,178	1,903	497	193	0	0
- Lignite	1,888	407	306	313	84	0	0
- Gas	6,520	6,103	5,489	4,466	2,696	1,137	0
- Oil	812	538	382	88	34	0	0
- Diesel							
- Nuclear	2,764	2,113	1,521	923	281	151	0
- Hydrogen	0	28	170	847	1,382	2,149	2,683
- of which renewable H2							
- Renewables (w/o renewable hydrogen)	6,633	15,571	27,629	48,795	61,267	69,740	76,368
- Hydro	4,315	4,334	4,900	5,462	5,893	6,044	6,485
- Wind	1,304	4,257	8,278	15,402	20,953	25,960	29,814
- PV	568	5,490	11,240	20,141	23,324	24,069	23,730
- Biomass (& renewable waste)	348	867	760	874	876	943	1,012
- Geothermal	86	271	676	1,214	1,964	2,417	3,242
- Solar thermal power plants	12	244	1,436	4,370	6,173	7,252	8,106
- Ocean energy	1	81	169	486	702	907	1,297
Distribution losses	1,597	2,076	2,453	3,452	3,820	4,084	4,375
Own consumption electricity	1,564	2,076	2,532	3,452	3,820	4,084	4,229
Electricity for hydrogen production	0	294	1,278	4,577	7,088	9,573	10,784
Electricity for synfuel production	0	0	82	364	1,118	1,436	1,533
Final energy consumption (electricity)	**23,869**	**25,463**	**30,884**	**43,238**	**48,709**	**51,849**	**55,447**
Variable RES (PV, Wind, Ocean)	1,873	9,828	19,687	36,029	44,979	50,935	54,841
Share of variable RES	6.9%	32.9%	52.9%	65.4%	69.7%	71.7%	71.8%
RES share (domestic generation)	24.5%	52.1%	74.2%	88.6%	94.9%	98.2%	100.0%
CHP share (of total generation)	16%	10%	8%	6%	6%	5%	5%

Fig. 1 Global electricity generation—OECM 1.5 °C

© The Editor(s) (if applicable) and The Author(s) 2022
S. Teske (ed.), *Achieving the Paris Climate Agreement Goals*,
https://doi.org/10.1007/978-3-030-99177-7

[GW]	2019	2025	2030	2035	2040	2045	2050
Power plants	**7,657**	**10,866**	**16,077**	**25,346**	**29,209**	**31,915**	**34,199**
- Hard coal (& non-renewable waste)	1,285	868	271	29	0	0	0
- Lignite	356	54	36	34	17	0	0
- Gas (incl. H₂)	1,934	1,604	1,461	1,336	1,007	767	650
- Oil	836	429	297	53	12	0	0
- Diesel							
- Nuclear	429	322	232	141	43	23	0
- Biomass (& renewable waste)	77	198	174	200	200	215	231
- Hydro	1,569	1,419	1,576	1,726	1,830	1,845	1,980
- Wind	617	1,582	2,979	5,327	7,026	8,424	9,644
of which wind onshore	617	1,350	2,528	4,393	5,733	6,728	7,620
of which wind offshore	0	233	451	934	1,293	1,696	2,024
- PV	537	4,197	8,212	14,093	15,658	16,599	16,950
- Geothermal	12	37	92	165	267	329	441
- Solar thermal power plants	5	113	657	1,979	2,770	3,223	3,603
- Ocean energy	1	44	91	262	379	490	701
Combined heat and power plants	**1,191**	**870**	**863**	**927**	**979**	**936**	**883**
- Hard coal (& non-renewable waste)	494	169	110	70	39	0	0
- Lignite	44	27	26	29	0	0	0
- Gas (incl. H₂)	542	518	507	460	322	144	0
- Oil	61	44	38	24	17	0	0
- Biomass (& renewable waste)	49	111	175	304	520	608	668
- Geothermal	1	1	6	8	10	12	17
- Hydrogen (fuel cells)	0	0	0	32	70	172	199
CHP by producer							
- Main activity producers	240	230	240	258	242	241	259
- Autoproducers	952	640	623	669	737	695	625
Total generation	**8,306**	**11,218**	**16,433**	**25,813**	**29,866**	**32,708**	**35,083**
- Fossil	5,010	3,186	2,183	1,333	717	262	0
- Hard coal (& non-renewable waste)	1,779	1,037	381	100	39	0	0
- Lignite	399	82	61	63	17	0	0
- Gas (w/o H₂)	1,934	1,595	1,404	1,093	632	262	0
- Oil & Diesel	897	473	336	77	30	0	0
- Diesel							
- Nuclear	429	322	232	141	43	23	0
- Hydrogen (fuel cells, gas power plants, gas CHP)	0	9	57	275	445	677	849
- Renewables	2,868	7,701	13,962	24,064	28,661	31,746	34,234
- Hydro	1,569	1,419	1,576	1,726	1,830	1,845	1,980
- Wind	617	1,582	2,979	5,327	7,026	8,424	9,644
of which wind offshore	0	233	451	934	1,293	1,696	2,024
- PV	537	4,197	8,212	14,093	15,658	16,599	16,950
- Biomass (& renewable waste)	127	309	349	503	720	823	899
- Geothermal	12	37	98	173	277	341	457
- Solar thermal power plants	5	113	657	1,979	2,770	3,223	3,603
- Ocean energy	1	44	91	262	379	490	701
Variable RES (PV, Wind, Ocean)	1,155	5,823	11,283	19,683	23,063	25,513	27,295
Share of variable RES	13.9%	51.9%	68.7%	76.3%	77.2%	78.0%	77.8%
RES share (domestic generation)	**34.5%**	**69%**	**85%**	**93%**	**96%**	**97%**	**100%**

Fig. 2 Global installed power generation capacities—OECM 1.5 °C

[PJ/yr]	2019	2025	2030	2035	2040	2045	2050
District heating plants	**5,691**	**5,884**	**5,884**	**5,884**	**5,884**	**5,884**	**5,884**
Coal	1,289	1,283	588	294	147	0	0
Lignite	174	173	173	59	0	0	0
Gas	3,301	3,465	3,465	2,942	1,471	588	0
Oil	479	478	478	353	235	0	0
- Fossil fuels	5,243	5,399	4,705	3,648	1,854	588	0
- Biomass	399	427	1,121	2,177	3,972	5,237	5,826
- Solar collectors	46	55	55	55	55	55	55
- Geothermal	3	4	4	4	4	4	4
Heat from CHP [1]	**10,974**	**5,885**	**5,641**	**6,279**	**6,897**	**6,977**	**6,926**
Coal	5,374	1,191	775	487	267	0	0
Lignite	472	192	183	204	0	0	0
Gas	4,403	2,369	2,255	2,028	1,511	691	0
Oil	207	84	74	41	30	0	0
- Fossil fuels	10,456	3,836	3,287	2,760	1,808	691	0
- Biomass	507	2,043	2,349	3,514	5,084	6,281	6,920
- Geothermal	10	5	5	5	6	5	5
- Hydrogen	0	0	0	0	0	0	0
Direct heating	**143,429**	**152,044**	**154,751**	**143,274**	**136,757**	**132,518**	**129,272**
Coal	28,032	15,544	8,648	4,813	0	0	0
Lignite	1,371	1,192	0	0	0	0	0
Gas	52,371	51,972	42,881	34,024	25,293	11,391	0
Oil	24,033	14,388	8,542	5,694	5,943	7,224	9,340
- Fossil fuels	105,808	83,096	60,071	44,531	31,235	18,615	9,340
- Biomass	31,340	42,248	57,164	39,154	30,837	27,481	21,900
- Solar collectors	1,243	7,891	13,093	17,308	19,778	22,137	26,125
- Geothermal	0	3,078	5,006	7,501	10,206	13,231	16,864
- Heat pumps [2]	292	8,674	11,947	29,211	39,297	45,820	50,029
- Electric direct heating	4,746	7,057	7,470	5,570	5,403	5,234	5,014
- Hydrogen							
Total heat supply [3]	**160,094**	**163,813**	**166,276**	**155,437**	**149,538**	**145,379**	**142,082**
Coal	34,696	18,017	10,012	5,594	414	0	0
Lignite	2,018	1,557	357	263	0	0	0
Gas	60,075	57,806	48,601	38,994	28,274	12,670	0
Oil	24,719	14,951	9,094	6,088	6,208	7,224	9,340
- Fossil fuels	121,507	92,332	68,063	50,939	34,897	19,894	9,340
- Biomass	32,246	44,717	60,634	44,845	39,893	39,000	34,646
- Solar collectors	1,246	7,895	13,097	17,311	19,782	22,140	26,129
- Geothermal	57	3,138	5,066	7,561	10,267	13,291	16,924
- Heat pumps [2]	292	8,674	11,947	29,211	39,297	45,820	50,029
- Electric direct heating	4,746	7,057	7,470	5,570	5,403	5,234	5,014
- Hydrogen	0	0	0	0	0	0	0
RES share (including RES electricity)	21.7%	39.0%	56.1%	64.7%	75.1%	85.7%	93.4%

[1] public CHP and CHP autoproduction

[2] heat from ambient energy and electricity use

[3] incl. process heat, cooking

Fig. 3 Global final heat demand—OECM 1.5 °C

[PJ/yr]	2019	2025	2030	2035	2040	2045	2050
road	**91,900**	**79,050**	**65,342**	**42,582**	**35,736**	**28,978**	**24,845**
- fossil fuels	85,703	72,255	57,294	16,522	7,546	825	0
- biofuels	3,804	3,990	4,158	5,600	6,849	6,857	6,209
- synfuels	0	0	0	0	0	0	0
- natural gas	2,149	851	710	629	284	0	0
- hydrogen	0	541	1,284	5,637	6,025	6,468	5,587
- electricity	244	1,413	1,896	14,193	15,032	14,828	13,049
rail	**2,872**	**2,661**	**3,110**	**3,249**	**3,444**	**3,707**	**4,001**
- fossil fuels	1,679	1,099	676	232	43	10	0
- biofuels	0	74	130	234	254	142	0
- synfuels	0	0	0	0	0	0	0
- hydrogen	0	0	0	0	0	0	0
- electricity	1,193	1,488	2,303	2,784	3,147	3,556	4,001
navigation	**2,453**	**11,461**	**11,853**	**11,958**	**12,079**	**12,251**	**12,424**
- fossil fuels	2,441	11,209	8,001	8,072	1,570	0	0
- biofuels	6	252	2,963	2,392	6,281	6,738	6,212
- synfuels	6	0	296	598	2,416	3,553	4,100
- hydrogen	0	0	593	897	1,812	1,960	2,112
- electric	0	0	0	0	0	0	0
- sail-hybrid							
aviation	**5,805**	**9,106**	**8,088**	**7,122**	**5,954**	**5,058**	**4,295**
- fossil fuels	5,805	8,924	7,400	2,849	595	101	0
- biofuels	0	182	687	3,561	3,751	3,338	2,878
- synfuels	0	0	0	712	1,608	1,618	1,417
- electric	0	0	0	0	0	0	0
total (incl. pipelines)	**105,978**	**102,279**	**88,392**	**64,912**	**57,213**	**49,993**	**45,566**
- fossil fuels	97,777	94,337	74,080	28,304	10,039	936	0
- biofuels (incl. biogas)	3,810	4,499	7,939	11,787	17,135	17,074	15,299
- synfuels	6	0	296	1,310	4,023	5,171	5,517
- natural gas	2,149	851	710	629	284	0	0
- hydrogen	0	541	1,877	6,534	7,837	8,428	7,700
- electricity	1,437	2,901	4,199	16,977	18,179	18,384	17,050
total RES	4,169	6,551	13,229	34,670	46,249	48,724	45,566
RES share	3.9%	6.4%	15.0%	53.4%	80.8%	97.5%	100.0%

Fig. 4 Global transport final energy demand—OECM 1.5 °C

{PJ/yr}	2019	2025	2030	2035	2040	2045	2050
Total (incl. non-Energy use)	**400,278**	**400,693**	**403,245**	**379,724**	**375,313**	**369,850**	**373,805**
Total Energy use 1)	360,974	362,934	364,237	341,125	336,184	330,262	333,337
Transport	**105,978**	**102,279**	**88,392**	**64,912**	**57,213**	**49,993**	**45,566**
- Oil products	97,777	94,337	74,080	28,304	10,039	936	0
- Natural gas	2,149	851	710	629	284	0	0
- Biofuels	3,810	4,499	7,939	11,787	17,135	17,074	15,299
- Synfuels	6	0	296	1,310	4,023	5,171	5,517
- Electricity	1,437	2,901	4,199	16,977	18,179	18,384	17,050
RES electricity	*353*	*1,510*	*3,117*	*15,039*	*17,253*	*18,051*	*17,050*
- Hydrogen	0	541	1,877	6,534	7,837	8,428	7,700
RES share Transport	**3.9%**	**6.4%**	**15.0%**	**53.4%**	**80.8%**	**97.5%**	**100.0%**
Industry	**120,884**	**119,361**	**132,074**	**140,330**	**147,271**	**161,951**	**177,417**
- Electricity	34,511	38,826	56,655	76,699	96,013	112,104	128,790
RES electricity	*8,470*	*20,212*	*42,046*	*67,945*	*91,122*	*110,072*	*128,790*
- Public district heat	6,075	5,321	5,883	6,548	7,192	7,880	8,529
RES district heat	*478*	*439*	*1,179*	*2,488*	*4,926*	*7,092*	*8,529*
- Hard coal & lignite	28,989	15,786	10,479	5,786	0	0	0
- Oil products	11,890	4,800	3,051	928	0	0	0
- Gas	31,263	33,350	23,957	21,671	17,421	9,319	0
- Solar	17	3,129	6,054	7,460	8,459	10,426	12,540
- Biomass	8,139	16,898	23,228	15,559	7,826	6,929	10,003
- Geothermal	0	1,173	1,989	3,562	4,546	6,024	7,524
- Hydrogen	0	78	778	2,118	5,815	9,268	10,032
RES share Industry	**14.1%**	**35.1%**	**57.0%**	**70.6%**	**83.3%**	**92.5%**	**100.0%**
Other Sectors	**134,111**	**141,294**	**143,771**	**135,883**	**131,700**	**118,318**	**110,354**
- Electricity	46,315	51,709	53,072	66,661	69,171	65,566	64,159
RES electricity	*11,367*	*26,919*	*39,386*	*59,052*	*65,648*	*64,377*	*64,159*
- Public district heat	6,044	6,413	6,771	6,631	6,351	6,065	5,717
RES district heat	*476*	*529*	*1,357*	*2,520*	*4,351*	*5459*	*5717*
- Hard coal & lignite	6,299	4,482	0	0	0	0	0
- Oil products	16,139	9,808	3,472	1,140	329	21	0
- Gas	30,183	27,592	25,636	17,116	10,930	3,131	0
- Solar	1,229	4,762	7,040	9,848	11,320	11,710	13,585
- Biomass	27,903	34,622	44,763	30,549	27,938	24,618	15,515
- Geothermal	0	1,905	3,017	3,939	5,660	7,206	9,340
- Hydrogen	0	0	0	0	0	0	2,038
RES share Other Sectors	**30.6%**	**48.6%**	**66.5%**	**77.9%**	**87.3%**	**95.8%**	**100.0%**
Total RES	**62,248**	**117,216**	**184,067**	**239,709**	**283,859**	**311,906**	**333,337**
RES share	**17.2%**	**32.3%**	**50.5%**	**70.3%**	**84.4%**	**94.4%**	**100.0%**
Non energy use	39,304	37,760	39,008	38,599	39,129	39,588	40,468
- Oil	28,798	27,898	28,806	28,363	28,739	28,932	32,097
- Gas	8,302	7,433	7,757	7,753	7,938	8,110	8,371
- Coal	2,205	2,429	2,445	2,483	2,453	2,546	0

Fig. 5 Global final energy demand—OECM 1.5 °C

[PJ/yr]	2019	2025	2030	2035	2040	2045	2050
Total (incl. non-energy-use)	**564,549**	**536,105**	**513,593**	**488,696**	**473,300**	**461,183**	**465,433**
- Fossil (excluding on-energy use)	**418,757**	**330,140**	**235,489**	**136,409**	**72,398**	**24,567**	**0**
- Hard coal	138,615	80,288	33,924	13,234	3,000	359	0
- Lignite	20,955	5,724	3,278	3,069	704	0	0
- Natural gas	121,586	117,698	104,028	84,014	55,245	23,249	0
- Crude oil	137,600	126,431	94,259	36,093	13,450	959	0
- Nuclear	30,156	24,148	17,219	10,350	3,119	1,664	0
- Renewables	**76,332**	**144,057**	**221,876**	**303,338**	**358,654**	**395,363**	**424,966**
- Hydro	15,534	15,601	17,640	19,665	21,214	21,757	23,345
- Wind	4,694	14,626	26,764	44,573	58,591	70,709	81,707
- Solar	3,433	30,123	68,644	134,896	166,698	181,253	192,489
- Biomass	52,300	79,302	100,722	90,526	92,287	96,388	94,159
- Geothermal	366	4,113	7,499	11,931	17,338	21,994	28,596
- Ocean energy	4	293	608	1,748	2,525	3,264	4,669
Total RES incl. electr. & synfuel import	**76,329**	**144,057**	**221,876**	**303,338**	**358,654**	**395,363**	**424,966**
RES share	15.4%	30.4%	48.5%	69.0%	83.2%	94.1%	100.0%
Non-energy use	**39,304**	**37,760**	**39,008**	**38,599**	**39,129**	**39,588**	**40,468**
Coal	2,205	2,429	2,445	2,483	2,453	2,546	0
Gas	8,302	7,433	7,757	7,753	7,938	8,110	8,371
Oil	28,798	27,898	28,806	28,363	28,739	28,932	32,097

Fig. 6 Global primary energy demand—OECM 1.5 °C

[Mt/CO2/yr]	2019	2025	2030	2035	2040	2045	2050
Condensation power plants	**10,147**	**6,997**	**3,636**	**1,696**	**840**	**301**	**0**
- Hard coal (& non-renewable waste)	5,333	3,758	1,124	116	0	0	0
- Lignite	1,864	263	164	150	71	0	0
- Gas	2,294	2,059	1,759	1,330	747	301	0
- Oil + Diesel	657	918	588	100	22	0	0
Combined heat and power plants	**3,610**	**1,696**	**1,405**	**1,178**	**678**	**215**	**0**
- Hard coal (& non-renewable waste)	2,472	792	523	337	187	0	0
- Lignite	258	137	130	147	0	0	0
- Gas	830	731	717	669	471	215	0
- Oil	50	37	35	24	20	0	0
CO2 emissions power and CHP plants	**13,758**	**8,693**	**5,040**	**2,874**	**1,518**	**516**	**0**
- Hard coal (& non-renewable waste)	7,805	4,549	1,647	454	187	0	0
- Lignite	2,121	399	294	297	71	0	0
- Gas	3,125	2,790	2,476	1,999	1,218	516	0
- Oil + Diesel	707	955	623	124	42	0	0
CO2 intensity (g/kWh) without credit for CHP heat							
- CO2 intensity fossil electr. generation	780	711	624	536	505	454	0
- CO2 intensity total electr. generation	509	291	135	52	24	7	0
CO2 emissions by sector	**35,225**	**25,045**	**17,395**	**9,924**	**5,031**	**1,573**	**0**
- Industry	6,685	4,328	3,133	2,378	1,113	522	0
- Other sectors	7,620	5,239	3,951	2,823	1,714	464	0
- Transport	7,490	7,095	5,565	2,127	754	71	0
- Power generation	13,430	8,383	4,746	2,596	1,450	516	0
Population (Mill.)	7,713	8,184	8,548	8,888	9,199	9,482	9,735
CO2 emissions per capita (t/capita)	**4.6**	**3.1**	**2.0**	**1.1**	**0.5**	**0.2**	**0.0**

Fig. 7 Global energy-related CO_2 emissions—OECM 1.5 °C

Index

Printed in the United States
by Baker & Taylor Publisher Services